SEA ICE ANALYSIS AND FORECASTING

This book provides an advanced introduction to the science behind automated prediction systems, focusing on sea ice analysis and forecasting. Starting from basic principles, fundamental concepts in sea ice physics, remote sensing, numerical methods and statistics are explained at an accessible level. Existing operational automated prediction systems are described and their impacts on information providers and end clients are discussed. The book also provides insight into the likely future development of sea ice services and how they will evolve from mainly manual processes to increasing automation, with a consequent increase in the diversity and information content of new ice products. With contributions from world-leading experts in the fields of sea ice remote sensing, data assimilation, numerical modelling, verification and operational prediction, this comprehensive reference is ideal for students, sea ice analysts and researchers, as well as decision-makers and professionals working in the ice service industry.

TOM CARRIERES has over 30 years of experience in leading operational ice modelling activities for the Canadian Ice Service. These activities include development, implementation and testing of automated prediction systems, iceberg drift and deterioration models, general circulation models and extended-range statistical models. Tom co-chairs International Ice Charting Working Group Data Assimilation Workshops and Canadian Sea Ice Model Working Group meetings to facilitate collaboration among Canadian and international experts.

MARK BUEHNER is a research scientist working for Environment and Climate Change Canada, and is considered an international expert in data assimilation. He is the scientific lead in Canada for the development of data assimilation systems for both operational sea ice prediction and global deterministic weather prediction. Mark is also a member of the Data Assimilation and Observing Systems Working Group of the World Meteorological Organization.

JEAN-FRANÇOIS LEMIEUX is a research scientist working for Environment and Climate Change Canada. His work includes the development of numerical algorithms, physical parameterizations and verification methods for sea ice modelling. He is one of the leading developers of the McGill sea ice model and also contributes to the development of the CICE sea ice model. Jean-François is strongly involved in the development and implementation of automated ice–ocean prediction systems.

LEIF TOUDAL PEDERSEN has over 30 years of experience in sea ice remote sensing activities at the Technical University of Denmark and the Danish Meteorological Institute. These activities include development, implementation and testing of sea ice information retrieval algorithms for a number of sea ice variables such as sea ice concentration, sea ice drift and sea ice type. Leif co-chairs International Ice Charting Working Group Data Assimilation Workshops and has for many years worked to facilitate collaboration among sea ice experts worldwide.

SEA ICE ANALYSIS AND FORECASTING

Towards an Increased Reliance on Automated Prediction Systems

Edited by

TOM CARRIERES
MARK BUEHNER
JEAN-FRANÇOIS LEMIEUX
Environment and Climate Change Canada

and

LEIF TOUDAL PEDERSEN
Earth Observation Laboratory (eolab.dk)

CAMBRIDGE
UNIVERSITY PRESS

CAMBRIDGE
UNIVERSITY PRESS

University Printing House, Cambridge CB2 8BS, United Kingdom

One Liberty Plaza, 20th Floor, New York, NY 10006, USA

477 Williamstown Road, Port Melbourne, VIC 3207, Australia

314-321, 3rd Floor, Plot 3, Splendor Forum, Jasola District Centre, New Delhi - 110025, India

79 Anson Road, #06-04/06, Singapore 079906

Cambridge University Press is part of the University of Cambridge.

It furthers the University's mission by disseminating knowledge in the pursuit of education, learning and research at the highest international levels of excellence.

www.cambridge.org
Information on this title: www.cambridge.org/9781108417426
DOI: 10.1017/9781108277600

First published 2017

A catalogue record for this publication is available from the British Library

Library of Congress Cataloging in Publication data
Names: Carrieres, Tom, editor.
Title: Sea ice analysis and forecasting : towards an increased resilience on automated prediction systems / edited by Tom Carrieres, Mark Buehner, Jean-François Lemieux, and Leif Toudal Pedersen.
Description: Cambridge, United Kingdom ; New York, NY: Cambridge University Press, 2017. | Includes bibliographical references and index.
Identifiers: LCCN 2017024710 | ISBN 9781108417426
Subjects: LCSH: Sea ice – Measurement. | Sea ice – Forecasting.
Classification: LCC GB2405 .S43 2017 | DDC 551.34/30287–dc23
LC record available at https://lccn.loc.gov/2017024710

ISBN 978-1-108-41742-6 Hardback

This is for Annette, Erin, Meaghan, Alessandra and Lola.

Contents

Colour plate section can be found between pages 78 and 79

Contributors

Lars Axell
Swedish Meteorological and Hydrological Institute, SE-601 76 Norrköping, Sweden

Laurent Bertino
Nansen Environmental and Remote Sensing Center, Thormøhlens gate 47 N-5006, Bergen, Norway

Sylvain Bouillon
Nansen Environmental and Remote Sensing Center, Thormøhlens gate 47 N-5006, Bergen, Norway

Mark Buehner
Meteorological Research Division, Environment and Climate Change Canada, 2121 TransCanada Highway, Dorval, Québec, H9P 1J3, Canada

Tom Carrieres
Canadian Ice Service, Environment and Climate Change Canada, 373 Sussex E-3, Ottawa, Ontario, K1A0H3, Canada

Barbara Casati
Meteorological Research Division, Environment and Climate Change Canada, 2121 TransCanada Highway, Dorval, Québec, H9P 1J3, Canada

Alain Caya
Meteorological Research Division, Environment and Climate Change Canada, 2121 TransCanada Highway, Dorval, Québec, H9P 1J3, Canada

Frédéric Dupont
Meteorological Service of Canada, Environment and Climate Change Canada, 2121 TransCanada Highway, Dorval, Québec, H9P 1J3, Canada

Gregory Flato
Canadian Centre for Climate Modelling and Analysis, Environment and Climate Change Canada, c/o University of Victoria, PO Box 1700, Stn CSC, Victoria, BC, V8W 2Y2, Canada

Patrick Heimbach
University of Texas at Austin, Institute for Computational Engineering and Sciences and Institute for Geophysics, 201 East 24th Street, POB 4.232, Austin, TX 78712, USA

Sean Helfrich
U.S. National Ice Center, 4251 Suitland Road, NSOF, Washington, DC, 20395, USA

Georg Heygster
Institute of Environmental Physics, University of Bremen, P.O. Box 330440, D28334, Bremen, Germany

Nick Hughes
Forecasting Division of Northern Norway, Postboks 6314, 9293 Tromsø, Norway

Natalia Ivanova
Nansen Environmental and Remote Sensing Center, Thormøhlens gate 47 N-5006, Bergen, Norway

Juha Karvonen
Finnish Meteorological Institute, PB 503, FI-00101, Helsinki, Finland

Stefan Kern
Integrated Climate Data Center, University of Hamburg, Grindelberg 5, D-20144, Hamburg, Germany

Viatcheslav Kharin
Canadian Centre for Climate Modelling and Analysis, Environment and Climate Change Canada, c/o University of Victoria, PO Box 1700, Stn CSC, Victoria, BC, V8W 2Y2, Canada

Thomas Lavergne
Norwegian Meteorological Institute, PO Box 43, Blindern, NO0313, Oslo, Norway

Jean-François Lemieux
Meteorological Research Division, Environment and Climate Change Canada, 2121 TransCanada Highway, Dorval, Québec, H9P 1J3, Canada

Martin Losch
Alfred Wegener Institute, Helmholtz Centre for Polar and Marine Research, PO Box 120161, 27515 Bremerhaven, Germany

Arne Melsom
Norwegian Meteorological Institute, PO Box 43, Blindern, NO0313, Oslo, Norway

E. Joseph Metzger
Naval Research Laboratory, Oceanography Division (Code 7322), Building 1009, Room C132, Stennis Space Center, MS 39529, USA

Leif Toudal Pedersen
Earth Observation Laboratory (eolab.dk), Markmandsgade 14, 4 TV, DK-2300 Copenhagen S, Denmark and Technical University of Denmark, DTU-Space, Building 348, DK-2800 Lyngby, Denmark

Pam Posey
Naval Research Laboratory, Oceanography Division (Code 7322), Building 1009, Room C132, Stennis Space Center, MS 39529, USA

Pierre Rampal
Nansen Environmental and Remote Sensing Center, Thormøhlens gate 47 N-5006, Bergen, Norway

Michael Sigmond
Canadian Centre for Climate Modelling and Analysis, Environment and Climate Change Canada, c/o University of Victoria, PO Box 1700, Stn CSC, Victoria, BC, V8 W 2Y2, Canada

Greg Smith
Meteorological Research Division, Environment and Climate Change Canada, 2121 TransCanada Highway, Dorval, Québec, H9P 1J3, Canada

Adrienne Tivy
Canadian Ice Service, Environment and Climate Change Canada, 373 Sussex E-3, Ottawa, Ontario, K1A0H3, Canada

Rasmus Tonboe
Danish Meteorological Institute, Lyngbyvej 100, DK-2100, København Ø, Denmark

Louis-Bruno Tremblay
Department of Atmospheric and Oceanic Sciences, Room 945, Burnside Hall, McGill University, 805 Sherbrooke Street West, Montréal, Québec, H3A 0B9, Canada

Martin Vancoppenolle
Sorbonne Universités, UPMC Paris 6, CNRS/IRD/MNHN, Laboratoire d'Océanographie et du Climat Institut Pierre-Simon Laplace, Paris, France.

Timothy Williams
Nansen Environmental and Remote Sensing Center, Thormøhlens gate 47 N-5006, Bergen, Norway

Preface

We are in the midst of a major transformation in the way ice services are provided to the marine community from a labour-intensive manual process to a more automated approach that relies on the output of a numerical prediction system. This book is designed to elaborate on these ideas by briefly reviewing current operational practices, examining in detail the components of automated ice prediction systems and then proposing ideas about how these will lead to the evolution of sea ice services. The four main components of automated prediction systems reviewed here include numerical sea ice models, observations, data assimilation systems and objective evaluation procedures. Starting with basic concepts of the relevant science, each topic will be developed to a moderately advanced level. It is our intention neither to provide a thorough literature review nor to provide detail at the level of scientific journal papers.

The target audience includes sea ice researchers, graduate and undergraduate students and national ice service personnel as well as the marine and weather prediction communities. The book is intended to bridge the gap between researchers and operational personnel as well as to provide an overview of the field to the marine and modelling communities. It is expected that it will appeal to non-experts as well as those who have expertise in more focused aspects of automated prediction systems.

Acknowledgements

We would like to thank the many contributors who have provided invaluable information and suggestions. We would also like to thank our external reviewers John Falkingham, Dirk Notz and David Hebert, who all helped ensure that this book consistently addressed the needs of our audience.

Abbreviations

3DVar	Three-Dimensional VARiational data assimilation
4DVar	Four-Dimensional VARiational data assimilation
AARI	Arctic and Antarctic Research Institute
AATSR	Advanced Along-Track Scanning Radiometer
ACC	Anomaly Correlation Coefficient
ACNFS	Arctic Cap Nowcast/Forecast System
AD	Algorithmic Differentiation
AIDJEX	Arctic Ice Dynamics Joint EXperiment
AMSR	Advanced Microwave Scanning Radiometer
APS	Automated Prediction System
ASCAT	Advanced SCATterometer
ASI	Artist Sea Ice (algorithm)
ASPeCt	Antarctic Sea ice ProcEsses and ClimaTe
AVHRR	Advanced Very High Resolution Radiometer
CanSIPS	CANadian Seasonal to Inter-annual Prediction System
CCCma	Canadian Centre for Climate Modelling and Analysis
CCMEP	Canadian Centre for Meteorological and Environmental Prediction
CDI	Close Drift Ice
CERSAT	Centre ERS d'Archivage et Traitement (IFREMER)
CI	Confidence Interval
CICE	Community Ice CodE (Los Alamos)
CIS	Canadian Ice Service
CLS	Collecte Localisation Spatiale
CMCC	Continuous Maximum Cross Correlation
CMEMS	Copernicus Marine Environment Monitoring Service
CRREL	Cold Regions Research and Engineering Laboratory
DMI	Danish Meteorological Institute
EB	Elasto-Brittle (ice rheology)
EM	ElectroMagnetic
EnKF	ENsemble Kalman Filter
ESA	European Space Agency

ESD	Error Standard Deviation
ESMR	Electrically Scanning Microwave Radiometer
EUMETSAT	EUropean organization for the exploitation of METeorological SATellites
EVP	Elastic–Viscous–Plastic (ice rheology)
FDD	Freezing Degree Days
FESOM	Finite-Element Sea-ice Ocean circulation Model
FI	Fast Ice
FMI	Finnish Meteorological Institute
FSD	Floe Size Distribution
FSTD	sea ice Floe Size and Thickness Distribution
FY	First Year
GD	Giops weekly Delayed model analysis
GDPS	Global Deterministic Prediction System
GIOPS	Global Ice Ocean Prediction System
GIS	Geographic Information System
GOFS	Global Ocean Forecast System
GR	Giops Real-time weekly analysis
HH	Horizontal transmit–Horizontal receive
HSS	Heidke Skill Score
HyCOM	HYbrid Coordinate Ocean Model
IA	Ice Analyst
IMB	Ice Mass Balance buoy
IMS	Interactive multisensor snow and ice Mapping System
IR	InfraRed
ISRO	Indian Space Research Organization
JAXA	Japanese Aerospace eXploration Agency
ITD	sea Ice Thickness Distribution
ITP	Ice Tethered Profiler
LIM	Louvain-la-neuve sea Ice Model
MAE	Mean Absolute Error
MAGIC	MAp-Guided Ice Classification (algorithm)
MCA	Maximum Covariance Analysis
MCC	Maximum Cross-Correlation
MEMLS	Microwave Emissivity Model for Layered Snowpacks
MITgcm	Massachusetts Institute of Technology General Circulation Model
MIZ	Marginal Ice Zone
MLR	Multiple Linear Regression
MODIS	MODerate-resolution Imaging Spectroradiometer
MOS	Model Output Statistics
MSE	Mean Square Error
MY	Multi-Year

NASA	National Aeronautics and Space Administration
NAVGEM	NAVy Global Environmental Model
NAVOCEANO	NAVal OCEANographic Office
NCODA	Navy Coupled Ocean Data Assimilation
NEMO	Nucleus for European Modelling of the Ocean
neXtSIM	NEXT-generation Sea Ice Model
NIC	National Ice Center
NIIS	National Ice Information Service
NPP	Noaa Polar Platform
NRL	Naval Research Laboratory
NSCAT	Nasa SCATterometer
NSIDC	National Snow and Ice Data Center
NT2	Nasa Team 2 (algorithm)
NWP	Numerical Weather Prediction
ODI	Open Drift Ice
OI	Optimal Interpolation
OIB	Operation Ice Bridge
OPA	Océan PArallélisé
OSCAT	Oceansat-2 SCATterometer (ISRO)
OSISAF	Ocean and Sea Ice Satellite Application Facility
OVP	Ocean Vertical Profiles
OW	Open Water
PCT	Proportion Correct Total
pdf	Probability Density Function
PET	Proportion Error Total
PM	Passive Microwave
PP	Perfect Prognosis
QuikSCAT	QUIcK SCATterometer
RDPS	Regional Deterministic Prediction System
RGPS	RADARSAT Geophysical Processing System
RHS	Right-Hand Side
RIPS	Regional Ice Prediction System
RMSE	Root Mean Square Error
SAR	Synthetic Aperture Radar
SEEK	Singular Evolutive Extended Kalman (filter)
SICCI	Sea Ice Climate Change Initiative
SLA	Sea Level Anomalies
SLAR	Side-Looking Airborne Radar
SMAP	Soil Moisture Active Passive
SMHI	Swedish Meteorological and Hydrographic Institute
SMMR	Scanning Multi-channel Microwave Radiometer
SMOS	Soil Moisture and Ocean Salinity

SS	Skill Score
SSM/I	Special Sensor Microwave Imager
SSMIS	Special Sensor Microwave Imager/Sounder
SST	Sea Surface Temperature
TIROS	Television InfraRed Observation Satellite
TOPAZ	Towards an Operational Prediction system for the north Atlantic european coastal Zones
ULS	Upward-Looking Sonar
UTC	Universal Time (Coordinated)
VCDI	Very Close Drift Ice
VIIRS	Visible Infrared Imaging Radiometer Suite
VODI	Very Open Drift Ice
VP	Viscous–Plastic (rheology)
VV	Vertical transmit–Vertical receive
WHOI	Woods Hole Oceanographic Institute
WMO	World Meteorological Organization

1
Introduction

Tom Carrieres

1.1 Overview

Sea ice is most popularly recognized as an important indicator of climate change due to media reports of recent record low amounts of Arctic sea ice and future ice climate scenarios that include the complete loss of summer ice in the Arctic. Perhaps less widely appreciated is the day-to-day importance of sea ice to those involved in marine activities. Whether it's considered an essential platform for either transportation or research or it represents a serious hazard to shipping or resource extraction, up-to-date information about current and future sea ice conditions is essential to many different components of society.

While sea ice information is still largely produced manually, ongoing advances in the development of automated computer-based prediction systems for sea ice provide opportunities for new or alternative types of ice information. Such systems are complex and consist of a number of essential components. They are quite distinct from sea ice models used purely for research purposes since an automated prediction system (APS) is required to produce high-quality information for clients reliably, often in near-real time. In this sense, an APS for sea ice is, in many ways, similar to other environmental prediction systems, such as those for numerical weather prediction (NWP). And as with NWP, the term "prediction" is used here to refer to all of the required components of the system. Principally, this consists of collecting and processing sea ice observations that are then used, through data assimilation, to obtain a complete and accurate estimate of the current sea ice conditions. This estimate is then used to initialize a sea ice forecast model to produce the forecast of future sea ice conditions. Thus, the outline of this book bears strong similarities to an equivalent book on weather prediction, and lessons learned from the introduction of NWP would also echo the discussions here.

Sea ice models have been in development for many years but have mainly been used in climate models and process studies. To adapt these models for the shorter forecast periods and higher spatial resolution forecasts that are needed to support a national ice information service (NIIS), model initialization and certain physical processes take on greater importance. Examples include land-fast ice formation and short-timescale ocean processes such as tides and wave–ice interactions. The conservation of mass and heat that is crucial for climate modelling is not essential for operational forecasting where the model state is repeatedly reinitialized by the data assimilation system. Chapter 2 reviews relevant sea ice physics and numerical modelling techniques.

One might expect that the automated use of satellite observations within an APS should be straightforward, since they are routinely used in current NIIS operations. However, humans have sophisticated pattern recognition ability, and analysts generally have to undergo lengthy training before they are able to prepare operational products of sufficiently high quality. The appearance of sea ice features depends not only on quantities such as concentration and thickness but may also be affected by snow cover, surface melting, surface temperature, salinity content, roughness, surface emissivity, etc. In addition, observations from a particular sensor may depend on the surface footprint, solar illumination, incidence angle and atmospheric interference. Unlike the analyst, who has knowledge and experience, automated systems must rely on a sometimes ambiguous mathematical relationship between the observation and the desired ice feature as well as the related uncertainty in both the observation and the mathematical relationship. Chapter 3 provides a review of sea ice observations and their utility.

Sea ice data assimilation systems should use as many independent observations as possible for computing an estimate of the current sea ice conditions. The complexity of data assimilation systems can span a wide range. The simplest approach uses data insertion to replace values predicted by the forecast model with those obtained directly from a single type of observation. More sophisticated methods, similar to those currently used for NWP, account for the spatial–temporal and multivariate statistical relationships of the errors in a short-term model forecast and observations when assimilating a wide range of diverse observation types. The choice of which data assimilation approach to employ by a particular NIIS can depend on the computational and scientific resources available and the final use of the APS output. Chapter 4 reviews data assimilation techniques and some specific challenges when applying them to sea ice prediction.

The integration of advances in forecast models, observations and data assimilation within a sea ice APS is taking place to varying degrees at a number of centres. The output from these systems may have NIIS as clients while others may provide a direct service to the marine community. Chapter 5 reviews key characteristics of a number of existing operational sea ice APS.

The objective evaluation of both manually generated products and products generated from the output of an APS to provide a measure of accuracy is crucial to their optimal use. Consistency has not yet been achieved in either evaluation methodology or the data used for evaluation for both types of products available from a single NIIS, let alone from different NIISs. Different approaches are required to evaluate various types of predicted variables, whether they are categorical, scalar or vector. Chapter 6 provides a review of objective evaluation techniques including examples used to measure the value of APS outputs.

The concept of using output from an APS, either exclusively or in combination with other data sources, when generating ice products will often be referred to as numerical guidance. Chapter 7 focuses on current ice services and how the integration of numerical guidance can lead to effective improvements in operational procedures and new products.

1.2 A Brief Overview of Ice Services

Every day, a large volume of highly detailed ice information is provided to a wide variety of clients ranging from resource planners to ship crews navigating their vessels through extremely hazardous ice conditions. This information is considered operational because it must be reliably available in near-real time, seven days per week. Operational sea ice products are a crucial data source for safe and efficient operations in ice-covered waters. A suite of observations, numerical modelling systems and dedicated infrastructure and personnel working within specialized centres in numerous countries supports the generation of operational products. Highly trained ice analysts (IA) prepare these products using an essentially manual process that may not always be well documented, especially in scientific literature. To help understand how this operational production might evolve in the future, it is useful to begin with a background on operational ice services.

Ice occurring in the marine environment either forms in situ, referred to here as sea ice, or originates from land to form icebergs. While the focus of this book is on sea ice, it was an iceberg that sank the RMS Titanic in 1912, resulting in heightened general awareness of the threats posed by ice at sea. Monitoring of marine ice at that time was limited to ship and shore reports, with little or no information being widely distributed. By the 1950s and 1960s, visual aerial reconnaissance and radio facsimile communications made it feasible to provide near-real-time ice information in support of icebreaking, marine transportation and fishing operations.

Since the monitoring and reporting of sea ice conditions requires very specialized expertise and infrastructure, a number of countries have set up their own operational ice services. Some of the currently operational NIIS and organizations that provide similar services are: Argentine Naval Hydrographic Service; Canadian Ice Service (CIS); Danish Meteorological Institute (DMI); Finnish Meteorological Institute (FMI); German Federal Maritime and Hydrographic Agency; Japan Hydrographic and Oceanographic Department; Norwegian Meteorological Institute; Polish Hydrological Forecasting Office; Russian Arctic and Antarctic Research Institute (AARI); Swedish Meteorological and Hydrological Institute (SMHI); and, the US National Ice Centre (NIC). Other countries prepare sea ice information products as required, including Australia, Chile and Kazakhstan. Each NIIS provides information about past, present and future ice conditions, usually in the form of spatial information superimposed on maps or as text bulletins. In areas of geographic overlap, many of the NIIS coordinate their products to ensure consistent information is provided to clients. Coordination of ice services is facilitated by the World Meteorological Organization (WMO) and more detailed information on NIIS of the world may be found in the publication 'Sea-Ice Information Services in the World' (WMO, 2010).

To date, many of the NIIS innovations have been due to improvements in sea ice observations and information technology. Aircraft equipped with side-looking airborne radar (SLAR) went into operation in the 1970s and 1980s. Meanwhile, the NIISs were early adopters of using imagery from weather satellites for sea ice interpretation, starting in the 1970s. The widespread availability of operational satellites equipped with synthetic aperture radar (SAR) instruments in the 1990s has greatly improved the reliability and also the affordability of existing ice services, since it greatly reduced the need for observations from aircraft. These and other satellite observations combined with tailored analysis workstations incorporating geographic information systems (GIS), have allowed ice analysts to produce more accurate sea ice charts along with a large variety of derived products.

NIIS products may be categorized according to their use: (1) tactical products that are prepared in near-real time; and (2) strategic ice products that are more relevant to longer time periods. This book generally focuses on the first type of product but both are described briefly below.

1.2.1 Tactical Ice Products

Tactical ice products are provided to clients in near-real time and include mainly graphical charts and text bulletins. The preparation of these products involves manually interpreting and reconciling a wide variety of mainly satellite observations, while taking into account expected ice, weather and oceanographic conditions. Satellite observations do not provide ice information directly. Therefore, expert interpretation plays a significant role in their effective use. Similarly, forecasting ice conditions involves expert use of weather and oceanographic information along with an advanced knowledge of sea ice physics.

Clients typically use the resulting ice products to guide their daily activities such as navigating through or around ice-covered waters or identifying optimum locations for fishing. All of these products must adhere to operational deadlines, often resulting in a limited amount of time for the analysts to synthesize available data. In general, ice charts follow the guidance provided by WMO (2015), although regional adaptations are common. Ice charts depict areas of uniform ice conditions that are described in terms of the total ice concentration and partial concentrations of various ice types and corresponding predominant floe sizes. An example of an ice chart is shown in Figure 1.1 while a description of the WMO 'egg' code used in the chart is shown in Figure 1.2. Daily charts are prepared for operationally active marine areas whether or not current observations cover all areas. Since observations are rarely made at the valid time of the chart, the analysts at some centres simply stitch together the most recent observations and adjust areas of intersection for spatial continuity. At other centres, the more difficult procedure of predicting the evolution of ice conditions between the observation time and the valid time of the chart is used when producing the chart. Additional information may also be provided on the chart related to ice thickness measurements, sea surface temperatures and a description of the data sources used for a given area.

Since not all ships are equipped with the communications and reception equipment necessary to receive the graphical ice charts, ice bulletins made up of only text are often used. Ice bulletins contain at the very least a description of total ice concentration. They

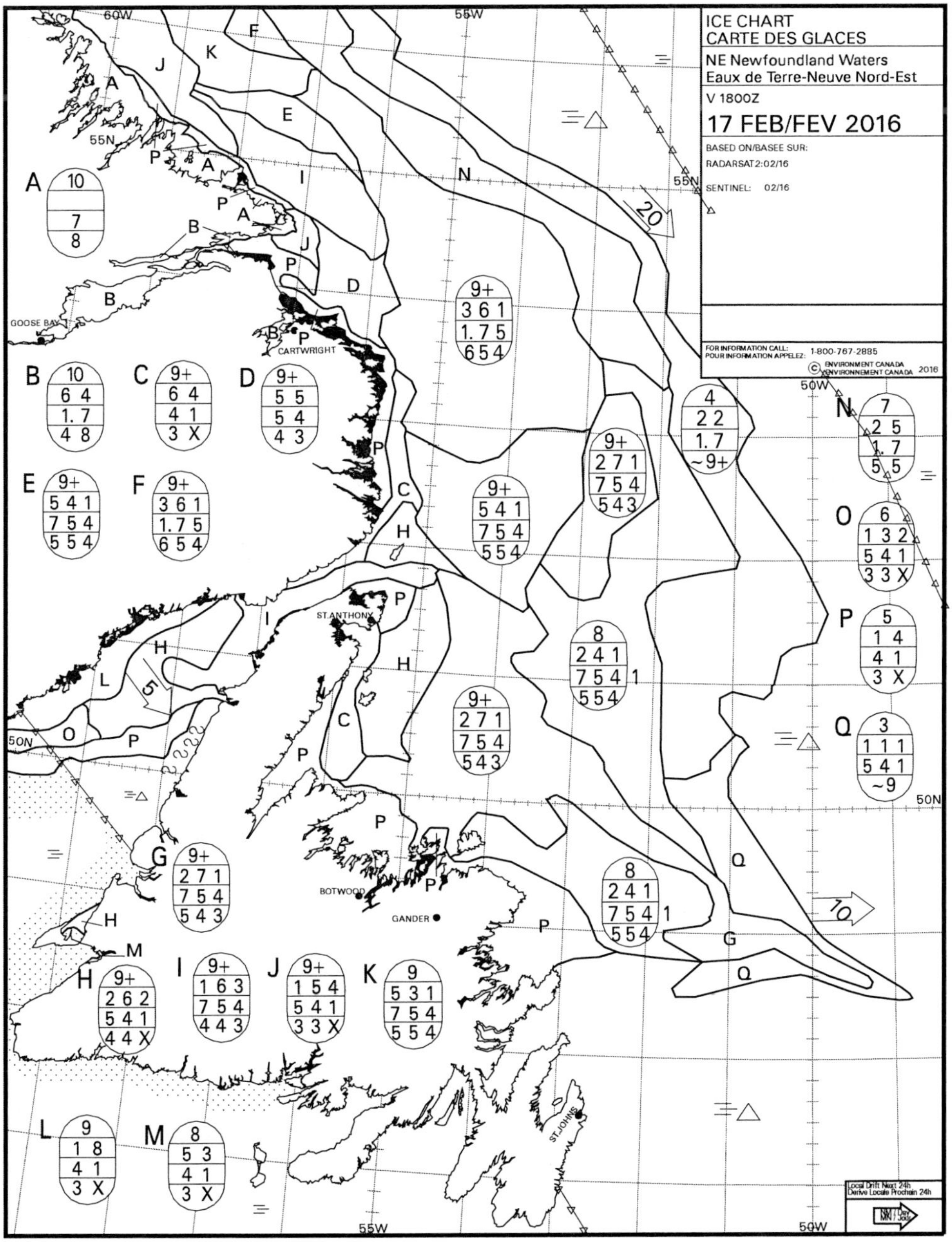

Figure 1.1: An example of a daily ice chart as prepared by the Canadian Ice Service. Conditions are those expected at the chart valid time and are presented as delineated areas of relatively uniform ice conditions described using the egg code format, described in Figure 1.2. The 24-hour forecast ice drift direction is shown as arrows with net drift distance inserted as a number with units of nautical miles. The iceberg limit is depicted as a line with overlaid triangles. Areas of ice free water are identified as (=) and bergy water as (△). Open water areas (i.e. between 0 and 1 tenth of ice) are dotted. © Her Majesty the Queen in Right of Canada, as represented by the Minister of the Environment Canada, 2017.

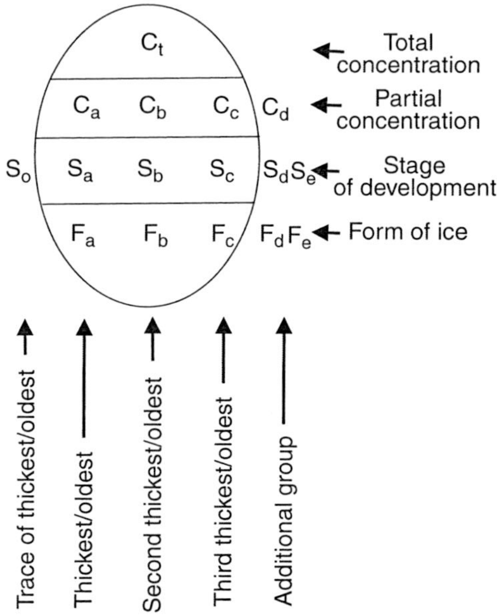

Figure 1.2: The WMO egg code format. Concentrations are given in tenths while the stage of development and form of ice are numerical codes related to the age or thickness of the ice and the size or form of the ice, respectively. © Her Majesty the Queen in Right of Canada, as represented by the Minister of the Environment Canada, 2017.

may also provide information on ice edge location, ice stage of development and a short-term forecast of expected changes.

Other tactical products are linked more closely to observations, including the satellite images themselves and image analysis charts. Image-based products are portions of satellite images or whole images that are often compressed to accommodate the limited communication bandwidth available in many areas at sea. Full resolution SAR and optical satellite image products may provide the high spatial resolution suitable for navigation purposes, although clients must have the expertise necessary to interpret the data. To assist in this area, expert analysts interpret the satellite imagery and provide image analysis charts. These charts are based on an analysis of a single satellite image or image swath and they are valid at the time of the satellite pass. Usually they are prepared using SAR or optical imagery, although other image sources and meteorological information can be used to assist in the image interpretation. Since ice conditions can evolve rapidly, every effort is made to prepare and transmit these products as quickly as possible.

Clients for tactical ice products include Coast Guard icebreaking and vessel routing, marine transportation, offshore oil and gas industry, national defence, fishing industry, aboriginal/local transportation, port authorities and tourists/adventurists.

1.2.2 Strategic Ice Products

Given its chaotic nature, the predictability of the atmosphere is about two weeks, resulting in a direct limitation on the predictability of sea ice. However, anomalies in the large-scale ocean and sea ice conditions relative to climatological averages (e.g. sea surface temperature, ice area, ice thickness) can provide longer-term predictability of large-scale characteristics. Hence, forecasts from beyond several days up to a season provide information more related to the general evolution of conditions or departures from normal.

Similar configurations of a physically-based APS could be used for short-term and seasonal forecasts. However, NIIS often rely on statistical or analogue techniques for predicting the evolution of the sea ice cover on monthly to seasonal timescales and sea ice APS are not yet widely used for guidance. Forecasts beyond the coming ice season are not normally issued by NIIS, although they may provide general trends in ice cover at very coarse scales.

Climate products are another type of strategic ice product. They are quite diverse and can include, for example, the normal extent of ice for a given area and time of year or extreme ice conditions at a point location. They are typically used as input to the design of structures and vessels, planning new ship routes or to help monitor climate change. They are usually based on a 30-year period as that is considered long enough to capture interannual variability while short enough to represent current normal conditions. Typical products include the frequency of occurrence of ice and each ice stage of development on a monthly basis. It is crucial that the source products used to generate such an ice climatology do not introduce artificial trends such as may result from a change in input data or a change in preparation techniques.

Clients for these products include long-range planners for marine transportation routing and vessel deployment, marine planners, naval architects, climate research scientists and policy makers.

1.3 The Components of an Automated Prediction System

In some cases, current daily ice chart and bulletin production may fall short of new and evolving client information requirements. In addition, climate change may result in year-round Arctic shipping being feasible, thus significantly increasing the demand for ice services. In other cases, new products, such as site-specific information and probabilistic information, are required to support local communities and risk-based decision makers. Opposing these expanding requirements is the financial pressure on governments to limit increases to the budgets of the centres that provide these services. While a large number of observations should lead to more accurate ice products, it takes considerable time and expertise to analyse and integrate the information. As a result, the most accurate products require the greatest amount of time to prepare. While the current, mostly manual procedures are becoming more efficient, the provision of a significant increase in ice information would still require a roughly proportional increase in the number of ice analysts.

Of course, these issues have been faced by the much larger weather service community. Advances in NWP have allowed weather forecasters to focus on short-range, high-impact weather, while at the same time more accurate and longer-range forecasts are automated. This automation requires investments in forecast model and data assimilation research as well as the supporting information technology and communications. For the weather community, these investments are providing increasingly accurate and reliable weather forecasts.

The sea ice community is going through a similar transition, but with the confidence that longer-term benefits will be achieved, as experienced with NWP. The purpose of this book is to provide an advanced introduction to some of the scientific and technological advances in automated sea ice prediction systems that enable this transformation. It is also suggested that, as the accuracy of these systems improves, they can increasingly be used to provide input either to a manual or fully automated process for the generation of ice products. This will facilitate the expansion and increased tailoring of ice services while remaining financially affordable. It should also be noted that, to further improve the accuracy of weather forecasts, NWP systems are being developed that increasingly incorporate more components of the complete Earth system, including representations of the ice and ocean. This ongoing evolution can also benefit centres responsible for operational ice services, as it provides an excellent opportunity for coordinated development of coupled sea ice, ocean and weather prediction systems to ensure that the resulting systems meet the combined requirements of ice, ocean and weather information services. Hence, large investments in coupled prediction systems that only meet the needs of NIIS are not required. More modest investments into improving automated sea ice prediction, combined with close collaboration with weather and ocean APS developers, should result in significant ongoing benefits.

Figure 1.3 depicts the components of an APS. Some parts of this system emulate the manual ice analysis and forecasting process. One starts with an initial estimate of current conditions, also referred to as a background state, which in this case is provided by a short-term forecast from a numerical model. A correction to the background state is obtained by synthesizing observations from a large variety of sources using a data assimilation system. This corrected background state, referred to as the analysis state, then serves as the starting point to predict future conditions using the forecast model and information about other relevant environmental conditions. As the forecast model increases in complexity to include ocean and atmospheric components, less reliance on external forcing is required. Initially, manual intervention could be used to correct deficiencies in the system, although the goal would be to minimize or eliminate such intervention altogether. A crucial component of this system is routine objective evaluation in order to identify and correct problem areas as well as to quantify expected accuracy. Some systems could also include post-processing that would correct errors in APS outputs using a variety of statistical techniques.

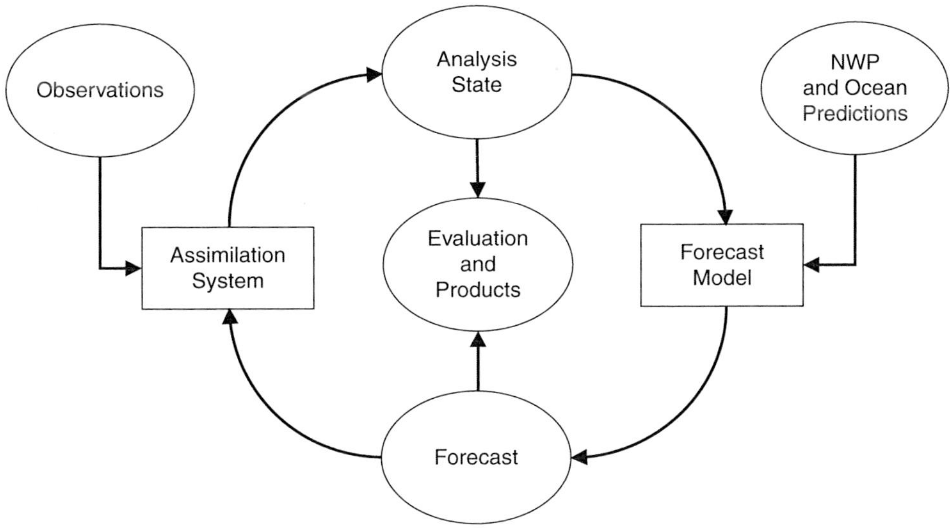

Figure 1.3: A schematic of the main components of an automated prediction system

1.4 Chapter Summary

A NIIS provides information about past, present and future ice conditions to a large variety of clients both in near-real time and for non-time-critical operations. Although these services rely on manual approaches, they will continue to adapt due to the availability of new technologies and evolving client requirements. With the availability of operational APS outputs and the ongoing developments to improve these systems, it is expected that ice services will rely more and more on APS products for guidance.

The components (described in the following chapters) of an ice APS include a sea ice model, observations, a data assimilation system and objective evaluation and post-processing, all aimed at providing new and improved ice information products.

References

WMO (2010). *Sea-Ice Information Services in the World*, WMO No. 574. World Meteorological Organization.

WMO (2015). *WMO Sea-Ice Nomenclature, Volume III – International System of Sea-Ice Symbols*, WMO No. 259. World Meteorological Organization.

2

Sea Ice Physics and Modelling

Jean-François Lemieux, Sylvain Bouillon, Frédéric Dupont, Gregory Flato, Martin Losch, Pierre Rampal, Louis-Bruno Tremblay, Martin Vancoppenolle, Timothy Williams

The purpose of sea ice automated prediction systems (APS) is to forecast the evolution of the sea ice cover in a particular region over a certain period of time. To produce these forecasts, a sea ice model requires information about the current sea ice conditions and the atmospheric and oceanic forcing that largely govern the evolution of sea ice. An estimate of the current sea ice conditions, referred to as the analysis state, can be obtained by applying a data assimilation procedure (see Chapter 4) that employs a large number of diverse sea ice observations (see Chapter 3). Forcing from the atmosphere and the ocean can be obtained from atmospheric and oceanic models that continually interact with the sea ice model in a coupled framework or from precomputed atmospheric and oceanic forecasts with an offline approach.

This chapter describes the important physical processes for sea ice forecasting. It discusses how these dynamic and thermodynamic processes are represented in the sea ice models used in large-scale operational sea ice prediction systems. We define large-scale prediction systems here as systems producing, for example, pan-Arctic forecasts or forecasts over a large region (e.g. Baffin Bay, Hudson Bay, Beaufort Sea). The models used for these applications are referred to as large-scale sea ice models. This chapter also introduces how the equations for dynamics and thermodynamics are typically implemented and solved numerically. Recent model developments are also presented.

In this book, the timescales of interest vary from a few hours (short-term forecasting) to a few months (seasonal forecasting). Over a few days, there could be significant changes in sea ice conditions in the marginal ice zone (MIZ), i.e. the region that marks the transition between the pack ice and the open ocean. The skill of a short-term sea ice forecast therefore strongly depends on the quality of the initial sea ice conditions (e.g. sea ice concentration) and ocean conditions (e.g. sea surface temperature) in the MIZ and regions nearby. The forcing from surface winds and from surface ocean currents is also of primary importance for predicting the movement of the ice edge. Over longer timescales, surface winds and ocean currents remain important for accurate forecasts. Moreover, the proper representation of many physical processes becomes crucially important for longer-term forecasts when feedbacks associated with thermodynamic and dynamic processes can lead to significant biases. For example, the vertical growth/melt of thick ice is irrelevant for short-term forecasts but becomes crucial on seasonal timescales.

Many of the ideas for sea ice modelling presented below originated from the Arctic Ice Dynamics Joint EXperiment (AIDJEX), a joint project during the 1970s funded by the United States and Canada. For more detailed review of sea ice modelling, the reader is referred to the work of Notz (2012) and Hunke *et al.* (2010). The review article of Feltham (2008) is also relevant for the representation of rheology in sea ice models, and the article by Hunke *et al.* (2011) is of interest for recent model developments associated with sea ice thermodynamics.

2.1 Introduction to Sea Ice Physics

As opposed to icebergs, which are created from the calving of glaciers from land ice (e.g. Greenland, Antarctica), sea ice forms from seawater that freezes. For sea ice to form, the seawater must be cooled to its freezing point. The freezing point of seawater is a function of salinity; it decreases as the salt content increases. A freezing point of −1.8°C is a typical value for seawater in the polar oceans. As seawater cools, it becomes denser. This denser seawater sinks and triggers convective mixing, bringing warmer water (e.g. remnant solar heating from the previous summer) up from below to the surface mixed-layer. As a consequence, this convective process delays the initial formation of sea ice. This is very different than the formation of ice in a lake; the density of freshwater also increases as freshwater cools but only until a temperature of 4°C, below which the density decreases, thus inhibiting pre-freezing convection and ventilation of warmer deeper water. This is the reason why the water temperature below the thermocline in a lake is around 4°C.

Because of vertical ocean stratification, the convective mixing does not usually affect the whole water column. For instance, in the Arctic Ocean, there is usually a well-defined vertical salinity gradient at the base of the mixed-layer,[1] referred to as the cold halocline, which separates the surface mixed-layer from the warmer and saltier water below. For this reason, the convective process only takes place between the surface mixed-layer and the halocline waters (i.e. the first several tens of metres of the water column). Once this upper layer reaches the freezing point temperature, sea ice starts to form. Then, the ocean heat fluxes associated with surface convection is negligible since the halocline waters that are brought up to the surface are also at the freezing point temperature. This is the main reason for the presence of a relatively thick perennial ice cover (two to six metres) in the Arctic Ocean. In the Southern Ocean, both the salinity and temperature increase at the base of the mixed-layer (i.e. there is no cold halocline) and surface convection ventilates a significant amount of heat from below the mixed-layer, keeping the sea ice relatively thin (about one metre) and therefore seasonal.

In the first stage of development of sea ice, needle-shaped ice crystals suspended in cold seawater are formed. This mixture is referred to as frazil ice. With further cooling, a mushy layer called grease ice develops. Then, with the action of wind and waves, this mushy layer

[1] Note that around the freezing point temperature of seawater, variations in density are almost entirely defined by changes in salinity rather than temperature.

can be consolidated to form plates of diameters of a few tens of centimetres to a few metres, which are referred to as pancakes. Once this thin layer of ice exists, additional growth occurs at the base of the ice layer. For this to happen, the ocean mixed-layer must continue to lose heat to the atmosphere. This heat flux is in fact latent heat released when liquid seawater changes phase into solid ice.

The ability of the ice to insulate the relatively warm ocean surface from the relatively cold atmosphere depends on its thickness. Consequently, the conductive heat flux through the ice and the growth rate both decrease as the sea ice thickness increases. For instance, typical ice growth rate in winter over open water during the winter is approximately ten centimetres per day and of the order of one centimetre per day over typical ice thickness of three metres. When snow covers the ice surface, ice growth is further reduced for a given ice thickness, because of the significantly lower thermal conductivity of snow compared to ice. This is another factor explaining the presence of a relatively thinner ice cover in the Southern Ocean, as snow precipitation in this region is, on average, more abundant than in the Arctic.

Sea ice floats at the surface of the ocean because it is less dense than seawater. For typical sea ice and seawater densities (917 kg m^{-3} and 1035 kg m^{-3}, respectively), roughly 10 per cent of the ice thickness lies above the water line (the freeboard) and the remaining 90 per cent is underwater (the draft). This follows from Archimedes' principle. When a snow cover is present on top of the ice, the ice surface is further depressed below sea level. If the snow cover is sufficiently heavy, it can push the ice upper surface below the water line. This floods the base of the snow with seawater and can lead to snow-ice formation, i.e. snow is converted to sea ice. Again, because of higher snow precipitation, this snow-ice formation process is more common in the Southern Ocean than in the Arctic.

As sea ice forms, it has an important impact on ocean properties and stratification. Indeed, part of the salt contained in the seawater is expelled during the formation of the ice. This process is called brine rejection. Because the cold and salty brine is denser than the water it originates from, it sinks, enhances mixing in the ocean surface layers and feeds the cold halocline layer. The remaining salt in the ice is in the form of liquid brine contained in brine pockets. These brine pockets adjust their volume and salinity to maintain their freezing point at the local temperature. The amount of brine (per unit volume) sets the bulk ice thermal properties (thermal conductivity, specific heat and energy of melting).

Melting can occur both at the base and at the top surface of sea ice. Lateral melting also occurs at the edges of ice floes. The basal melt is determined by the conductive heat flux through the ice and the turbulent ocean heat flux from below. The surface melt depends on the net energy balance (turbulent and radiative heat fluxes) at the ice or snow surface – these heat fluxes in turn depend on the surface temperature at the interface between the ice or snow and the atmosphere. A crucial factor for the surface melt is the albedo of the ice or snow. This unit-less parameter defines the fraction of downward shortwave (solar) radiation that is reflected by the surface. The albedo strongly depends on the characteristics of the surface (ridges, melt pond, snow presence and crystal structure, etc.). While lateral melt is usually a small contributor to the total melt, it directly impacts the surface area covered by sea ice and this impact is amplified by the ice-albedo feedback. That is, due to its much lower albedo than sea ice, as more of the ocean is exposed to solar radiation, more heat is absorbed by the ocean, thus enhancing the melt.

During summer, a significant fraction of the melt water can accumulate at the surface and form melt ponds. As water has a lower albedo than ice or snow, melt ponds can strongly affect the surface radiative balance and lead to further melting. The surface melt water that percolates through the ice together with the basal melt water increase surface stratification. This creates a lid of relatively fresh (light) water at the ocean surface, called the seasonal halocline, which limits vertical mixing between the surface water and winter mixed-layer. Ice that disappears completely during the melt season is called seasonal (or first-year) ice, while ice that survives is referred to as perennial (or multi-year). Multi-year ice that goes through many seasonal cycles becomes progressively less saline due to different desalination processes such as gravity drainage, flushing and surface melt water percolation.

The temporal evolution of the sea ice cover is governed not only by thermodynamic processes, but also by dynamic processes. This is mostly due to the action of the surface winds but also of ocean currents in certain regions. Sea ice typically moves horizontally with velocities between 0 and 1 m s^{-1}. An important implication of sea ice dynamics is that it constitutes a freshwater transport when ice forms in one location, drifts and then melts in another location. For example, sea ice transported southward from the Arctic to the northern North Atlantic is a source of freshwater as it melts. This increases the stratification and has the potential to reduce deep water formation (linked with the global thermohaline circulation) in regions such as the Greenland, Icelandic, Norwegian and Labrador seas.

The dynamic processes of sea ice also have important implications for ice–ocean–atmosphere interactions. In regions of compact ice, the wind stress acting on the ice cover creates mechanical stresses inside the ice, referred to as internal ice stresses. Unlike a fluid such as water, sea ice can resist some internal stresses without undergoing significant deformation (elastic deformation and viscous creep are present, but are negligible). However, when critical internal stresses are reached, the ice either fails in tension, compression and/or shear.

For instance, the wind field can create openings in the ice pack when the ice fails in tension. Large openings, referred to as polynyas (pools of open water in an otherwise fully ice covered ocean), are common along the Eurasian coastline. Surface winds can also create long openings or cracks in the ice pack that are referred to as leads. Satellite observations reveal that leads can extend over distances of hundreds of kilometres and have widths ranging from a few metres to a few kilometres. In winter, leads are important sites for the formation of sea ice. Indeed, as the ocean surface water is exposed to the cold atmosphere, it loses heat to the atmosphere and new sea ice can rapidly form. In summer, leads are also important for the ocean radiative balance as the exposed open water can absorb a large fraction of the incident solar radiation due to its low albedo, causing enhanced melting, as already mentioned.

Similarly, in the presence of strong convergent winds, the compressive internal stress might reach the maximum value that can be sustained by the sea ice, causing the ice to fail in compression. Such events can give rise to spectacular features, referred to as ice pressure ridges. Sea ice ridges are formed as two sections of ice are pressed together, causing the forces at the interface to increase. These forces result in local bending, fracturing the ice into blocks that are pushed both upward and downward (against gravity and buoyancy respectively) creating a structure which has a 'sail' protruding upward and a much larger 'keel' protruding down into the ocean. The overall thickness of ridged ice may be many times (~10 times) the thickness of the level ice from which it is formed, and therefore is more likely to survive summer melt. In fact, a significant fraction of the sea ice that survives the summer melt to become multi-year ice begins as ridged ice. Sea ice ridges in the Arctic are mainly formed north of the Canadian Arctic Archipelago where the dominant winds blow from the Eurasian coastline towards the Canadian coastline. This ice in turn is recirculated in the central Arctic as part of the Beaufort Gyre. This explains the large spatial extent of the perennial ice cover in the Arctic in the twentieth century before the recent rapid retreat. Due to their high thicknesses, pressure ridges are a significant hazard for ships navigating in ice infested waters.

In shallow coastal regions with depths up to about 20 m, large pressure ridges can reach the sea floor where they become anchored. These grounded ridges might then act as anchor points to stabilize and maintain a coastal sea ice cover in place. This immobile coastal sea ice is referred to as land-fast ice. Grounded ridges acting as anchor points for land-fast ice have been observed, for example, along the coast of Alaska and in the Laptev Sea. The grounded ridges at the boundary between the land-fast ice and the drifting ice are called stamukhi. Land-fast ice is also observed in some regions that are too deep for pressure ridges to reach the sea floor. In this case, the ice can stay in place due to the lateral propagation of internal ice stresses that originate where the ice is in contact with the shore. For example, it is thought that the land-fast ice cover in the Kara Sea forms primarily through static arching; these horizontal arches take footing on islands offshore between which they form. The fact that sea ice can form arches is an indication that it has some resistance to tensile stresses. Land-fast ice held in place by its contact with the shore in narrow channels, such as those found in the Canadian Arctic Archipelago, is sometimes referred to as landlocked ice.

In the MIZ, waves can have a significant impact on sea ice. Indeed, waves can cause sea ice to drift by transferring momentum as they travel into the ice. Waves can also enhance melting both through increased ocean mixing and by causing the ice to break up into smaller floes that experience increased lateral melting due to their larger ratio of lateral to horizontal surface area. In the winter, the breakup of ice due to waves also affects the ice strength and therefore the motion of sea ice. With the recently observed decrease in the Arctic summer extent of sea ice, wave-ice interactions are becoming more important as the fetch over which waves can be generated is increasing (e.g. in the Beaufort Sea). The combination of large waves and broken ice floes is one of the main hazards for ships operating in polar areas.

2.2 Sea Ice Models for Large-Scale Applications

Aerial pictures of sea ice reveal that the ice cover is composed of many ice floes with dimensions ranging between a few metres to a few kilometres. In winter, in the middle of the pack, ice floes tend to be larger and with diamond shapes, while in summer or in the MIZ, smaller and more rounded floes are usually observed. Even though one could model the evolution of the ice cover as a collection of particles (i.e. the individual ice floes), this approach is complex and computationally expensive for large-scale applications. Furthermore, providing accurate initial conditions for this type of model is still difficult as observations at such high resolution are not frequent and are not yet available at all locations.

This is why current large-scale operational sea ice prediction systems do not track individual ice floes but rather treat the ice as a continuum. To this end, these models simulate the evolution of a sea ice thickness distribution (ITD). The ITD defines the fractional area covered by ice of different thicknesses. The integral of the ITD over all non-zero thicknesses is equal to the well known quantity called sea ice concentration. Compared to the discrete ice floe approach, modelling the evolution of the ITD allows for the much simpler continuum mathematics to be used and leads to models that are more computationally efficient. The Los Alamos Community Ice CodE (CICE) model and the Louvain-la-Neuve sea Ice Model (LIM) are two commonly used continuum-based large-scale sea ice models.

It is worth noting that models that additionally describe the evolution of a floe size distribution (FSD) have been recently proposed. These are important particularly in the MIZ where wave action has an impact on the floe size and therefore on the mechanical properties of sea ice and its subsequent motion under the action of winds and ocean currents. These models should not be viewed in a Lagrangian sense as tracking the exact positions of ice floes. Rather, they describe the evolution of a distribution that provides the fraction of a certain area covered by floes with a particular horizontal size.

As explained below, there are four fundamental equations used by sea ice models to describe the behaviour of sea ice: the conservation of momentum (Section 2.3), mass (Section 2.4), energy (Section 2.5) and salt (briefly discussed in Section 2.6). All of the processes described below can be related to these equations.

More specifically, the sea ice momentum equation describes how sea ice is set in motion due to the action of various forces. How the deformations (divergence/convergence and shear) of the ice cover under the action of external loads are simulated is also discussed. Given the simulated ice drift and deformations, this chapter also introduces how the transport of sea ice and the ridging/opening processes are represented in these models. Finally, the formulations of lateral melt and vertical growth and melt (intrinsically linked with the conservation of energy and salt) in sea ice models are presented.

As is discussed in Section 2.7, the set of coupled equations for sea ice are complex and analytical solutions are not possible. This means that the evolution of prognostic variables cannot be described by known mathematical functions. It is, however, possible to approximate the evolution of these prognostic variables in a forecast by using temporal and spatial discretizations. Details are given in Section 2.7, but for the moment it is important to know that the approximate solutions are obtained at discrete times (time levels) and at specific locations on the model domain. The interval between time levels is referred to as the time step. The domain is decomposed in small areas called grid cells. The distance between these grid cells depends on the spatial resolution.

2.3 Conservation of Momentum

As the ratio between the horizontal scale (1000 km) and the vertical scale (1–10 m) is large, sea ice dynamics is considered to be a two-dimensional (2-D) problem. The sea ice drift obeys the 2-D momentum equation, given by

$$m\frac{D\mathbf{u}}{Dt} = -\hat{\mathbf{k}} \times mf_c\mathbf{u} + A\boldsymbol{\tau}_a + A\boldsymbol{\tau}_w + \nabla \cdot \boldsymbol{\sigma} - mg_e\nabla H_o, \tag{2.1}$$

where $\mathbf{u} = u\hat{\mathbf{i}} + v\hat{\mathbf{j}}$ is the horizontal sea ice velocity vector, m is the combined mass of ice and snow per unit area, $\hat{\mathbf{i}}, \hat{\mathbf{j}}, \hat{\mathbf{k}}$ are unit vectors aligned with the x, y, z axes of the coordinate system, f_c is the Coriolis parameter, $\boldsymbol{\tau}_a$ is the wind stress, $\boldsymbol{\tau}_w$ is the water stress, A is the sea ice concentration, $\boldsymbol{\sigma}$ is the internal stress tensor with $\nabla \cdot \boldsymbol{\sigma}$ referred to as the rheology or ice interaction term, g_e is the acceleration due to gravity and H_o is the sea surface height.

Equation 2.1 is Newton's second law of motion ($m\mathbf{a} = \mathbf{F}$) applied to sea ice. Note, however, that the units of the different terms are not Newtons (N) but rather N m^{-2}. The internal stresses are vertically integrated stresses such that their units are N m^{-1}. The term on the left-hand side of the equation is the inertial term. On the right-hand side, the terms respectively represent the Coriolis force due to the rotation of the Earth, the wind forcing, the friction due to the underlying water, the mechanical term for ice–ice interactions and the force due to the slope of the ocean surface. The dominant momentum balance is usually between the rheology term, the wind stress and the water stress, with the other terms usually constituting a small residual. Figure 2.1 displays some of these stresses that are applied to a certain area of sea ice.

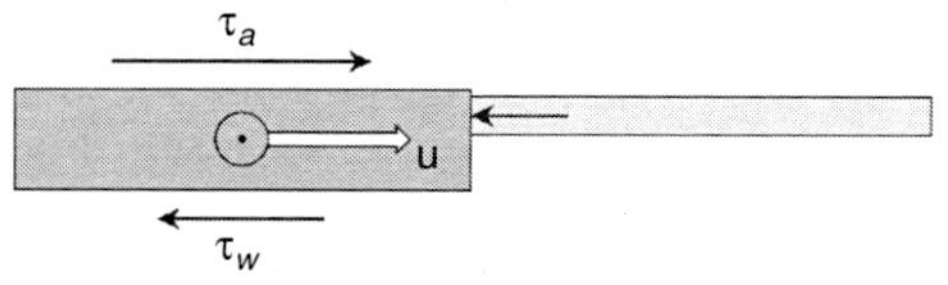

Figure 2.1: Schematic of forces per unit area (stresses) applied on sea ice (the dark grey rectangle on the left). The wind stress sets the ice in motion toward the right with velocity u (white arrow). The ice is slowed down by the water stress and by the presence of thinner ice on the right (pale grey). With the ice going toward the right, the Coriolis force, shown by the black dot inside the grey circle, is pointing out of this page (assuming the ice is in the Northern Hemisphere). The term due to the slope of the ocean surface is assumed to be zero.

Note that in most sea ice models, the advection of momentum is neglected such that the material derivative D/Dt becomes $\partial/\partial t$. This can be justified because the timescales of changes in forcing are smaller than those associated with the advection. At higher resolution, this practice may have to be revisited.

The momentum equation can in fact be viewed as two equations: one for the u component and one for v. In order to solve these equations, the terms must be defined with known quantities (such as the wind stress obtained from an atmospheric model forecast) and/or written as a function of u and v. In this case, there are two equations and two unknowns (u and v), assuming m and A are known. For simplicity, we will sometimes give the example of the 1-D momentum equation:

$$m\frac{\partial u}{\partial t} = A\tau_a + A\tau_w + \frac{\partial \sigma}{\partial x}, \tag{2.2}$$

which is obtained by setting $v = 0$ and $\partial/\partial y = 0$ in the 2-D sea ice momentum (Equation 2.1).

We discuss in more details the wind and water stresses in Section 2.3.1 below and the sea ice rheology term in Section 2.3.2.

2.3.1 Wind and Water Stresses

The wind and water stresses are usually expressed by quadratic relations. Hence, the wind stress is given by

$$\boldsymbol{\tau}_a = \rho_a C_{da}|\mathbf{u}_a - \mathbf{u}|[(\mathbf{u}_a - \mathbf{u})cos\,\theta_a + \hat{k} \times (\mathbf{u}_a - \mathbf{u})sin\,\theta_a], \tag{2.3}$$

where ρ_a is the air density, C_{da} is called the air drag coefficient, $\mathbf{u}_a$ is the surface (or close to the surface) 2-D wind and θ_a is a turning angle to represent the rotation of the wind vector within the planetary boundary layer. When using the surface wind (as done in most sea ice models), the wind turning angle is zero and the above expression is much simpler. As $\mathbf{u}$ is typically much smaller than $\mathbf{u}_a$, it is usually neglected in the formulation of the wind stress.

Similarly, a quadratic expression is used for the water stress. It is given by

$$\boldsymbol{\tau}_w = \rho_w C_{dw}|\mathbf{u}_w - \mathbf{u}|[(\mathbf{u}_w - \mathbf{u})cos\,\theta_w + \hat{k} \times (\mathbf{u}_w - \mathbf{u})sin\,\theta_w], \tag{2.4}$$

where ρ_w is the seawater density, C_{dw} is the water drag coefficient, $\mathbf{u}_w$ is the horizontal ocean current vector and θ_w is a turning angle to represent rotation of the ocean current vector in the oceanic surface boundary layer (i.e. the Ekman spiral). Similarly, the turning angle is set to zero if surface currents are used. Note that the length scale for the Ekman spiral is much smaller than that of the atmosphere and a small but non-zero water turning angle may be necessary unless the vertical resolution of the model is very high (less than 1 m).

These stress terms should represent the effect of both the skin drag and the form drag. The skin drag relates to the tangential stress over a 'flat' surface characterized by only a small-scale roughness. On the other hand, the form drag is related to the normal stress (or pressure) when a fluid flows over larger obstacles such as pressure ridges, floe edges and melt ponds edges.

In the simplest implementations, the air and water drag coefficients are constant in space and in time and only represent the effect of the skin drag. More sophisticated models define the air drag coefficient based on a small-scale roughness length and take into account the stability of the atmosphere. Some models now include the effect of both the skin drag and the form drag.

2.3.2 Rheology

When ice is drifting in the MIZ, its dynamical behaviour is mostly determined by the wind stress and the friction due to the underlying water with the Coriolis and sea surface tilt terms being of less importance. However, in zones of higher sea ice concentration, the effect of ice–ice interactions, as represented by the rheology term, is also important for the momentum balance. Basically, this term describes the internal stresses applied to the ice in a grid cell by the ice in surrounding grid cells. How to represent the rheology in sea ice models is an area of very active research. Despite recent scientific papers that have introduced new sea ice rheologies, most large-scale sea ice models are still based on Hibler's viscous–plastic rheology (Hibler, 1979). Many of the ideas that led to the development of the viscous–plastic rheology by Hibler can be traced back to the AIDJEX project in the mid-1970s.

Contrary to a fluid (such as water or air), where the deformations are linearly related to the applied stresses, sea ice is considerably more complex. Observations made during AIDJEX suggested that for low internal stresses, sea ice is rigid and does not deform significantly while, once critical stresses are reached, the ice can deform in tension (divergence), or compression (convergence) and/or shear. Based on observations that the patterns of leads are similar to fractures in other granular materials simulated with success using plasticity theory (e.g. sand, clay), Max Coon and co-authors proposed to model sea ice as a plastic material in the AIDJEX model (Coon *et al.*, 1974). A perfect plastic material does not deform until the applied stress reaches a critical value called the yield stress. Once the material has deformed, these plastic deformations are permanent. An example is the formation of pressure ridges. For a ridge to form, a critical compressive stress must be reached, after which the ice fails in compression. Once the internal stress decreases (e.g. due to a reduction in wind stress), the ridge is maintained and the deformation is permanent.

From a numerical point of view, the perfect plastic approach is unsuitable as it leads to a mathematical singularity (a division by zero). To resolve this issue, one has to describe how the ice behaves for subcritical stresses. To address this, Coon and his co-authors proposed to model sea ice as an elastic–plastic material (Coon *et al.*, 1974). For subcritical stresses, the ice deforms elastically, i.e. the deformation is reversible and there is no loss of energy in the elastic regime. Only when the stress reaches the yield criterion can the ice deform permanently. Even though the elastic–plastic model represented a major leap forward in the development of sea ice models, this approach has some limitations. As the elastic strain is reversible, one has to keep track of the history of the strain. This seriously complicates the numerical implementation of this approach. From a physical point of view, it is also not clear that an elastic–plastic approach is the most appropriate. Indeed, even though sea ice can behave as an elastic material in laboratory experiments, it is not obvious that in reality a collection of floes separated by various cracks would lead to an elastic behaviour.

This consideration and the difficulty in implementing the elastic–plastic model led Hibler to propose an alternative: the Viscous–Plastic (VP) rheology (Hibler, 1979). With the VP approach, the plastic regime is the same as in the AIDJEX model. However, for subcritical stresses, Hibler's idea is to describe sea ice as a very viscous fluid (or creep flow). Viscosity characterizes the resistance to deformation of a fluid when a stress is applied to it. As an example, honey is much more viscous than water. If one puts a teaspoon of honey beside the same volume of water on an inclined plane, the honey would flow (i.e. deform) much more slowly than the water would do. Compared to the elastic–plastic model, the approach proposed by Hibler considerably simplifies the numerical implementation.

Figure 2.2 is a representation of the VP rheology in 1-D. By convention, compressive stresses are negative, while tensile stresses are positive. The critical compressive and tensile stresses are respectively defined by σ_c and σ_t. If the ice is subjected to a compressive stress with a magnitude $\sigma > -\sigma_c$ (at point a in Figure 2.2), it exhibits a very small deformation $\dot{\varepsilon} = \partial u/\partial x$. If the wind stress increases such that σ reaches the critical stress $-\sigma_c$, the ice fails in compression and can exhibit large deformations. This case characterizes the formation of a ridge in the model. Note that this process cannot go on indefinitely. Indeed, as the ice converges, the ice thickness increases and so does the ice strength. This means that the ice (locally) can resist larger compressive stresses. A state of stress with $\sigma < -\sigma_c$ is impossible.

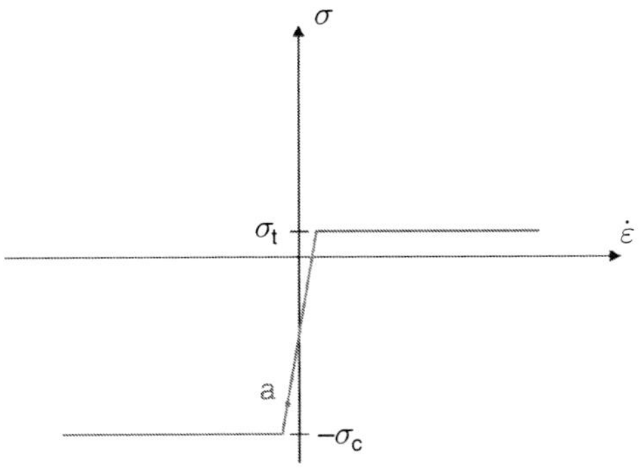

Figure 2.2: Viscous–plastic stress-strain rate relation in 1-D. The viscous regime is characterized by the linear regime between applied stress and resulting deformation – applicable for the range $-\sigma_c < \sigma < \sigma_t$. The plastic regime is applicable for $\sigma = -\sigma_c$ and $\sigma = \sigma_t$.

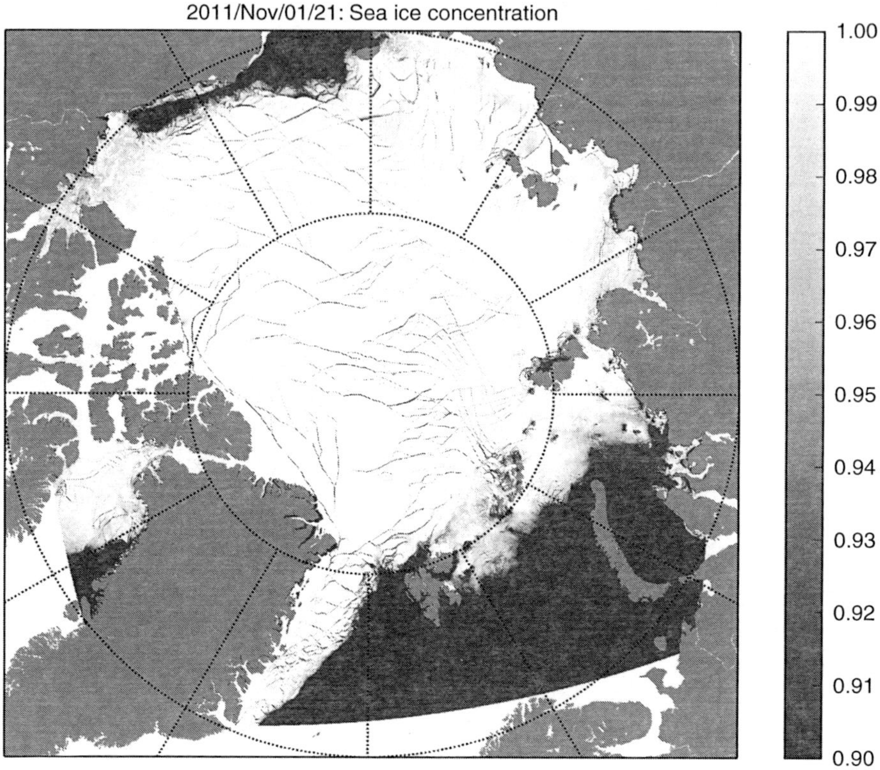

Figure 2.3: Sea ice concentration on 21 November 2011 as simulated by the MITgcm. The spatial resolution is 1 km. (A black and white version of this figure will appear in some formats. For the colour version, please refer to the plate section.)

In contrast, if the forcing creates tensile stresses within the ice, the ice in the VP model has a relatively small tensile strength σ_t compared to the strength in compression. This is justified by the fact that geophysical sea ice has cracks oriented in different directions such that it offers little resistance in tension. When the ice fails in tension, the ice velocity field is divergent. This is how a lead or a polynya can form in a 1-D VP model. Note that σ_c and σ_t are not constants; they depend on the sea ice thickness distribution (open water fraction and fraction covered by sea ice of different thicknesses) at a certain location. They are respectively $\sigma_c = P(\alpha + 1)/2$ and $\sigma_t = P(\alpha - 1)/2$, where $\alpha \sim 1.12$ is a constant which is a function of the ratio of the compressive to shear strength of the material and P is called the ice strength. We will discuss later in this chapter how the ice strength is parameterized in current large-scale sea ice models.

While the critical stresses in the 1-D rheology are defined by two points ($-\sigma_c$ and σ_t), the critical stresses that mark the transition between the viscous and the plastic regimes are defined by a curve (or yield curve) in the more realistic 2-D framework. One might argue that the yield curve could be obtained by subjecting pieces of sea ice to normal and shear stresses in a laboratory experiment. However, the yield curve for sea ice should not represent the failure modes of a single piece of solid ice (or ice floe), but rather should describe the geophysical behaviour of the sea ice cover over the spatial scale of several kilometres, which typically consists of many floes, ridges and cracks. Due to this complexity and combined with the fact that in situ ice stress measurements are difficult, the most appropriate yield curve for sea ice dynamics remains the subject of intense research. Hibler proposed, for its mathematical simplicity, to represent the yield curve as an ellipse (not shown). With this yield curve, simulated sea ice has strong resistance in compression, moderate resistance in shear and small resistance in tension. Figure 2.3 shows an example of a high-resolution simulation using a 2-D VP model (the MITgcm). The displayed sea ice concentration field shows that many leads are simulated. These leads are manifestations of failures of the ice cover. They are a consequence of a divergent ice velocity field (most of the time associated with a shear strain rate).

2.4 Conservation of Mass

Sea ice thickness ranges from a few millimetres for thin newly formed ice to several metres for pressure ridges. These variations in thickness arise from two competing processes: deformation, which creates both openings and ridges, and thermodynamic growth and melt which, over an annual cycle, tends to increase the thickness of thin ice and decrease the thickness of thick ice. In regions of deformed ice, the thickness can vary significantly over horizontal distances as small as a few metres. This is many orders of magnitude smaller than the typical horizontal spatial resolutions (1 km–20 km) of current large-scale sea ice forecasting systems. Even with the expected continued increase in computational power over the next decade, it will remain impossible to explicitly resolve such small scales in sea ice models.

Nevertheless, it is important for a sea ice model to represent processes that strongly depend on the ice thickness, such as thermodynamic growth or its mechanical strength, as previously mentioned. This can be done by defining a subgrid-scale sea ice thickness distribution (ITD) function (Thorndike *et al.*, 1975). The ITD function is defined by the fractional area covered by ice of different thicknesses (similar to a histogram or sample distribution function – see Flato (1998) for a more complete discussion of the properties and underlying assumptions). An example of a thickness distribution function based on electromagnetic sensor observations (see Chapter 3 for more detail) in the Beaufort Sea is provided in Figure 2.4.

One can define the ITD function, denoted $g(x, y, t, h)$, more precisely by letting $g(x, y, t, h)dh$ be the fraction of area in some region (centred on location x, y) with thickness between h and $h + dh$ at time t. We henceforth refer to the thickness distribution function as $g(h)$, noting that it is a function of location and time. The function $g(h)$ is normalized, like a probability density function, such that its integral (area under the curve) equals 1. Thorndike *et al.* (1975) derived an evolution equation for $g(h)$:

$$\frac{\partial g(h)}{\partial t} = -\nabla \cdot \left(g(h)\mathbf{u}\right) + \psi - \frac{\partial\left(fg(h)\right)}{\partial h} + L, \qquad (2.5)$$

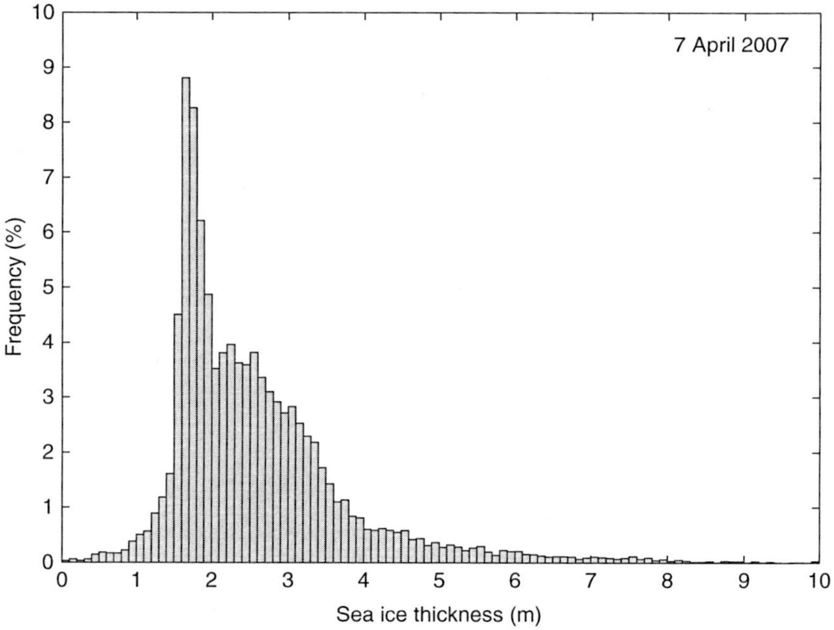

Figure 2.4: Observationally based sea ice thickness distribution histogram illustrating the frequency of occurrence of ice of different thicknesses based on airborne electromagnetic induction measurements. The data was collected along a transect in the Beaufort Sea on 7 April 2007. The spatial resolution of the sensor is 50 m and the observational error is 10 cm for level ice. Data provided by Stefan Hendricks, Alfred Wegener Institute.

that describes the change in $g(h)$ in response to advection, deformation and redistribution, vertical growth or melt and lateral melt, indicated by the four terms on the right-hand side of this equation. The advection term is simply the horizontal transport of ice by the velocity field **u** (obtained by solving the momentum equation). This will be further discussed in Section 2.4.2. The redistribution term, ψ, is more complex and represents transfers of ice from one ice thickness to another when mechanical redistribution (ridging, lead opening) is present. This will be discussed in Section 2.4.3. The third term represents the changes in $g(h)$ due to vertical growth and melt, with rate f. When f is positive, ice is growing (getting thicker) and when f is negative ice is melting (getting thinner). Since f has the same units as velocity (m s^{-1}), this term can be thought of as advection of ice in the 'thickness' dimension, with growth or melt transporting ice area from one thickness to another. The last term, L, represents the effect of lateral melt which removes ice area, typically as a function of thickness. This occurs in response to ocean heat fluxes (e.g. heat gain in open water areas within the ice cover or convected upward from the deeper ocean). The vertical growth and melt and the lateral melt are thermodynamic processes that will be further discussed in Sections 2.4.4 and 2.4.5, respectively. It should be noted that in the above formulation, the fraction of area covered by open water is naturally represented as $g(0)$.

Some common sea ice quantities, like mean thickness and concentration, are defined in terms of integrals of $g(h)$. In particular: the grid-cell total ice concentration is defined as $A = \int_{\varepsilon}^{\infty} g(h)dh$, (where ε is an arbitrarily small, non-zero number); and grid-cell mean thickness (ice volume per unit total area) is defined as $\overline{h} = \int_{0}^{\infty} hg(h)dh$.

In the context of numerical models of sea ice, Equation 2.5 above is discretized both in space (using some form of horizontal grid) and in thickness (typically by defining a number of fixed thickness categories, although Lagrangian approaches have also been applied, e.g. Lipscomb, 2001). The function $g(h)$ is discretized in N_c categories by specifying the fractional areas $a_1, a_2, \ldots\ a_{N_c}$ and thicknesses $h_1, h_2, \ldots\ h_{N_c}$ for each of the categories $(1, 2, \ldots\ N_c)$. The fractional area a_0, which corresponds to the open water, must also be specified such that $\sum_{k=0}^{N_c} a_k = 1$. Typically, the number of categories N_c in sea ice models is set between 5 and 10.

2.4.1 Two-Category Versus Multi-Thickness Category Models

Models based on the ITD, as described above, are often referred to as 'multi-category' models (e.g. CICE, LIM3) because they represent the ice cover as a discrete number of ice thickness categories. In such models, the thermodynamic calculations (growth, melt, heat conduction) are carried out separately for each thickness category. A simplification, employed in earlier models such as that of Hibler (1979), and still used in some models (e.g. MITgcm and neXtSIM), instead represents the sea ice cover in terms of two state variables, A and $\overline{h}$, defined above. In this case, only one thermodynamic calculation over ice is required, making use of the corresponding 'physical' ice thickness, $\overline{h}/A$. Such models are sometimes referred to as 'two-category' models since they effectively consider an open water, or zero-thickness, category, along with a category representing the mean thickness. Two-thickness category models are simpler and less computationally expensive, but their representation of ridging and growth/melt is rather crude. The evolution equations for A and $\overline{h}$ in a two-category model are

$$\frac{\partial A}{\partial t} = -\nabla \cdot (A\mathbf{u}) + S_A \text{ and } \frac{\partial \overline{h}}{\partial t} = -\nabla \cdot \left(\overline{h}\mathbf{u}\right) + S_h, \tag{2.6}$$

where S_A and S_h are thermodynamic source terms (Hibler, 1979), and A is limited to being equal to or smaller than 1.0 (Hibler, 1979).

Resolving multiple thickness categories allows for a better representation of the non-linear processes associated with ice thickness. As far as ice thermodynamics are concerned, there are two key non-linear effects, linked to the growth and melt of sea ice, respectively.

First, the ice growth rate rapidly decreases with increasing ice thickness because insulation increases with thickness. This creates a negative feedback between changes in growth and thickness. In practice, multi-category models exhibit more ice growth as compared with their two-category counterparts. This is explained by the fact that the mean growth rate of an explicitly resolved thickness distribution is larger than the growth of a single ice column

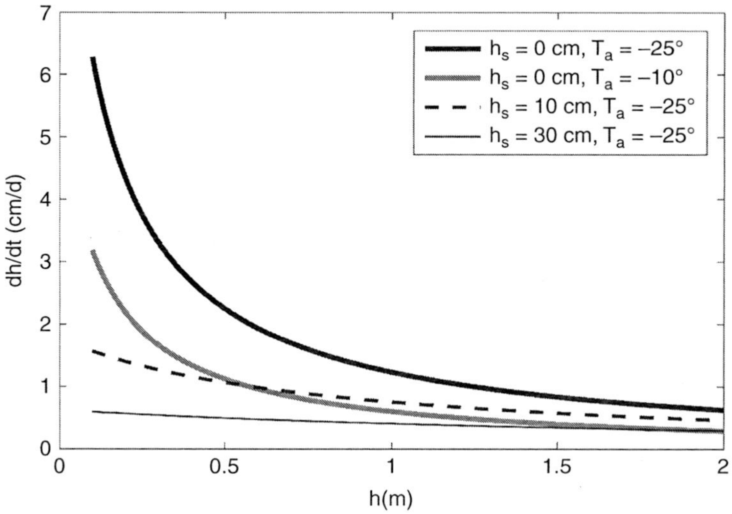

Figure 2.5: Idealized calculations showing the relationship between the sea ice growth rate (in cm per day) and the ice thickness for various snow depth (h_s) and air temperature combinations (T_a). After Maykut (1986).

with the same mean thickness. Two processes slightly reduce this effect: the presence of snow, which increases insulation, and the fact that thin ice is typically warmer than thick ice during the growth season. Figure 2.5 shows the results of simple computations illustrating the fact that sea ice and snow can strongly reduce the sea ice growth rate because they are good insulators (especially true for snow).

Second, the rate of melting is nearly independent of thickness because the conductive fluxes are small during the melt season. However, if resolved, the complete melt of thin ice rapidly increases the open water fraction in the early melt season. The extra absorbed solar radiation warms the upper ocean, enhancing the basal and lateral melting. Due to this mechanism, multi-category models tend to melt more sea ice than two-thickness category models.

Both of these non-linear processes amplify the sea ice seasonal cycle, which makes predicted sea ice conditions from multi-category models generally more consistent with observations.

2.4.2 Horizontal Transport

This section explains how the ITD evolves due to the transport of sea ice, the first term on the right in Equation 2.5. Horizontal transport, also referred to as advection, is a dynamical term that can lead to changes in the ice volume at a given location. It depends on the sea ice velocity **u**, obtained from the integration of the momentum equation, Equation 2.1. This is one way in which the continuity (mass) equation is coupled to the momentum equation.

There are two general formulations for the horizontal advection. The transport can either be formulated based on an Eulerian or a Lagrangian framework. With the Eulerian formulation, an advection scheme calculates the amount of ice that is imported or exported in a fixed grid cell during a certain time interval. With the Lagrangian approach, the nodes (or corners) of the grid cells are advected following the ice velocity field. There are also hybrid schemes that combine elements of Lagrangian and Eulerian formulations. One that has been implemented in sea ice modelling is the 'particle-in-cell' (PIC) scheme in which the momentum equation is solved on a fixed Eulerian grid, but advection is accomplished by assigning scalar quantities (like ice volume and area) to individual particles whose location is computed via Lagrangian advection. A sea ice model based on the PIC formalism was described by Flato (1993), and developed further for operational implementation by Sayed and Carrieres (1999).

Most large-scale sea ice models use Eulerian grids that are fixed in space and time. With this framework, the term $-\nabla \cdot \left(g(h)\mathbf{u}\right)$ represents the fractional area of ice of thickness h that enters in a given grid cell minus the fractional area that leaves the same grid cell (per unit of time). For multi-category models, this advection operator needs to be applied for each category. On the other hand, for two-thickness category models, $-\nabla \cdot \left(\overline{h}\mathbf{u}\right)$ and $-\nabla \cdot (A\mathbf{u})$ are calculated for the $\overline{h}$ and A equations, respectively.

Figure 2.6 shows a simple example that explains the advection operator in an Eulerian framework. A convergent ice velocity field is assumed with $v = 0$ and $h = h_1$ is considered spatially uniform before the advection and redistribution. This means that the ice is contained in one single category (i.e. $a_k = 1$). The pale grey shape on the far right shows the net volume of ice $[u(x) - u(x + \Delta x)]\Delta t \Delta y h_1$ that is advected into the grid cell during the time interval Δt (i.e. the model time step). With an initial concentration of 1 and a convergent ice velocity field, the term $-\nabla \cdot (a_k\mathbf{u})$ alone would lead to a concentration larger than 1, which is unphysical. The next section explains how this 'additional surface' is redistributed to other categories in order to maintain the concentration at a value of 1.

The numerical implementation of the horizontal transport poses a challenge because sea ice can develop sharp gradients across leads, pressure ridges and near the ice edge. It is useful to select high-order stable schemes in order not to erode large gradients in the ice fields numerically. To avoid unphysical over- and undershoot (e.g. negative ice thickness), flux-limited schemes (that introduce extra diffusion) are used. For example, CICE uses a second-order conservative remapping scheme with a limiting function to preserve monotonicity. Figure 2.7 displays the simulated ice thickness field $(\overline{h})$ using different advection schemes in a two-category model. These results demonstrate that the choice of an advection scheme can have a significant impact on the preservation of strong gradients; simple schemes such as the first-order upwind approach introduce a large amount of numerical diffusion.

The advection of snow also has to be represented by the model. In reality, snow is advected with the ice but can also drift relative to the ice depending on the wind forcing and the surface conditions (e.g. roughness, larger obstacles). Large-scale sea ice models, however, generally consider only the snow motion due to ice drift. In this case, the advection is treated in the same way as the transport of ice. This ad hoc simplification

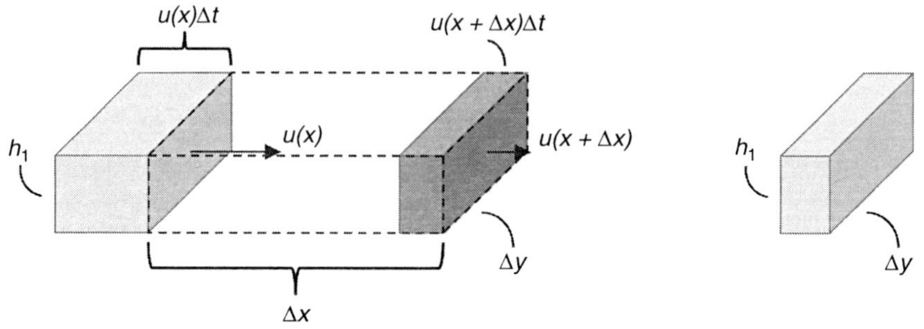

Figure 2.6: Schematic explaining the (Eulerian-upwind) advection of ice in a model grid cell. It is assumed that $v = 0$, that the u component exhibits convergence (i.e. $u(x)>u(x + \Delta x)$) and that $h = h_1$ is spatially uniform before the advection (and redistribution). The volume defined by the dashed lines (left side of the figure) shows a model grid cell of dimensions Δx, Δy covered by ice of thickness h_1. The volume (with a length $u(x)\Delta t$) of ice that comes into the grid cell during a time interval Δt is shown in pale grey (on the left) while the volume (with a length $u(x + \Delta x)\Delta t$) of ice that leaves the grid cell during the same Δt is shown in dark grey (on the right). The net volume of ice advected into the grid cell, shown in the right panel, is equal to the volume in (pale grey) minus the volume out (dark grey).

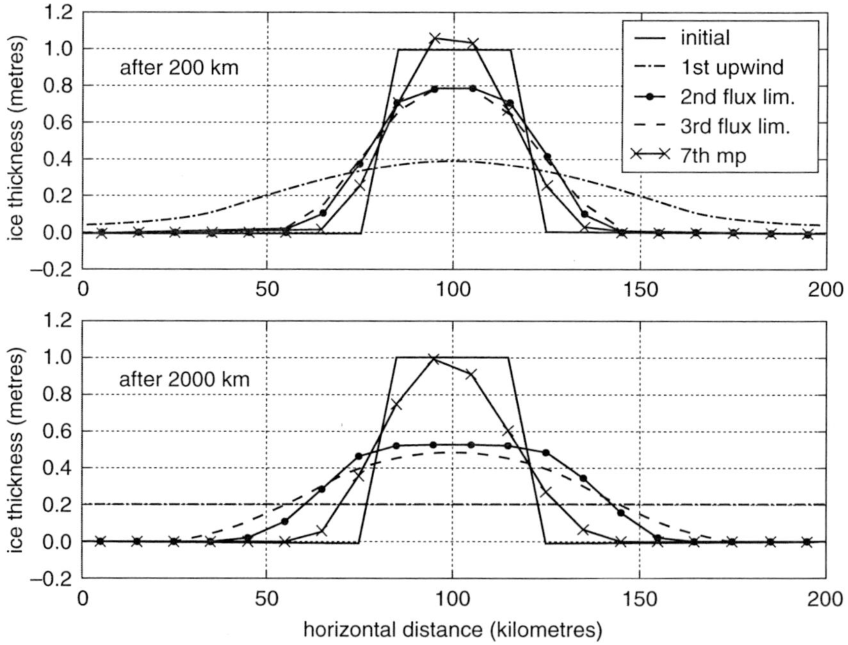

Figure 2.7: Illustration of the effects of different advection schemes in a re-entrant channel of 200 km length. The original box-like feature ('ice floe', solid black line) is 1 m thick and 50 km wide and is advected from left to right; the ice velocity is 1 m s^{-1}, the spatial resolution is 10 km and the time step is 2500 s. The two-thickness category model MITgcm was used for these idealized numerical experiments. The top panel corresponds to a full cycle, the bottom panel to ten full cycles through the domain. The first-order upwind advective scheme smooths the feature so that after 2000 km no gradients are left. With higher-order schemes (second-order central differences with Superbee flux limiter, third-order direct space time method with flux limiter, seventh-order monotonicity preserving scheme) the shape is also smoothed because of the numerical diffusion introduced in these schemes, but the numerical diffusion of the original shape is much smaller.

assumes implicitly that the large-scale distribution of snow is negligible and snow drift only has a role in the snow distribution within a particular grid cell. Blowing snow models exist but they have not yet been integrated in operational large-scale sea ice models (see Section 2.8.7 for a further discussion).

2.4.3 Rafting, Ridging and the Redistribution Function

In this section, we discuss how the redistribution function ψ (the second term on the right in Equation 2.5) is parameterized in multi-category models. This dynamical term is related to the advection term discussed in the previous section. Redistribution does not change the volume of ice at a certain grid cell, contrary to advection. The redistribution term only rearranges the ice in the different categories depending on the local ITD and the deformation fields (divergence/convergence and/or shear).

Figure 2.6 showed the transport of volume (and surface) of ice in a grid cell during a time interval Δt. With this example, the advection term alone would lead to a concentration larger than 1. It is the redistribution term that rearranges the ice in different categories to

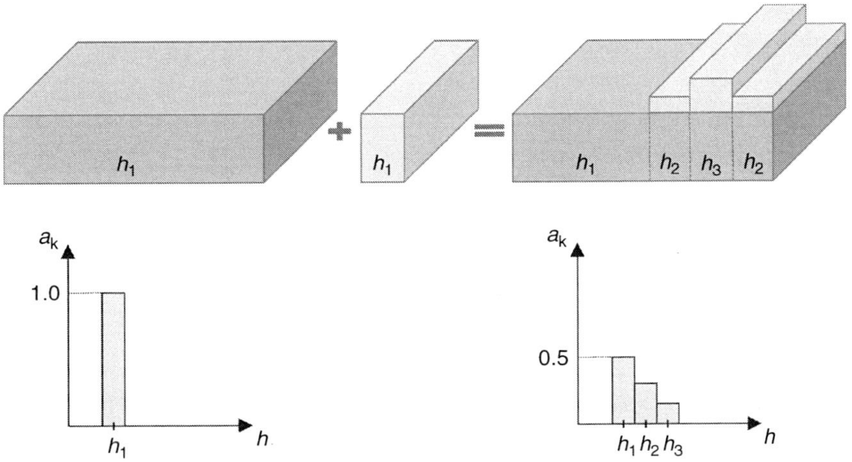

Figure 2.8: Schematic explaining ridging and the redistribution function in a simple example. It is assumed that $v = 0$, $A = 1$ and that h is spatially uniform ($h = h_1$) before the advection/redistribution. This means that, initially, all the ice is in a single category (lower left panel). The pale grey volume in the middle of the upper panel is the net volume of ice transported into this grid cell. This volume is calculated by the advection term (see Figure 2.6). Because of the convergent motion, this volume (with $h = h_1$) is redistributed to larger thickness categories by the redistribution function ψ (lower right panel). The total concentration after the redistribution is still 1.0 but the total volume (or mean thickness) has increased. Note that the upper panel is just a cartoon to help to understand the redistribution process and should not be confused with what is really considered by the model and what happens in the real world. For simplicity, a single ridge is shown on the right of the grid cell. In the model, there is no concept of the exact location of ridges in the grid cell but rather that the ITD has changed in the grid cell due to this simulated ridging event. In the real world, many different ice thicknesses are created when a ridge is formed and most of the volume of the ridge is under water (i.e. the keel).

ensure that the concentration stays bounded between 0 and 1. This process is illustrated in Figure 2.8. The pale grey shape in the middle represents the net volume of ice advected into the grid cell as calculated by the advection term (the same volume of ice shown on the right side of Figure 2.6). The function ψ then redistributes the thin ice to thicker categories while ensuring conservation of volume and that the concentration does not exceed 1.

In the case of divergence, the redistribution term provides a source of open water area $g(0)$. Pure shear deformation is area-preserving and so does not formally imply either convergence or divergence. However, in an ice pack comprised of a mixture of ice floes of various thicknesses, shear deformation generally involves simultaneous local convergence and divergence, and hence simultaneous ridging and opening.

Parameterizations of redistribution determine how much ice is ridged and the thickness of the ridged ice. This process remains a source of uncertainty in sea ice models since the process itself is difficult to observe directly, and many factors are at play. For example, in a given region (e.g. a model grid cell) there will be a wide range of ice thicknesses present and the variations in the two-dimensional strength that arises from this thickness variability will affect the small-scale deformation field and hence the ridging behaviour. It is generally assumed that thinner ice is more likely to participate in ridging than thicker ice, and this is built into most parameterizations. Observations (e.g. Amundrud *et al.*, 2004) indicate some relationship between the thickness of the ice participating in the ridging process and the resulting ridge thickness, and so this too is generally represented in the parameterizations. However, the detailed formulation of such parameterizations is beyond the scope of this chapter and can be found, for example, in Lipscomb *et al.* (2007).

Finally, there is a connection between ridging and the two-dimensional strength P that enters the rheology (see Section 2.3.2). The mechanical work needed to create ridges is related to the ability of the ice pack to resist convergent motion, or more commonly referred to as the sea ice compressive strength. Early parameterizations of large-scale ice strength were based on equating the work done by deformation to the change in gravitational potential energy associated with building a ridge (Rothrock, 1975). Empirical evidence based on comparing modelled ice drift to observations (Flato and Hibler, 1995), and theoretical evidence based on detailed simulations of individual ridges (Hopkins, 1994), indicate that although scaling with potential energy seems to hold, the actual energy used in ridge building is perhaps a factor of 10 or more larger due to friction and other factors.

2.4.4 Thermodynamic Source Terms: Vertical Processes

In this section we introduce the thermodynamic source term associated with vertical growth and melt ($-\partial(fg(h))/\partial h$, the third term on the right in Equation 2.5). The representation in sea ice models of vertical sea ice thermodynamic processes has increased in complexity over time. Yet, the fundamental features of this representation were established in the 1970s by the AIDJEX group (see for instance, Maykut and Untersteiner, 1971; Coon *et al.*, 1974; Thorndike *et al.*, 1975; Hibler, 1979).

The vertical thermodynamic processes include freezing/melting at the base, surface melting, formation of new sea ice over open water and snow-ice formation. The change in thickness over a time interval Δt due to vertical processes must be calculated for all the thickness categories. Hence, the total change in thickness in a certain category is given by $\Delta h_T = \Delta h_{su} + \Delta h_b$ where Δh_{su} is the change at the surface due to surface melting and Δh_b is the change at the base due to freezing or melting. Figure 2.9 is a cartoon illustrating the process of basal melt and how it affects the ITD.

Once the change in thickness Δh_T due to all vertical growth and melt processes is computed, the growth rate f in Equation 2.5 is retrieved by dividing Δh_T by the time interval Δt. As described below, the change in mass (i.e. Δh_T) can be calculated by considering the equations for the conservation of energy (Section 2.5) and the conservation of salt (Section 2.6).

The formation of new ice over open water receives a particular treatment. Models generally follow the approach of Hibler (1979). With this approach, once the mixed-layer reaches the freezing point temperature of seawater, any additional oceanic heat loss is

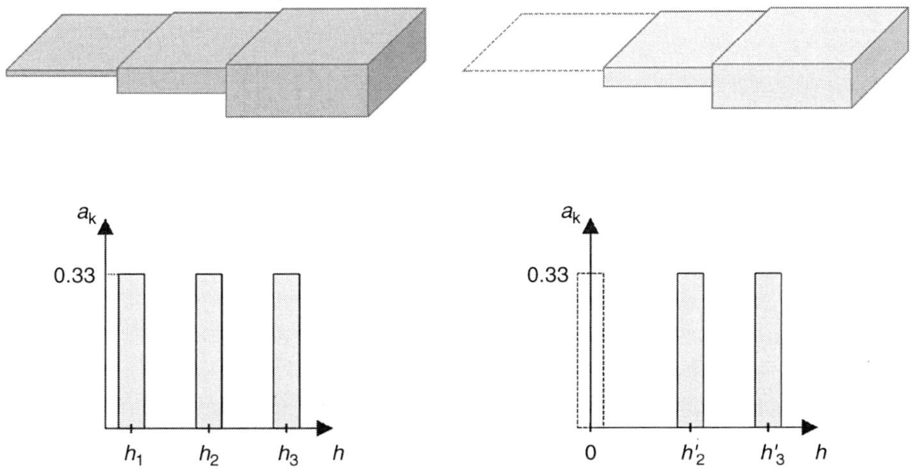

Figure 2.9: Schematic illustrating how sea ice melting reduces ice concentration in multi-category models, assuming basal melting only. The upper panels represent what happens in reality; the lower panels show how this is represented in the model in terms of ice categories. Before melting (on the left), the ice (in dark grey) in the grid cell is characterized by thin ice (h_1), ice of intermediate thickness (h_2) and thick ice (h_3). The three partial concentrations are 1/3 such that the total concentration is 1. The upper right panel shows the ice (in light grey) after a time interval Δt. It is assumed that all of the thin ice melts during this time interval. The surface that melted is represented by the dashed line. Note that the volume of ice that melted is roughly the same for all three thicknesses; as opposed to ice growth, ice melt is less dependent on ice thickness. The total concentration after the melt is reduced to 2/3 as all the ice in the thinnest category has melted. For simplicity, it is assumed that the thickest categories were not remapped to other categories but slightly shifted toward lower thicknesses with $h'_2 < h_2$ and $h'_3 < h_3$. The fractional area covered by open water is represented by the dashed category at $h = 0$ m in the lower right panel.

converted into a volume of sea ice v_0. The thickness (h_0) of the newly formed ice is the result of thermodynamic (growth) and dynamic (e.g. winds, waves) processes. These processes are too fast to be resolved in current sea ice models, and in any case they are not yet properly understood. For this reason, h_0 is simply a specified constant and the fractional area covered by the new ice is calculated from v_0. In multi-category models, the growth rate of new ice is assigned to $f(h_0)$. In two-thickness category models, new ice and its properties are aggregated onto pre-existing ice, if any.

2.4.5 Thermodynamic Source Terms: Lateral Melt

Because the horizontal distances spanned by sea ice are much larger than its thickness, ice growth and melt were initially assumed purely vertical. Research since then has mainly focused on the vertical processes and lateral processes have been generally overlooked. Lateral melting (L, the fourth term on the right in Equation 2.5) was not considered in early studies (e.g. Maykut and Untersteiner, 1971). The process modelling study of lateral melting by Steele (1992), proposed that lateral melting depends on floe size and becomes non-negligible only for ice floes smaller than ~100 m. As the floe size distribution (FSD) is still not well understood and just starting to be included in sea ice models, lateral melting is usually parameterized, typically following Bitz *et al.* (2001). In this formulation, the lateral melt rate is an empirical function of temperature difference between the ice and the water, and the effective decrease in ice concentration is inversely proportional to the prescribed mean floe size (Steele 1992).

Two-thickness category models include a melting term designed to decrease sea ice concentration when sea ice melts. Yet, rather than representing the proper lateral melt, this term parameterizes the melting of unresolved thin ice. In multi-category models, the melting of thin ice is explicitly resolved such that this extra term should not be included.

We expect further progress on the parameterization of lateral melt when FSD models, currently being developed (Horvat and Tziperman 2015; Zhang *et al.*, 2015), will be implemented in large-scale sea ice models. This could be important in the increasingly large MIZ where floe size is ~100 m and wave action is important (see Section 2.8.6).

2.5 Conservation of Energy

Because the growth (melt) of ice releases (takes) energy, the conservation of mass is tightly coupled to the conservation of energy through the thermodynamic source terms (see Sections 2.4.4 and 2.4.5). Ice growth and melt processes are typically formulated by converting any heat imbalance ΔF, for instance at the ice–atmosphere or ice–ocean interfaces into a mass change following:

$$\rho \frac{dh}{dt} = -\frac{\Delta F}{\Delta E}, \tag{2.7}$$

where ρ is the sea ice density, and $\Delta E = E_2 - E_1$ is the specific enthalpy change (J kg^{-1}) corresponding to the phase transition from a thermodynamic state 1 to state 2. Ice enthalpy E is generally defined as minus the energy required to melt a unit mass of sea ice and to bring it to a temperature of 0°C (see Bitz and Lipscomb, 1999 and Schmidt *et al.*, 2004 for details), so that ΔE is positive when ice melts. With time discretization, $dh/dt \sim \Delta h/\Delta t$ where Δh is either Δh_{su} or Δh_b as introduced in the previous section.

In this section, for simplicity, we focus on the thermodynamic representation of sea ice and only briefly discuss snow. Usually, the treatment of snow is simple (assuming constant thermal properties and density) and involves similar equations. The modelling of snow is further discussed in Section 2.8.7.

2.5.1 Thermodynamic Formulation and Enthalpy Definition

A first prerequisite for computing ice growth and melt is the formulation of ice enthalpy, which follows from the thermodynamic assumptions of the model. There are three classical thermodynamic formulations used in present-day sea ice models. All assume a constant ice density.

The simplest one is the so-called Semtner (1976) 0-layer formulation (SM0L), which neglects sensible heat storage and the effect of brine inclusions. There is no internal temperature computation, and the ice enthalpy is simply minus the latent heat of freezing $L_0 = 3.35 \times 10^5$ J kg^{-1}. This formulation has been widely used and is still used in some sea ice models nowadays.

The Semtner (1976) 3-layer formulation (SM3L) includes sensible heat storage, adding a temperature dependence to the ice enthalpy $E = E(T)$, but ignores brine inclusions. As with the SM0L formulation, this approach has been widely used.

The third one is the Bitz and Lipscomb (1999) formulation (BL99) that includes sensible heat storage and the thermal effect of brine inclusions, adding a salinity dependence: $E = E(S, T)$. In the standard implementation of BL99, the bulk salinity S can vary in the vertical but does not have a time dependence.

New approaches (see Section 2.8.8) accounting for a time-dependent salinity profile have been developed but have not yet been implemented in operational sea ice forecasting systems.

The BL99 formulation is at the base of many current sea ice models. In the BL99 formulation, sea ice is assumed to be made of gas-free pure ice and saline brine. This is justified physically since salt ions are not embedded in the ice crystalline lattice other than in trace amounts. Hence, the bulk ice salinity S is directly related to the brine fraction ϕ and the brine salinity, S_{br}:

$$S = \phi S_{br}. \tag{2.8}$$

At timescales larger than a few minutes, thermal equilibrium is maintained between brine inclusions and the solid freshwater ice, which is the second assumption of this formulation.

Hence, the volume and salinity of the brine inclusions adjust their size in order to maintain the temperature of the brine at its freezing temperature. The third assumption is that the relation between the freezing point temperature and salinity is linear. These last two assumptions imply that brine salinity and the local temperature T are related through:

$$T = -\mu S_{br}, \tag{2.9}$$

where μ is a constant equal to 0.054 °C (g/kg)$^{-1}$.

Combining Equation 2.8 and Equation 2.9 results in the following relationship for the brine fraction: $\phi = -\mu S/T$. Hence, from the three model assumptions, one gets that the sea ice phase composition (namely ϕ) can be specified from S and T, which therefore give a complete thermodynamic description of the system. The S and T dependence of the brine fraction is depicted in Figure 2.10. With these assumptions, the ice specific enthalpy is a function of S and T:

$$E(S,T) = c_{pi}(T + \mu S) - L_0\left(1 + \frac{\mu S}{T}\right) - c_{pw}\mu S, \tag{2.10}$$

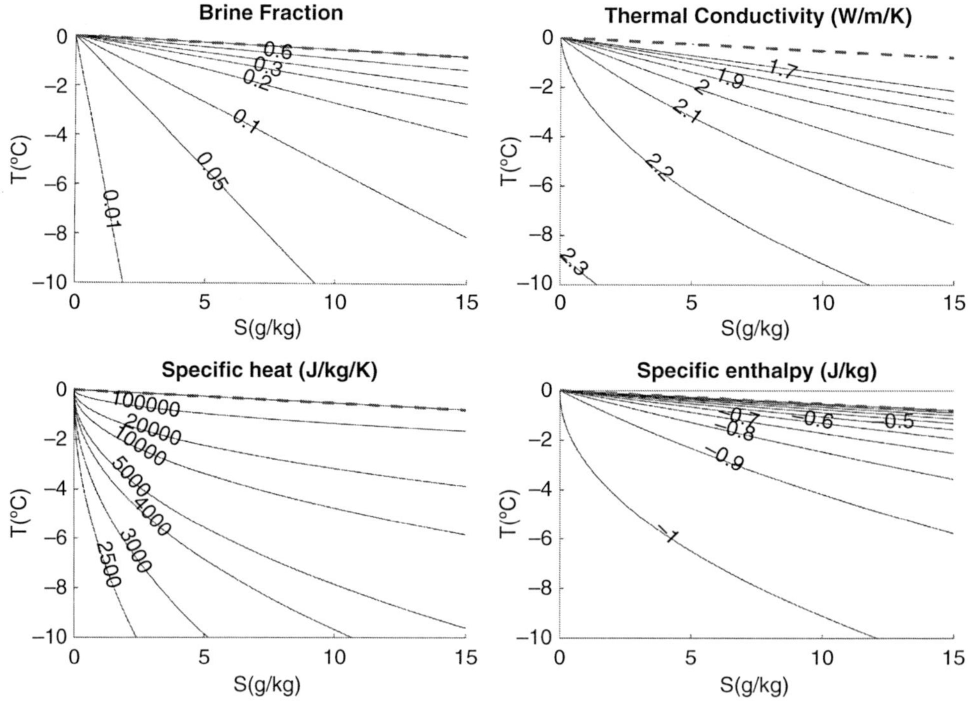

Figure 2.10: Dependence of sea ice thermal properties on bulk salinity and temperature as represented in the BL99 approach for sea ice thermodynamics: brine fraction ($\phi = -\mu S/T$), thermal conductivity k (Equation 2.14), specific heat c_p (Equation 2.13) and specific enthalpy divided by the pure ice latent heat (Equation 2.10).

where c_{pi} and c_{pw} are the pure ice and seawater specific heats (J (kg K)$^{-1}$). Assuming zero salinity, one retrieves the SM3L specific enthalpy. Assuming zero salinity and $c_{pi} = c_{pw} = 0$, one retrieves the SM0L formulation.

In order to conserve energy, all thermodynamic formulations must be consistent with the expression of the specific enthalpy (see Schmidt *et al.*, 2004).

2.5.2 Calculation of Ice Growth and Melt

We now describe how vertical ice growth and melt are computed from the corresponding energy inputs (see Figure 2.11 for a sketch of the heat fluxes), regardless of the enthalpy formulation.

At the ice surface in cold conditions, the heat conduction flux matches the net atmospheric heat loss. Hence, the sum of the surface internal heat conduction flux (F_c^{su}, positive upwards with the superscript *su* referring to the 'surface') and of the net atmospheric heat flux $F_{net}^{atm} = (1-\alpha)F_{SW} + \left(F_{LW}^{d} - F_{LW}^{u}\right) + F_s + F_l$, positive downwards) is zero. This condition is used to determine the surface temperature:

$$F_{net}^{atm}(T^{su}) + F_c^{su}(T^{su}) = 0. \tag{2.11}$$

If the net atmospheric surface heat flux is sufficient to bring T^{su} above 0°C (typically during the warm season), T^{su} is capped at 0°C and the excess of heat $\Delta F^{su} = F_{net}^{atm} + F_c^{su}$ is used to melt the ice. The thickness change Δh_{su} and the corresponding surface melt rate can be obtained from Equation 2.7 by dividing ΔF^{su} by the change in enthalpy $\Delta E(T, S)$.

The net solar radiation flux $(1-\alpha)F_{SW}$ is of major influence to the surface energy balance and the surface melt rate, so that even small changes can significantly affect the sea ice mass balance. A critical parameter that determines the net solar radiation balance is the surface

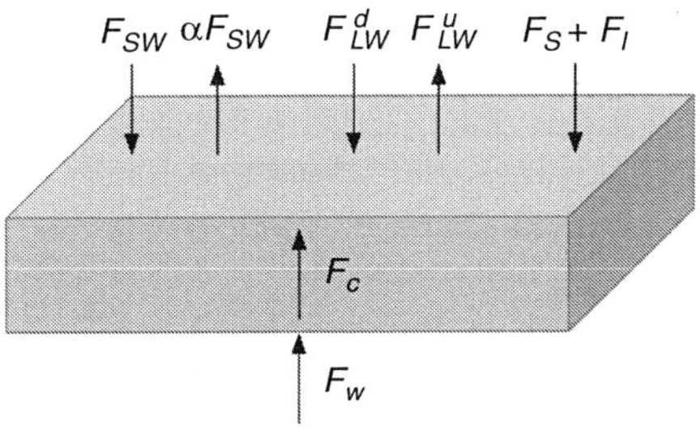

Figure 2.11: Schematic showing the different heat fluxes (W m^{-2}) used for vertical growth and melt computations. F_{SW} is the incoming shortwave radiation, α is the ice albedo, F_{LW}^{d} is the downwelling (*d*) longwave radiation (infrared) from the atmosphere, F_{LW}^{u} is the upwelling (u) longwave radiation emitted by the ice surface, F_s is the sensible heat flux, F_l is the latent heat flux, F_c is the vertical conductive heat flux through the ice cover and F_w is the ocean heat flux. Note that F_c is a function of the local temperature gradient and thermal conductivity – and therefore is a function of depth. For simplicity, we show one single conductive heat flux in the ice and assume the surface is snow-free.

albedo α. The simplest albedo formulation is to use prescribed constant values for ice and snow. A second approach is to specify the dependence of the albedo on ice thickness, snow depth, surface temperature and melt pond fraction. Typically, the albedo takes a single, broadband value. Multi-band schemes are, however, now available. The most advanced schemes explicitly resolve the radiative transfer equation, e.g. using the Delta-Eddington approach, and diagnose the sea ice optical properties, including the surface albedo. As snow and melt ponds strongly impact the surface albedo, recent developments in modelling the evolution of snow (Section 2.8.7) and melt ponds (see Section 2.8.9) will certainly improve the representation of the surface albedo in operational sea ice prediction systems.

At the ice base, it is assumed that the temperature is at the underlying seawater freezing point. The energy fluxes involved are the turbulent ice–ocean heat flux F_w and the internal conduction of heat F_c^b (with the superscript b referring to the 'base'). If the oceanic heat flux is smaller than the conductive heat flux, new ice forms at the ice base. Conversely, when the turbulent ice–ocean heat flux is larger than the conductive heat flux, the excess energy is used to melt ice. In other words, any negative (positive) heat imbalance $\Delta F^b = F_w - F_c^b$ between the heat conduction and the ocean heat flux (F_w is positive upward) is converted into ice growth (melt) using Equation 2.7 (i.e. Δh_b is calculated).

2.5.3 Diffusion of Heat and Thermal Properties

The vertical growth and melt rate depend on internal conduction fluxes, which themselves depend on the vertical temperature profile. To obtain the latter, a 1-D heat diffusion equation should ideally be solved (conservation of energy), that is

$$\frac{\partial}{\partial t}(\rho E) = \frac{\partial}{\partial z}\left(k\frac{\partial T}{\partial z}\right) + Q, \tag{2.12}$$

where z is the vertical coordinate, $T(z,t)$ is the internal ice temperature that needs to be solved, k is the bulk ice thermal conductivity (see Figure 2.10) and Q is the rate of warming associated with light absorption and scattering in the ice.

Equation 2.12 equates the change in ice internal energy (the left-hand side) to the convergence of heat fluxes associated with heat conduction and radiative transfer through sea ice (right-hand side).

BL99 approach: If brine inclusions are considered, the internal energy is that of saline ice, and the change of internal energy can be written as $\rho c_p(S,T)\partial T/\partial t$. The sea ice *effective* specific heat

$$c_p = c_{pi} + \frac{L_0 \mu S}{T^2} \tag{2.13}$$

includes the contributions of the warming of pure ice and brine and internal melting around brine inclusions. The second term in the effective specific heat drastically increases near the freezing point as the brine volume increases in order to maintain internal phase equilibrium (see Figure 2.10).

The most consistent framework is to formulate the bulk sea ice thermal conductivity as a weighted mean of pure ice and brine contributions (Hunke *et al.*, 2011):

$$k(S,T) = [1 - \phi(S,T)]k_i + \phi(S,T)k_{br}, \tag{2.14}$$

where k_i and k_{br} are the pure ice and brine thermal conductivities, respectively (see Figure 2.10). As shown in Figure 2.10, the dependence of k on S and T is rather weak, though, as the conductivities of pure ice and the one for brine have similar values. With such expressions for c_p and k, the heat diffusion equation is non-linear and requires iterative procedures to be solved (see Dupont *et al.*, 2015).

SM3L and SM0L approaches: Assuming $S = 0$ in the equations above, one retrieves the SM3L approach with a classical, linear heat diffusion equation. In the SM0L case, $\partial E/\partial t = 0$ and there is no heat equation to solve. In this case, the conduction flux is equal for the lower and upper ice interfaces, and takes the following form: $k\Delta T/h$ where k is the effective thermal conductivity and ΔT is the surface-bottom temperature difference. The effective thermal conductivity may account for the impact of snow (in the 0L model), or of the unresolved ITD (in the 3L model). Neglecting brine in the SM0L approach leads to a shift in the phase (a few weeks) and a reduction in the amplitude (10–20 per cent) of the ice thickness seasonal cycle (Semtner, 1984). In the SM3L, brine inclusions are often parameterized using a latent heat reservoir storing the penetrating solar flux.

2.6 Conservation of Salt

Observations indicate large spatio-temporal variations in ice salinity. As the ice–ocean salt exchanges and the sea ice thermal properties (see Figure 2.10) directly respond to changes in ice salinity, sea ice models should include an equation describing the conservation of salt.

Until recently, there were no available formulations for brine dynamics, which is why the sea ice models used in sea ice forecasting systems represent the conservation of salt in very simple terms only. As far as the computation of sea ice thermal properties is concerned, most forecasting systems follow the BL99 approach, prescribing a vertical salinity profile $S(z)$ following observations from Central Arctic multi-year ice. Such ice has a relatively low ice salinity, particularly near the surface, which significantly differs from first-year sea ice, which is more saline, particularly near the surface. In Arctic sea ice simulations, such an approach leads to underestimate sea ice growth and overestimate melting, so that the resulting simulated sea ice volume is significantly underestimated (Vancoppenolle *et al.*, 2009b).

In addition, using a similar $S(z)$ formulation in the computation of the ice–ocean exchanges of salt would break the global conservation of salt. This is because the ice would form (at the base) with a relatively higher salinity than the salinity at which it would melt (at the surface). Hence the salinity difference would not be traceable and induce a loss of salt for the global ocean. This is why a vertically-constant salinity must be imposed in the computation of ice–ocean exchanges if temporal changes in ice salinity are not represented.

With the recent development of brine dynamic schemes, a full account for the conservation of salt has been introduced in several sea ice models. These developments, discussed in Section 2.8.8 are yet to be implemented in operational sea ice forecasting systems, which would improve their physical realism.

2.7 Numerical Implementations

In geophysical models, the set of coupled physical equations is usually highly non-linear. This prevents the derivation of analytical solutions. In other words, it is not possible to express the evolution of a certain prognostic variable with known mathematical functions. This is also the case for sea ice as modelled with the equations given above.

It is nevertheless possible to approximate the evolution of these prognostic variables in a model by discretizing the governing equations. With discretization, an approximate solution is obtained at discrete times and at specific positions on the domain. There are many different techniques to spatially and temporally discretize coupled equations such as those for sea ice. Here, we only give a brief introduction on discretization and on numerical solution techniques. More sophisticated methods are presented at the end of this chapter.

Consider the 1-D sea ice momentum equation (Equation 2.2) introduced earlier. With a temporal discretization, we wish to solve the momentum equation at specific times referred to as time levels. These time levels are $\Delta t, 2\Delta t, 3\Delta t, \ldots$ where Δt is called the time step. The solutions that we wish to obtain are the velocity fields at each time level. These are concisely written as $u^1, u^2, u^3, \ldots$ with the superscript $n = 1, 2, 3, \ldots$ referring to the time levels. The number of time levels performed depends on Δt and on the length of the forecast. The field u^0 is referred to as the sea ice velocity initial condition. It could be obtained from a previous forecast or from an estimate obtained through the use of data assimilation.

Similar to the time discretization, the spatial discretization means that the equations are solved at specific locations on the domain. We consider a simple 1-D uniform Cartesian grid (shown in Figure 2.12). On this staggered grid, the velocities u are defined at the nodes while the tracers (i.e. A, m, σ, etc.) are positioned between the u-points at the T-points. The subscript $i = 1, 2, 3, \ldots, N+1$ refers to the position on the grid and we define the vector $\mathbf{u}^n = \{u_1^n, u_2^n \ldots u_{N+1}^n\}$, that is a vector that contains all the velocity components at time level n. This vector of size $N+1$ arises from the spatial discretization and should not be confused with the horizontal velocity vector $\mathbf{u}$ of Equation 2.1. The spatial resolution is defined by Δx, the distance between u_i and u_{i+1} that is assumed here to be spatially constant.

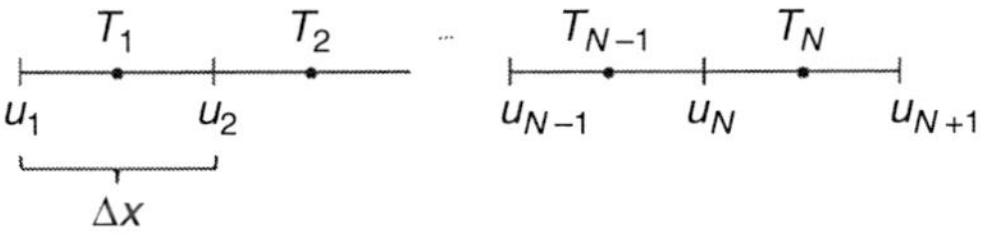

Figure 2.12: Schematic of a staggered 1-D grid with grid cells of size Δx.

The momentum equation involves partial derivatives in time (the acceleration term) and in space (the rheology term). Using a method called finite difference, $\partial u/\partial t$ at time level n and at position i on the domain can be approximated by $\left(u_i^n - u_i^{n-1}\right)/\Delta t$. This is a first-order approximation; that is, the error is $O(\Delta t)$ and increases linearly with the size of Δt. A similar approach can be used for the spatial derivatives in the rheology term. For the moment, we simplify the notation for all of the terms on the right-hand side (RHS) and write these as RHS $(\mathbf{u}^*)$, where $\mathbf{u}^*$ is $\mathbf{u}^{n-1}$ or $\mathbf{u}^n$ as explained below. Due to the spatial derivatives in the rheology term, the RHS is a function of the vector $\mathbf{u}^*$, i.e. not only of u_i^* (in fact it depends on u_{i-1}^*, u_i^* and u_{i+1}^*). As the previous time-level solution $\mathbf{u}^{n-1}$ is known, the most straightforward approach is to write

$$\frac{m\left(u_i^n - u_i^{n-1}\right)}{\Delta t} = RHS\left(\mathbf{u}^{n-1}\right), \tag{2.15}$$

from which the solution can easily be obtained as $u_i^n = u_i^{n-1} + \Delta t m^{-1} RHS\left(\mathbf{u}^{n-1}\right)$. This process is repeated at all of the $N+1$ locations on the grid in order to obtain the complete $\mathbf{u}^n$ vector (with the use of boundary conditions at i=1 and i=N+1).

This numerical approach for solving the equation is referred to as an explicit method: the approximate solution at time level n is calculated directly from the solution at $n-1$. Such explicit methods are easy to implement, but a severe drawback is that they can exhibit numerical instabilities. A numerical scheme is unstable when round-off errors grow during the integration such that the approximate solution can become highly unrealistic (e.g. sea ice velocities that are larger than the speed of light!). Such instabilities could occur with the VP rheology due to the very large viscous coefficients in zones of small deformations. To maintain stability, a very small time step (less than 1 s) must be used. This, however, is prohibitive in terms of the computational resources required to produce forecasts.

It is possible to alleviate this constraint on the time step by adding an artificial elastic term to the VP stress formulation. The artificial elastic waves generated by such an approach are damped during the time integration such that the solution approaches the VP solution. This approach, called the Elastic–VP (EVP, Hunke, 2001) method is computationally efficient and can be implemented in a way that scales very well with the number of computer processors used. It is now used in many sea ice models such as CICE and LIM. Some authors, however, pointed out that the artificial elastic waves remain and that the EVP solution does not always converge towards the VP solution (e.g. Konig and Holland, 2010). Note that a modified version of the EVP has been developed and that it converges to the VP solution provided stability requirements are satisfied and a sufficiently large number of iterations are performed (Kimmritz *et al.*, 2015).

For computational efficiency, i.e. to be able to take large time steps, another approach is to solve the momentum equation implicitly. For this, the momentum equation is written as

$$\frac{m\left(u_i^n - u_i^{n-1}\right)}{\Delta t} = RHS(\mathbf{u}^n), \tag{2.16}$$

where the RHS is a function of $\mathbf{u}^n$ (an unknown), instead of depending on the velocity field at the previous time level.

Equation 2.16 must be solved at all the u locations on the grid. This can be concisely written as $\mathbf{A}(\mathbf{u}^n)\mathbf{u}^n = \mathbf{b}$, where $\mathbf{A}(\mathbf{u}^n)$ is a $(N+1) \times (N+1)$ matrix, $\mathbf{b}$ is a $(N+1)$ vector (that contains terms such as the wind stress) and $\mathbf{u}^n$ is the desired solution. Notice that $\mathbf{A}$ is a function of $\mathbf{u}^n$, that is the problem is non-linear. The size of the problem, i.e. $(N+1) \times (N+1)$, increases as the spatial resolution is increased. In 2-D, with the velocity vector including all the u and v components of the grid, the size of problem increases by a factor of four when doubling the spatial resolution.

This type of non-linear system of equations arises in many fields of engineering and science. Finding the solution of such a non-linear system of equations is therefore an area of very active research in applied mathematics. For the sea ice momentum equation, Picard solvers (e.g. Zhang and Hibler, 1997) have first been implemented. To improve the convergence rate of the non-linear solution (i.e. to decrease the computational time), more efficient Newton methods have recently been developed (e.g. Lemieux *et al.*, 2010).

The fact that spatial and temporal discretizations are used, in order to obtain a numerical solution, has an impact on the processes that can be represented. Indeed, processes with a spatial scale smaller than the spatial resolution Δx are not captured by the approximate numerical solution. Unresolved spatial processes are often referred to as subgrid-scale processes. Similarly, the time step Δt limits the highest frequency of physical processes that can be resolved.

In order to resolve smaller-scale processes (e.g. sea ice deformations) and improve forecasts, one has to decrease Δx. However, this increases the size of the problem and therefore the number of mathematical operations required. With current operational sea ice models, the computational requirements, including memory, are too large to solve using a single processor computer. It is therefore crucial to develop parallel code, that is, to distribute the computational effort over many processors. To minimize the use of computational resources, a solver for finding the solution of $\mathbf{A}(\mathbf{u}^n)\mathbf{u}^n = \mathbf{b}$ should scale efficiently with the number of processors. An efficient scaling or linear scaling means that the computational time remains roughly the same when both the size of the problem and the number of processors is doubled. While parallelization for massively parallel computers is straightforward for explicit EVP schemes, implicit schemes require special care. Losch *et al.* (2014) demonstrated linear scaling for more than 1000 processors with a Newton (implicit) solver for the sea ice momentum equation.

2.8 Recent Developments and Future Directions

We list here several recent developments that we consider important. For some we also discuss what we envision as being future directions of research in the development of sea ice models.

2.8.1 New Sea Ice Rheologies

In recent years, some authors have criticized the VP rheological framework arguing that it does not take into account anisotropy, that is the orientation of subgrid-scale cracks (Tsamados *et al.*, 2013) and that it fails to represent the intermittency and heterogeneity of observed sea ice deformations (Girard *et al.*, 2009). This motivated some researchers to develop new rheological framework for sea ice dynamics (e.g. Girard *et al.*, 2011; Tsamados *et al.*, 2013).

An example of a new rheology is the Elasto-Brittle (EB) approach (Girard *et al.*, 2011). This rheology combines linear elasticity, a Mohr–Coulomb criterion for brittle failure under shear and compressive stress states, and a model for progressive damage (Girard *et al.*, 2011). Compared to classical models that calculate the ice strength based only on the ITD, models based on the EB rheology also contain a variable that describes the level of subgrid-scale damage. Bouillon and Rampal (2015) used the EB rheology as the dynamical core for a new sea ice model called neXtSIM and showed its ability to represent the multi-fractal spatial scaling of sea ice deformations. Sea ice deformations simulated with this model spontaneously localize along linear-like dynamical features (cracks and pressure ridges) separating essentially undamaged ice plates/floes. As observed from satellite imagery, these simulated deformations are transient features; they are active for a while before ceasing their activity depending on the refreezing kinetics, the evolution of the wind forcing and the internal sea ice dynamics. Rampal *et al.* (2016) compared sea ice trajectories simulated by the neXtSIM model to the ones of the IABP buoy dataset and found that neXtSIM accurately reproduces the observed mean and fluctuating parts of sea ice motion. This model has recently been implemented in the operational sea ice forecast platform neXtSIM-F for the Barents and Kara Seas (see example in Figure 2.13).

2.8.2 Spatial Discretization with Unstructured Grids

Alternative discretization methods include finite element or finite volume methods on unstructured grids (FESOM,[2] neXtSIM). These unstructured grids with adaptive grid cell sizes are attractive, because they allow an accurate geometrical representation of coast lines and topography in general. The unstructured mesh is usually made up of triangles and/or quadrangles of varying size. The irregularity of the grid requires different discretization techniques that are usually more computationally expensive than methods for regular grids,

[2] Finite element sea ice ocean circulation model from the Alfred Wegener Institute for Polar and Marine Research.

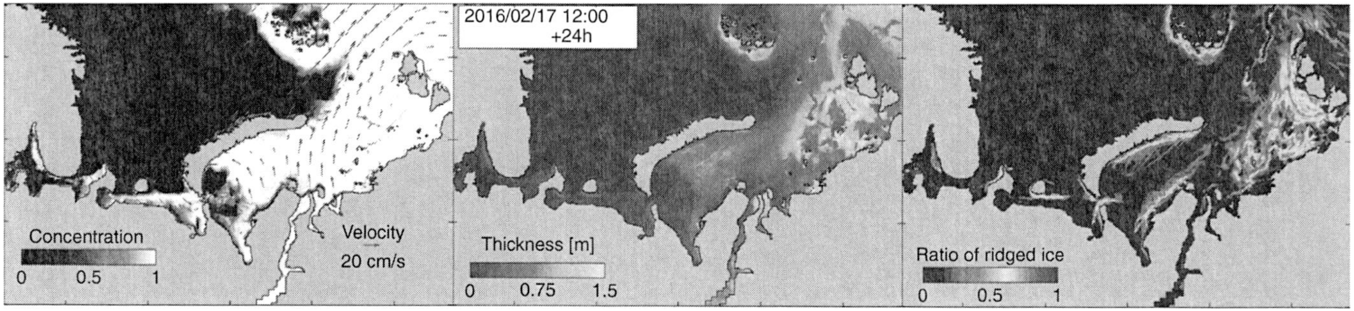

Figure 2.13: 24-hour forecast of sea ice concentration (left), drift (left), mean thickness (middle) with neXtSIM-F at 3 km resolution. The right panel displays the volume fraction of ridged ice in each grid cell of the model mesh. (A black and white version of this figure will appear in some formats. For the colour version, please refer to the plate section.)

but they also make possible the use of fully Lagrangian methods for advection that are particularly attractive for sea ice models. Figure 2.14 displays an example of an unstructured grid from the neXtSIM model.

2.8.3 Lagrangian Advection Schemes

As discussed in Section 2.4.2 (Horizontal transport), numerical advection of sea ice with many discontinuities introduces numerical diffusion that eventually leads to smearing out the discontinuities. Combining the Lagrangian and continuum approaches is an alternative to high order Eulerian schemes in order to preserve highly localized features like cracks, leads and ridges that are simulated by a sea ice model or initialized by the assimilation scheme (see Figure 2.14 for an example with the neXtSIM model).

When using a fully Lagrangian approach, the nodes of the model mesh belonging to ice-covered grid cells are moving with the simulated ice velocity **u**. This can lead to a highly distorted mesh in the locations where sea ice deforms, requiring the use of mesh adaptation and remapping technics. Using local mesh adaptation and remapping technics as in neXtSIM limits numerical diffusion to a small fraction of the elements. The local remapping also ensures strict conservation of physical quantities such as sea ice volume and area, which is required for long simulations. Conservation and continuity can also be preserved along open boundaries by treating specifically the elements touching the limits of the domain.

Having a moving mesh for the sea ice model implies some adaptations of the methods used for the coupling to other components, the post-processing and the parallelization of the code. Future comparisons of the performance between Lagrangian and Eulerian models will need to take these additional computational costs into account.

2.8.4 Form Drag

Many sea ice models only consider the skin drag for calculating the wind and water stresses. The skin drag relates to the stress over a ‘flat’ surface characterized by a small-scale roughness. However, the sails associated with pressure ridges, floe edges and melt pond edges are obstacles to the flow of air over sea ice. Similarly, ice keels and floe edges obstruct the flow of water under sea ice. Taking into account these form drag contributions in sea ice models is important for a correct representation of the air–ice and ice–ocean momentum exchanges.

Lüpkes *et al.* (2012) have introduced a parameterization, based on sea ice morphology, of the atmospheric drag coefficient. Tsamados *et al.* (2014) proposed a more complete approach for both atmospheric and oceanic form drag. The simulation results of Tsamados *et al.* (2014) indicate that form drag has a notable impact not only on the ice drift but also on the large-scale distribution of sea ice volume.

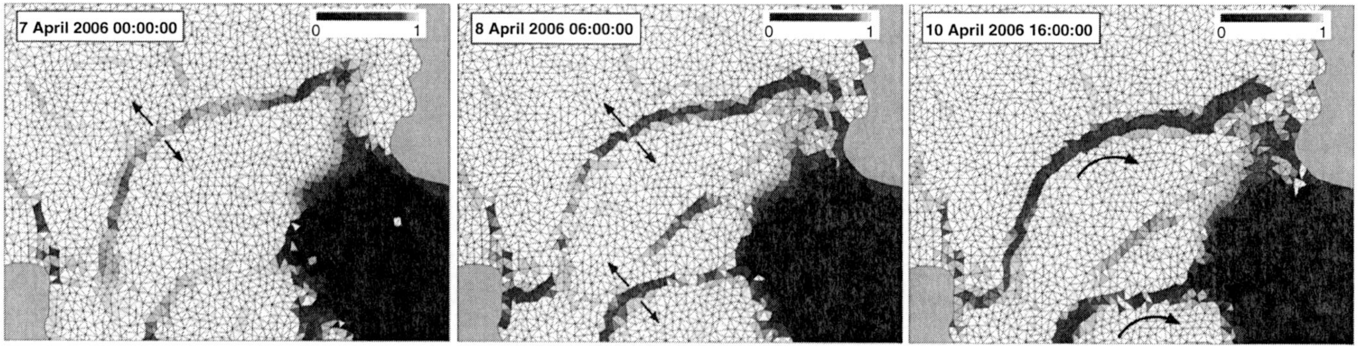

Figure 2.14: Example of a simulated sea ice concentration field and the underlying moving mesh obtained using a 10 km pan-Arctic neXtSIM configuration. The panels show a zoom on the Fram Strait, between the Northern East tip of Greenland (lower left corner) and the Spitzbergen Island (upper right corner). neXtSIM is a continuum-based sea ice model defined in a Lagrangian framework. As it uses an unstructured grid, it is discretized with the finite-element method instead of finite differences for structured grids. The panels illustrate how localized divergence generates discontinuities in the ice cover and how these discontinuities are preserved over time by the Lagrangian advection scheme. (A black and white version of this figure will appear in some formats. For the colour version, please refer to the plate section.)

2.8.5 Particle-Based Models

So far, continuum based sea ice models for large-scale applications have been presented in this chapter. However, when considering small scales (10 km^2 or less) and/or regions with a lower sea ice concentration such as in the MIZ, the discontinuous nature of sea ice may not be ignored with respect to sea ice mechanics and kinematics. In these conditions, a description in terms of an assembly of individual sea ice floes (e.g. Hopkins, 1996) in mechanical interaction should be favoured for an accurate description of the fine-scale structure of sea ice drift and internal forces. The numerical cost of these modelling approaches and the challenges posed by the modelling of floe-floe interactions (collisions) or how they can be combined with atmospheric as well as oceanic continuous models has resulted in them being much less developed so far than continuum mechanic models. These particle-based models consider simplified floe geometries and are based on a molecular dynamics scheme. Thus, these models estimate floe-floe contact forces based on an inter-penetration scheme which is a gross approximation if it is applied to a broad range of floe sizes and complex floe geometries.

Recently, Rabatel *et al.* (2015) proposed a different approach in which the dynamics of an assembly of individual ice floes of any shape and/or size is considered. As opposed to the molecular dynamic formulation, the model of Rabatel *et al.* (2015) is based on an event-driven algorithm with particular focus on collisions between floes, while avoiding inter-penetration. Between collisions, the motion of individual floes satisfies the linear and angular momentum conservation equations, with classical formulations for the Coriolis force and the wind and water stresses. The nature of the contacts is described through a constant coefficient of friction and a coefficient of restitution describing the loss of kinetic energy during the collision due to damage to and fracturing of the floes in the vicinity of the zone of contact.

A potentially interesting application of particle-based models would be to provide very short-term forecasts (a few hours) over a limited region. The boundary conditions could be provided by a large-scale continuum-based sea ice model. The assembly of floes for the initialization could be derived from high-resolution satellite observations, although automating this process will certainly be a challenge.

2.8.6 Wave-Ice Interactions

Historically, wave–ice interactions have mostly been studied from the viewpoint of the effect of ice on waves (see e.g. the review of Squire, 2007). Wave measurements have shown that wave energy decays exponentially with distance from the ice edge towards the ice pack (e.g. Wadhams *et al.*, 1988). With the recently observed decline in the Arctic summer sea ice extent, the fetch over which waves can be generated is increasing. As a consequence, there is growing interest in better understanding the effects of waves on ice. Recently, there has been a comprehensive field experiment in the Beaufort Sea with atmospheric, oceanographic and remote sensing measurements complementing in situ

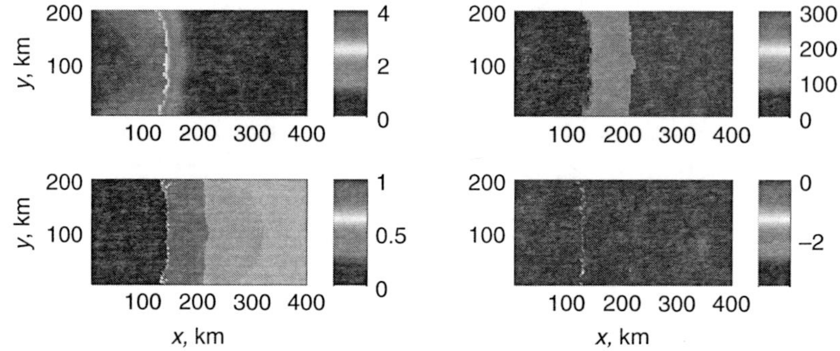

Figure 2.15: Waves breaking ice in an idealized experiment (the right-hand, upper, lower lines of grid cells correspond to land). The wave model, based on Williams *et al.* (2013a,b), is coupled to the neXtSIM sea ice model. The figure shows results after 48 h of steady pushing by a wave spectrum with significant wave height of 4 m and a peak period of 12 s, that is arriving from the left. Top left: significant wave height (m); top right: maximum floe size (m); bottom left: ice concentration; bottom right: $\log_{10}(1 - d)$, where d is the damage. Note the movement of and compression at the ice edge due to wave radiation stress as the waves continue to arrive at the ice. The ice has initial constant concentration of 0.7, thickness of 1 m, maximum floe size of 300 m, and $d = 0$. (A black and white version of this figure will appear in some formats. For the colour version, please refer to the plate section.)

wave measurements. One interesting observation made during this field campaign was the melting of a large field of pancake ice after waves passed through it.[3]

As waves can enhance ocean mixing, affect the drift of sea ice, change the floe size distribution (FSD) through break-up, and impact lateral melting, modellers are also interested in representing wave-ice interactions in sea ice models. Dumont *et al.* (2011) formulated the first parametrization of ice break-up due to waves, with the break-up producing more floes, and thus more floe edges where waves would be scattered. This in turn caused the waves to be attenuated more. Williams *et al.* (2013a, b) refined this model by removing one unrealistic breaking criterion and reformulating the wave energy transport equation into the type used by operational wave models.

The results of an idealized experiment using the Williams *et al.* (2013a, b) model are shown in Figure 2.15. After 48 h of simulations, the waves have broken the floes into smaller pieces over a distance of about 90 km. Due to the attenuation of the waves, momentum has been transferred to the ice, resulting in a movement of about 12 km. In practice this is done by adding a stress (the wave radiation stress) analogous to the wind or water stresses to the momentum equation (Equation 2.1). In the region of broken ice, the damage has increased to a high value of 0.999. Consequently, the ice stiffness and internal stress are very low in this area, leading to the ice becoming compacted near the ice edge. As well as the difficulties of modelling the dynamics of the broken ice, uncertainties in the amount of movement due to waves will partly come from the amount of attenuation that the waves experience – and particularly the shorter waves. Additionally, some of this lost momentum should be transferred to the ocean instead of the ice, which may also result in the ice moving less.

[3] A short movie about the October 2015 cruise in the Beaufort Sea (funded by the Office of Naval Research, USA) can be found at www.apl.washington.edu/project/project.php?id=arctic_sea_state.

Including wave–ice interactions in sea ice models implies that the FSD has to be represented and its evolution simulated. In the models of Dumont *et al.* (2011) and Williams *et al.* (2013a, b), the FSD is parameterized by a maximum floe size determined by the wavelengths present. Recently, Zhang *et al.* (2015) and Horvat and Tziperman (2015) both introduced models for the FSD that generalized the ITD approach of Thorndike *et al.* (1975). Zhang *et al.* (2015) considered the FSD independently from the ITD (although in a consistent way), while Horvat and Tziperman (2015) considered a joint floe size and thickness distribution (FSTD). The latter authors also went further in their discussions of the mechanisms affecting the FSTD, considering rafting and wave fragmentation, in addition to mechanical and thermodynamic processes (e.g. ridging or lateral melting).

2.8.7 Snow Modelling

We note that the representation of snow is still quite simple in most large-scale sea ice models, often consisting of a single layer with constant properties. Some authors (e.g. Lecomte *et al.*, 2014) have tested new approaches to represent snow in sea ice models. Their work illustrates that the representation of snow processes has a large impact on sea ice simulations. The development of more sophisticated modelling approaches for snow in models is likely to be a major theme of future research.

The presence of snow on the sea ice cover substantially modifies the ocean–atmosphere heat fluxes, the albedo of the surface, the moisture exchange between the surface and the atmosphere and the penetration of shortwave radiation through ice. Since all of these effects are directly related to the presence or the thickness of the snow cover, a good understanding of the processes involved in the snow cover evolution is important. The dominant source and sink terms in the snow mass budget have generally been associated with precipitation, melt and surface sublimation. However, recent modelling results suggest that other processes such as blowing snow effects (namely, blowing snow sublimation and airborne snow lost into leads and open waters) may also be important.

Precipitation, which occurs mainly in the form of snow, is an important term in the freshwater budget of the polar oceans. The state of the snow cover can influence the upper ocean, the sea ice and the lower atmosphere. Snow on sea ice reduces the ocean–atmosphere heat fluxes and consequently ice growth, through heat conduction. It also reflects shortwave radiation more effectively, owing to its relatively high albedo, and attenuates the penetration of light into the mixed-layer. Snow transported into open waters by winds or local precipitation (in winter), and snowmelt (in summer) affects the salinity of the surface waters and therefore the upper ocean stratification. While snow sublimation always provides moisture to the lower atmosphere, the latent heat required for the phase change comes from the ice/snow surface or the lower atmosphere, depending on whether surface or blowing snow sublimation is present. This in turn affects the temperature of the lower atmosphere, its stability and its ability to hold water vapour. Finally, since precipitation at high latitudes is expected to increase in a warmer world, a good understanding of the snow mass balance will become increasingly important.

2.8.8 Salt Dynamics

It has been known since the work of Malmgren (1927) that the ice salinity varies in space and time. New ice formation incorporates salt within brine inclusions. Snow-ice formation incorporates some salt as well, either from brine or seawater. As brine drainage expels salt from the ice, the ice salinity generally decreases with time. Brine drains from growing sea ice because of the temperature-driven unstable brine density gradient that promotes gravity-driven natural convection into the lowermost permeable sea ice. This process is referred to as gravity drainage. The percolation of meltwater through summer permeable ice, referred to as flushing, can drastically reduce the near-surface ice salinity in the Arctic. The most recent formulations of the dynamics of brine inclusions are based on the so-called mushy-layer theory, the theory of multi-phase materials (see Worster, 1992 and Hunke *et al.*, 2011).

Including salt dynamics requires a salt conservation equation. Based on the mushy-layer equations, Hunke *et al.* (2011) proposed the following 1-D formulation:

$$\frac{\partial S}{\partial t} = -w\frac{\partial S_{br}}{\partial z} + \frac{\partial}{\partial z}\left(D_S\frac{\partial S_{br}}{\partial z}\right) - \frac{\rho_i}{\rho_{br}}S_{br}\frac{\partial \phi}{\partial t}, \tag{2.17}$$

which states that ice salinity changes due to advection by brine motion, diffusion of salt, and the expulsion of salt when the brine fraction changes. D_S is the vertical diffusion coefficient for salt through the brine network. The vertical Darcy w velocity involves assumptions about the representation of gravity drainage and flushing. The salt budget must also account for the entrapment of salt during basal ice growth and snow-ice formation, assuming that all salt present originally in seawater is trapped in the ice. How salt dynamics should be represented is still a topic of active research.

An empirical approach to parameterize the salt budget in sea ice has been proposed by Vancoppenolle *et al.* (2009a). This is based on a single equation for bulk salinity, from which a profile shape is derived. Such a parameterization is sufficient to simulate the large-scale variations in ice salinity and their impact on ice thickness, producing salinity maps in line with those produced using Equation 2.17 as in Turner *et al.* (2015).

When salinity variations are included, the heat equation must be adjusted as well, by (1) using the correct expression for the time derivative of the sea ice enthalpy, accounting for salinity variations, and (2) advecting sensible heat due to brine motion.

Including salt dynamics has two implications on model results, as described by Vancoppenolle *et al.* (2009b). First, there are significant effects in ice growth and melt due to changes in the ice thermal properties. Second, the ice–ocean salt exchanges are affected as well, potentially changing the response of the ocean to sea ice growth and melt. The overall response is a slight increase in ice volume in the Arctic when salinity variations are incorporated.

2.8.9 Modelling of Melt Ponds

Melt ponds form when the meltwater produced by surface melting does not percolate fast enough through the porous ice and collects in depressions on the ice surface. As melt ponds significantly reduce the albedo, this leads to enhanced surface melt. It is therefore important for sea ice models to represent this positive feedback mechanism. Highly dependent on the surface topography of the ice, melt ponds are represented in large-scale sea ice models with varying degrees of complexity. The simplest parameterizations assume a prescribed melt pond fraction in the computation of melting ice albedo. More sophisticated approaches consider the sea ice topography for explicitly modelling the evolution of melt ponds. The model of Flocco *et al.*, 2010 derives a pseudo topography from the model ITD (and because of this their scheme is incompatible with two-category models) and accumulate meltwater starting from the lowest elevation. The model of Hunke *et al.*, 2013 follows the area and volume of melt ponds on level ice assuming a depth–area ratio for changes in melt pond volume. These approaches represent various processes related to melt ponds such as drainage through permeable sea ice and melt pond refreezing. Explicit melt pond schemes are starting to be introduced in sea ice forecasting systems.

2.9 Chapter Summary

This chapter gave an overview of the sea ice models used in large-scale sea ice prediction systems. These continuum-based models simulate the evolution of a sea ice thickness distribution. Dynamic processes of advection, deformation and redistribution were presented as well as thermodynamic processes of vertical growth/melt and lateral melt. Numerical methods for solving the sea ice equations were also introduced.

This chapter also presented some recent developments and possible future directions. For improved prediction of sea state and sea ice conditions in the MIZ, work is currently being done to improve processes such as wave-ice interactions and a better representation of the air–ice and ice–ocean momentum exchanges due to form drag. The modelling of snow, salt dynamics and the evolution of melt ponds are likely to be important themes of research for the years to come.

With the demand for higher-resolution forecasts and for a better definition of complicated geographical areas (e.g. narrow channels), sea ice models based on unstructured grids are being implemented. It is also anticipated that particle-based models (i.e. models that treat the ice as a collection of individual floes) will increasingly be used for forecast applications (probably in conjunction with continuum-based models for boundary conditions).

References

Amundrud, T.L., Melling, H. and Ingram, R.G. (2004). Geometrical constraints on the evolution of ridged sea ice. *Journal of Geophysical Research*, **109**, C06005. doi: 10.1029/2003JC002251.

Bitz, C.M. and Lipscomb, W.H. (1999). An energy-conserving thermodynamic model of sea ice. *Journal of Geophysical Research*, **104**, 15669–15677.

Bouillon, S. and Rampal, P. (2015). Presentation of the dynamical core of neXtSIM, a new sea ice model. *Ocean Modelling*, **91**, 23–37.

Coon, M.D., Maykut, G.A., Pritchard, R.S., Rothrock, D.A. and Thorndike, A.S. (1974). Modeling the pack ice as an elastic–plastic material. *AIDJEX Bulletin*, **24**, 1–105.

Dumont, D., Kohout, A. and Bertino, L. (2011). A wave-based model for the marginal ice zone including a floe breaking parameterization. *Journal of Geophysical Research*, **116**, C04001. doi: 10.1029/2010JC006682.

Dupont, F., Vancoppenolle, M., Tremblay, L.-B. and Huwald, H. (2015). Comparison of different numerical approaches to the 1D sea-ice thermodynamics problem. *Ocean Modelling*, **87**, 20–29. doi: 10.1016/j.ocemod.2014.12.006.

Feltham, D.L. (2008). Sea ice rheology. *The Annual of Fluid Mechanics*, **40**, 91–112.

Flato, G.M. (1993). A particle-in-cell sea-ice model. *Atmosphere-Ocean*, **31**, 3, 339–358.

Flato, G.M. and Hibler, W.D. (1995). Ridging and strength in modeling the thickness distribution of Arctic sea ice. *Journal of Geophysical Research*, **100**(C9), 18611–18626.

Flato, G.M. (1998). The thickness variable in sea-ice models. *Atmosphere-Ocean*, **36**(1), 29–36.

Flocco, D., Feltham, D.L. and Turner A.K. (2010). Incorporation of a physically based melt pond scheme into the sea ice component of a climate model. *Journal of Geophysical Research Oceans*, **115**, C08012. doi: 10.1029/2009JC005568.

Girard, L., Weiss, J., Molines, J.M., Barnier, B. and Bouillon, S. (2009). Evaluation of high-resolution sea ice models on the basis of statistical and scaling properties of Arctic sea ice drift and deformation. *Journal of Geophysical Research*, **114**, C08015. doi: 10.1029/2008JC005182.

Girard, L., Bouillon, S., Weiss, J., Amitrano, D., Fichefet, T. and Legat, V. (2011). A new modeling framework for sea-ice mechanics based on elasto-brittle rheology. *Annals of Glaciology*, **52**(57), 123–132.

Hibler, W.D. (1979). A dynamic thermodynamic sea ice model. *Journal of Physical Oceanography*, **9**, 815–846.

Hopkins, M.A. (1994). On the ridging of intact lead ice. *Journal of Geophysical Research*, **99**(C8), 16351–16360.

Hopkins, M.A. (1996). On the mesoscale interaction of lead ice and floes. *Journal of Geophysical Research*, **101**(C8), 18315–18326.

Horvat, C. and Tziperman, E. (2015). A prognostic model of the sea-ice floe size and thickness distribution. *The Cryosphere*, **9**, 2119–2134. doi: 10.5194/tc-9–2119-2015.

Hunke, E.C. (2001). Viscous–plastic sea ice dynamics with the EVP model: linearization issues. *Journal of Computational Physics*, **170**(1), 18–38. doi: 10.1006/jcph.2001.6710.

Hunke, E.C., Lipscomb, W.H. and Turner, A.K. (2010) Sea-ice models for climate study: retrospective and new directions. *Journal of Glaciology*, **56**(200), 1162–1172.

Hunke, E.C., Notz, D.K., Turner, A.K. and Vancoppenolle, M. (2011). The multi-phase physics of sea ice: a review for modellers. *The Cryosphere*, **5**, 989–1009. doi: 10.5194/tc-5–989-2011.

Hunke, E.C., Hebert, D.A. and Lecomte, O. (2013) Level-ice melt ponds in the Los Alamos sea ice model, CICE. *Ocean modelling*, **71**, 26–42, 10.1016/j.ocemod.2012.11.008.

Kimmritz, M., Danilov, S. and Losch, M. (2015). On the convergence of the modified elastic–viscous–plastic method for solving the sea ice momentum equation. *Journal of Computational Physics*, 296, 90–100. doi: 10.1016/j.jcp.2015.04.051.

König Beatty, C. and Holland, D.M. (2010). Modeling landfast sea ice by adding tensile strength. *Journal of Physical Oceanography*, **20**(1), 185–198. doi: 10.1175/2009JPO4105.1.

Lecomte, O. (2014). *Influence of Snow Processes on Sea Ice: A Model Study*. Université catholique de Louvain, Phd Thesis.

Lemieux, J.-F., Tremblay, B., Sedláček, J., Tupper, P., Thomas, S., Huard, D. and Auclair, J.-P. (2010). Improving the numerical convergence of viscous–plastic sea ice models with the Jacobian-free Newton Krylov method. *Journal of Computational Physics*, **229**, 2840–2852. doi: 10.1016/j.jcp.2009.12.011.

Lipscomb, W.H. (2001). Remapping the thickness distribution in sea ice models, *Journal of Geophysical Research*, **106**(C7), 13989–14000.

Lipscomb, W.H., Hunke, E.C., Maslowski, W. and Jakacki, J. (2007). Ridging, strength, and stability in high-resolution sea ice models. *Journal of Geophysical Research*, **112**, C03S91. doi: 10.1029/2005JC003355.

Losch, M., Fuchs, A., Lemieux, J.-F. and Vanselow, A. (2014). A parallel Jacobian-free Newton–Krylov solver for a coupled sea ice-ocean model. *Journal of Computational Physics*, **257**, 901–911. doi: 10.1016/j.jcp.2013.09.026.

Lüpkes, C., Gryanik, V.M., Hartmann, J. and Andreas, E.L. (2012). A parameterization, based on sea ice morphology, of the neutral atmospheric drag coefficients for weather prediction and climate models. *Journal of Geophysical Research*, **117**, D13112. doi: 10.1029/2012JD017630.

Malmgren, F. and Geofysisk Institutt. (1927). *On the Properties of Sea-Ice*. AS John Griegs Boktrykkeri.

Maykut, G.A. and Untersteiner, N. (1971). Some results from a time-dependent thermodynamic model of sea ice. *Journal of Geophysical Research*, **76**, 1550–1575.

Maykut, G.A. (1986). *The Surface Heat and Mass Balance: The Geophysics of Sea Ice*. Springer, 395–463.

Notz, D. (2012). Challenges in simulating sea ice in Earth system models. *WIREs Climate Change*. doi: 10.1002/wcc.189.

Rabatel, M., Labbé, S., and Weiss, J. (2015). Dynamics of an assembly of rigid ice floes. *Journal of Geophysical Research Oceans*, **120**, 5887–5909. doi: 10.1002/2015JC010909

Rampal, P., Bouillon, S., Ólason, E. and Morlighem, M. (2016). neXtSIM: a new Lagrangian sea ice model. *The Cryosphere Discussion*, **10**(3), 5885–5941. doi: 10.5194/tcd-9-5885-2015.

Rothrock, D.A. (1975). The energetics of the plastic deformation of pack ice by ridging. *Journal of Geophysical Research*, **80**(33), 4514–4519.

Sayed, M. and Carrieres T. (1999). Overview of a new operational ice model. *Proceedings of the Ninth (1999) International Offshore and Polar Engineering Conference*, Volume II, 622–627, Brest, France, May 30–June 4, 1999.

Schmidt, G.A., Bitz, C., Mikolajewicz and Tremblay, L.-B. (2004). Ice-ocean boundary conditions for coupled models. *Ocean Modelling*, **7**, 59–74. doi: 10.1016/S1463-5003(03)00030-1.

Semtner, A.J. (1976). A model for the thermodynamic growth of sea ice in numerical investigations of climate. *Journal of Physical Oceanography*, **6**, 379–389.

Semtner, A.J. (1984). On modelling the seasonal thermodynamic cycle of sea ice in studies of climatic change. *Climatic Change*, **6**, 27. doi: 10.1007/BF00141666.

Squire, V.A. (2007). Of ocean waves and sea-ice revisited. *Cold Regions Science and Technology*, **49**(2), 110–133.

Steele, M. (1992). Sea ice melting and floe geometry in a simple ice-ocean model. *Journal of Geophysical Research*, **97**(C11), 17729–17738.

Thorndike, A.S., Rothrock, D.A., Maykut, G.A. and Colony, R. (1975). The thickness distribution of sea ice. *Journal of Geophysical Research*, **80**(33), 4501–4513.

Tsamados, M., Feltham, D.L. and Wilchinsky, A.V. (2013). Impact of a new anisotropic rheology on simulations of Arctic sea ice. *Journal of Geophysical Research Oceans*, **118**, 91–107. doi: 10.1029/2012JC007990.

Tsamados, M., Feltham, D.L., Schroeder, D., Flocco, D., Farrell, S.L., Kurtz, N., Laxon, S.W. and Bacon, S. (2014). Impact of variable atmospheric and oceanic form drag on simulations of Arctic sea ice. *Journal of Physical Oceanography*, **44**(5), 1329–1353. doi: 10.1175/JPO-D-13-0215.1.

Turner, A.K. and Hunke, E.C. (2015). Impacts of a mushy-layer thermodynamic approach in global sea-ice simulations using the CICE sea-ice model. *Journal of Geophysical Research: Oceans*, **120**(2), 1253–1275. doi: 10.1002/2014JC010358.

Vancoppenolle, M., Fichefet, T., Goosse, H. and Bouillon, S. (2009a). Simulating the mass balance and salinity of Arctic and Antarctic sea ice. 1. Model description and validation. *Ocean Modelling*, **27**, 33–53. doi: 10.1016/j.ocemod.2008.10.005.

Vancoppenolle, M., Fichefet, T., Goosse, H. and Bouillon, S. (2009b). Simulating the mass balance and salinity of Arctic and Antarctic sea ice. 2. Importance of sea ice salinity variations. *Ocean Modelling*, **27**, 54–69. doi: 10.1016/j.ocemod.2008.11.003.

Wadhams, P., Squire, V.A., Goodman, D.J., Cowan, A.M., and Moore, S.C. (1988). The attenuation rates of ocean waves in the marginal ice zone. *Journal of Geophysical Research*, **93**(C6), 6799–6818.

Williams, T.D., Bennetts, L.G., Squire, V.A., Dumont, D. and Bertino, L. (2013a). Wave–ice interactions in the marginal ice zone. Part 1: Theoretical foundations. *Ocean Modelling*, **71**, 81–91. doi: 10.1016/j.ocemod.2013.05.010.

Williams, T.D., Bennetts, L.G., Squire, V.A., Dumont, D. and Bertino, L. (2013b). Wave–ice interactions in the marginal ice zone. Part 2: Numerical implementation and sensitivity studies along 1D transects of the ocean surface. *Ocean Modelling*, **71**, 92–101. doi: 10.1016/j.ocemod.2013.05.011.

Worster, M.G. (1992). The Dynamics of Mushy Layers. *Interactive Dynamics of Convection and Solidification*, NATO ASI Series, **219**, 113–138, Kluwer.

Zhang, J., and Hibler, W.D. (1997). On an efficient numerical method for modeling sea ice dynamics. *Journal of Geophysical Research*, **102**(C4), 8691–8702.

Zhang, J., Schweiger, A., Steele, M. and Stern, H. (2015). Sea ice floe size distribution in the marginal ice zone: Theory and numerical experiments. *Journal of Geophysical Research Oceans*, 120, 3484–3498. doi: 10.1002/2015JC010770.

3

Sea Ice Observations

Leif Toudal Pedersen, Rasmus Tonboe, Stefan Kern, Thomas Lavergne, Natalia Ivanova, Georg Heygster

Observations are crucial components in any automated prediction system (APS) for sea ice. Today, most sea ice observations are carried out by satellite-based instruments that measure emitted or scattered electromagnetic radiation. This chapter first describes the interaction between electromagnetic radiation and the physical components of the ice/ocean/atmosphere system. An understanding of these interactions is essential for appreciating how different types of satellite data can be used for retrieving information about sea ice. The most common instruments currently used for sea ice observations are then introduced. Next, a review is presented on how key sea ice variables, including ice concentration, ice thickness and ice drift, can be derived using observations from these instruments. Throughout, the focus is on the elements that are important for the quality of the observations. Finally, methods that are commonly used in ice analysis to integrate information from multiple sources are introduced (ice charts and observation operators). Within an automated prediction system for sea ice (such as those presented in Chapter 5), the information from a wide range of sea ice observations is combined with a short-term forecast from a sea ice model (described in Chapter 2) using data assimilation (described in Chapter 4).

This chapter is intended as an introduction to the topic of sea ice observations. For additional details and related mathematical equations, the reader should consult the references made throughout this chapter to more specialized books and scientific articles.

3.1 Sea Ice Electromagnetic Properties

Estimates of sea ice variables such as sea ice concentration and extent, ice drift, ice type and even ice thickness obtained from satellite observations have become a primary source of sea ice information both for operational monitoring and for climate monitoring. Various types of satellite-based instruments used for this purpose have been in orbit since the 1970s. However, such instruments do not measure sea ice variables directly. They measure electromagnetic radiation at various wavelengths and polarizations, and from these measurements sea ice information is inferred. The conversion of the electromagnetic signals measured by satellite-based instruments into physical information about the ice is carried out using automated retrieval algorithms or manually by ice analysts. Alternatively, satellite observations can be used directly within a data assimilation system (see Chapter 4) to obtain sea ice information by utilizing a forward model of the observation process (also referred to as an observation operator). A forward model computes the scattered or emitted electromagnetic radiation as a function of the physical state of the sea ice.

3.1.1 Basic Principles

In order to assess the performance of sea ice retrieval algorithms, it is important to understand the interaction between the electromagnetic radiation and physical properties of the ice and the ocean. Figure 3.1 shows the electromagnetic spectrum from long to short wavelengths covering microwaves (centimetre to millimetre wavelengths), infrared (micrometre wavelengths) and visible light (400–700 nanometres).

In vacuum, the relationship between wavelength (λ) and frequency (f) of electromagnetic radiation is via the speed of light c (3.0×10^8 m s^{-1})

$$\lambda = \frac{c}{f} \quad \text{or} \quad f = \frac{c}{\lambda}. \tag{3.1}$$

For electromagnetic propagation in snow and sea ice, the propagation speed is less than the speed of light in vacuum, but the relationship between frequency, speed and wavelength is still valid. In the following, we will be using frequency or wavelength according to the general tradition in the applications we present. In optical remote sensing, wavelength is often used, whereas in microwave remote sensing, frequency is more common.

Remote sensing of sea ice is carried out using a variety of satellite instruments operating at a wide range of electromagnetic wavelengths (Table 3.1). Instruments for remote sensing are often referred to as either *passive* or *active*. Passive instruments rely on some external source of radiation (such as the Sun or thermal emission) whereas active instruments (such as radars and lasers) provide their own source of radiation. In general, the microwave range is preferred due to the ability of microwave radiation to penetrate clouds and because this radiation can also be used for observations independent of solar illumination.

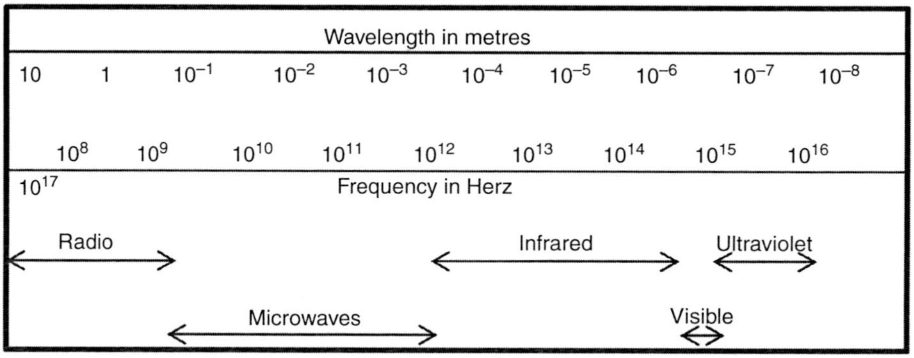

Figure 3.1: The electromagnetic spectrum. Frequency and wavelength.

Table 3.1: *Examples of some typical satellites and instruments and their operating wavelength range/frequency. The wavelengths given here are in air. Electromagnetic frequency is normally not used in the visible and infrared range but they are given here for comparison (Giga= G= 10^9 and Tera= T= 10^{12}). In microwave remote sensing wavelength or frequency is also often identified by letter-bands (such as L-band, C-band etc.). The definition of these bands originates from the infancy of radar technology and was standardized by IEEE Standard 521–2002. THIR: Thermal Infrared, NIR: Near infrared, VIS: Visible light.*

Frequency	Wavelength	Band	Example of instruments
1.4 GHz	21 cm	L	SMOS, SMAP, ALOS
6 GHz	5 cm	C	AMSR, Envisat ASAR, RADARSAT, Sentinel-1, ASCAT
10 GHz	3 cm	X	AMSR, TSX, CSK
12–19 GHz	2.5–1.6 cm	Ku	AMSR, SSMI(S), QuikSCAT, Envisat RA2, Cryosat
37 GHz	0.8 cm	Ka	AMSR, SSMI(S)
50–60 GHz	0.6–0.5 cm	V	SSMIS
85–91 GHz	0.35–0.33 cm	W	AMSR, SSMI(S)
25–30 THz	10–12 μm	THIR	AVHRR, MODIS, Landsat, NPP VIIRS, Sentinel-3, Envisat AATSR, ERS ATSR
200–430 THz	0.7–1.5 μm	NIR	AVHRR, MODIS, Sentinel-2 and 3, NPP-VIIRS
430–1000 THz	0.3–0.7 μm (300–700 nm)	VIS	AVHRR, MODIS Sentinel-2 and 3 NPP VIIRS

3.1.2 Atmospheric Interference

In order to be able to observe sea ice and other surface features, it is necessary to be able to see through the atmosphere. The clear sky atmosphere is transparent or semi-transparent in the visible and infrared part of the electromagnetic spectrum as well as at millimetre to centimetre wavelengths. However, the intervening atmosphere, including specific gases, clouds and precipitation, affects the satellite measurements to varying degrees in the whole range of wavelengths from ultraviolet to microwaves. The transparency of the atmosphere is limited in the ultraviolet by stratospheric ozone, in the visible primarily by clouds, and in the near infrared and thermal infrared by clouds and the greenhouse gases including water vapour. As some regions near the poles are completely dark 24 hours a day in winter, the usefulness of channels relying on sunlight for illumination is limited during this period.

The selection of wavelengths for a given application depends on the desired resolution and on the acceptable amount of atmospheric influence. In general, higher frequency means potentially better spatial resolution at the cost of higher sensitivity to the atmosphere. Since microwave radiation generally penetrates the atmosphere it allows observations of the Earth's surface during most weather conditions, whereas infrared and visible light cannot

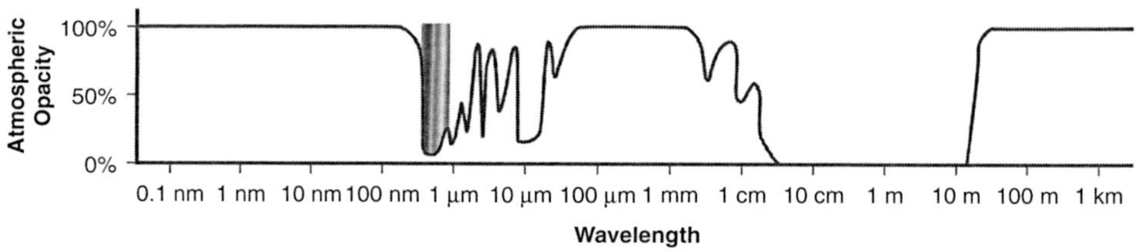

Figure 3.2: The opacity of the atmosphere. Wavelength bands with large transmissivity (small opacity) are transparent to radiation and are also referred to as atmospheric windows. Modified from NASA illustration; online, courtesy NASA.

penetrate clouds. Thus, visible and infrared wavelengths can only be used to gather information about sea ice under cloud-free conditions.

In the microwave region (1 mm to 1 m wavelength) the main atmospheric influence is due to attenuation by gases and by particles suspended in the air (water in clouds). The molecular attenuation by gasses generally increases towards higher frequencies and is associated with a number of absorption lines (molecular resonance frequencies) where specific gasses such as water vapour or oxygen attenuate the radiation significantly. The main absorption lines in the microwave spectrum are at 22.2 GHz (water vapour), 55–65 GHz (oxygen), 118 GHz (oxygen) and 183 and 325 GHz (water vapour). Liquid water in clouds also absorbs (and scatters) microwave radiation. The cloud attenuation effect is strongest at higher frequencies (above 15–20 GHz). The combined effect of absorption and scattering in the atmosphere, and in particular the absorption lines, lead to a situation where the atmosphere is fairly transparent in certain wavelength bands and almost opaque in others. The transparent parts of the spectrum are referred to as atmospheric windows and these are where surface properties can be measured from satellites.

3.1.3 Permittivity and Refractive Index

Sea ice and seawater are dielectric materials, meaning they interact with electromagnetic radiation. The permittivity ε (the square of the refractive index n) characterizes the dielectric properties. The permittivity of snow is determined by its density and liquid water content, while the permittivity of sea ice is largely driven by the brine- and the air-inclusion volume.

The permittivity determines the propagation of radiation, including the reflection and transmission at interfaces between different materials and the absorption inside the materials. Scattering of radiation in sea ice and the snowpack is detectable at wavelengths shorter than 3 cm. The permittivity of a material also controls the speed of light and the wavelength of electromagnetic radiation inside materials. The wavelength is reduced by the square root of the permittivity of the material where the radiation propagates

$$\lambda = \lambda_0/\sqrt{\varepsilon}, \tag{3.2}$$

where λ_0 is the wavelength in vacuum and λ is the wavelength inside the material. For example, in dry snow the wavelength and speed of light is about 80 per cent of the values in vacuum and in sea ice it is further reduced to about 60 per cent. The speed of light is modified such that the frequency remains the same. The permittivity in absorbing media, such as sea ice, is a complex number where the real part primarily describes the refractivity and the imaginary part the attenuation or loss factor ($n^2 = \varepsilon = \varepsilon' + j\varepsilon''$, where $j = \sqrt{-1}$).

The reflectivity, R, at the boundary between medium 1 and 2 for radiation incident at an angle θ is:

$$R_V = \left| \frac{n_1 \, cos\theta_i - \left(n_2 \sqrt{1 - \left(\frac{n_1}{n_2} \sin \theta_i \right)^2} \right)}{n_1 \, cos\theta_i + \left(n_2 \sqrt{1 - \left(\frac{n_1}{n_2} \sin \theta_i \right)^2} \right)} \right|^2,$$

$$R_H = \left| \frac{\left(n_1 \sqrt{1 - \left(\frac{n_1}{n_2} \sin \theta_i \right)^2} \right) - n_2 \, cos\theta_i}{\left(n_1 \sqrt{1 - \left(\frac{n_1}{n_2} \sin \theta_i \right)^2} \right) + n_2 \, cos\theta_i} \right|^2 \qquad (3.3)$$

where indexes V and H refer to vertical and horizontal polarization respectively and n_1 and n_2 are the refractive indexes of the two media. The relationship between incidence and transmission angle is given by Snell's law and is also a function of permittivity or refractive index

$$\frac{\sin \theta_1}{\sin \theta_2} = \frac{n_2}{n_1}. \qquad (3.4)$$

Radiation incident on a surface will either be reflected at the interface, absorbed by the surface, or transmitted through the surface. Hence, the reflectivity, absorptivity and transmissivity are related by $|R| + |\alpha| + |T| = 1$ where α is the absorptivity, R is reflectivity and T is transmissivity. For opaque surfaces $|T| = 0$ and hence $|R| + |\alpha| = 1$. According to Kirchhoff's law, absorptivity is equal to emissivity (e) in thermodynamic equilibrium, and thus

$$e = 1 - |R|. \qquad (3.5)$$

The emissivity of a material is its effectiveness in emitting energy as thermal radiation. The reflectivity and transmissivity between natural layers inside the air–snow–ice system are closely related to the emissivity, reflectivity, interface roughness and incidence angle. The reflectivity of visible light is also referred to as the albedo (whiteness) of a surface.

The difference between interaction of radiation with matter at vertical polarization (V) and horizontal polarization (H) (Equation 3.3) is the rationale for many instruments that carry out measurements at these two linear polarizations. The polarization ratio or polarization difference is important in many applications such as sea ice concentration retrieval.

3.1.3.1 Thermal Emission

Materials in thermal equilibrium (at any given instant this is true for snow and sea ice) emits and absorbs equal amounts of thermal radiation. For idealized blackbody materials (perfect emitters and absorbers) the emission is described by Planck's radiation law. In the thermal infrared part of the spectrum (10–12 microns) dry and wet snow are almost perfect blackbodies whereas at longer wavelengths dry snow is not.

Earth surface materials such as water, snow and ice are in general not perfect emitters of microwave radiation, and hence we characterize them by their emissivity (e) (Equation 3.5), which is the emitted radiation (at the same wavelength) as a fraction of the radiation of a perfect blackbody (at the same physical temperature). A blackbody has an emissivity of 1 and a perfect reflector an emissivity of 0. The relationship between emissivity and reflectivity (R) is given by Equation 3.5 according to Kirchhoff's radiation law.

We define the brightness temperature of a material (T_B) as

$$T_B = eT_P. \tag{3.6}$$

The brightness temperature (in Kelvin) is a measure of the energy emitted by the medium and T_P is the effective physical temperature. Because of microwave penetration into snow and ice across a sometimes steep temperature gradient, T_P is not synonymous with the surface temperature but rather is an integrated temperature across the emitting layer. It varies with microwave frequency due to differences in penetration depth.

3.1.3.2 Permittivity of Water

The permittivity of water is a function of salinity and temperature (Figure 3.3). In the microwave range, for sea water (with salinity 35 psu) it is generally an order of magnitude larger than the permittivity of ice and snow. This explains why the reflection coefficient at the air-water interface is quite large and consequently why the emissivity of water is relatively small (see Equation 3.6). The large loss factor is the reason why the penetration depth in water is very small. The permittivity of water is largest at longer microwave wavelengths and thus the emissivity is smaller at larger wavelengths.

3.1.3.3 Permittivity of Snow

The permittivity of dry snow is nearly independent of both temperature (when it is below freezing) and electromagnetic frequency and is mainly a function of snow density. Typical values of dry snow permittivity are between 1 and 2. Due to the larger permittivity of water

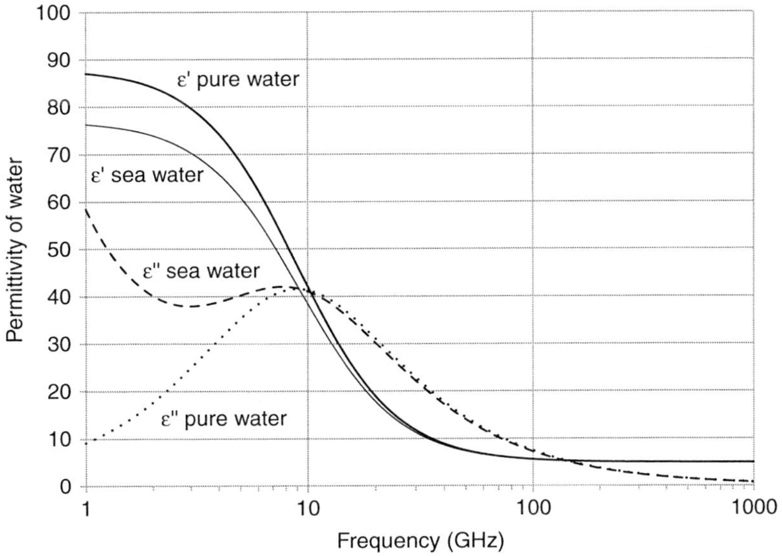

Figure 3.3: The permittivity and loss factor of pure and sea water. For more information see also Meissner & Wentz (2012).

than that of pure ice (e.g. snow crystals) across a wide range of microwave frequencies, the permittivity of wet snow is primarily a function of its volumetric water content (Grenfell *et al.*, 1986).

3.1.3.4 Permittivity of Sea Ice

The permittivity of sea ice at microwave frequencies is closely related to the salinity or brine volume of the ice. The salinity and volume of brine embedded as inclusions in ice is related to the ice temperature since the freezing temperature depends on salinity. In general, the permittivity of a mixture is a function of the volume fraction and the shape and distribution of the inclusions and the permittivity of these and the background material. The appropriate permittivity model to use is dependent on the ice type and structure. Several empirical models for snow and sea ice permittivity have been proposed, and these have to be selected for the specific type of snow or sea ice and frequency where they are applicable. Large differences exist between different models, especially for simulating the loss factor (ε''). In such models, the dielectric properties of the background material and the inclusions are combined in so-called mixing formulas. Some recommendations are given in Table 3.2. First-year (FY) sea ice permittivity is often simulated as a dielectric mixture between pure ice and liquid brine pockets where the brine volume is determined by the sea ice salinity and temperature, as described in Section 2.5.1. Different assumptions are then made about the shape and orientation of the brine pockets for computing the microwave interaction with these as scatterers and absorbers (see Table 3.2). The appropriate

Table 3.2: *The appropriate model for computing/modelling permittivity of different sea ice types using mixing models, as recommended by Shokr (1998).*

Ice type	Model type
First-year frazil ice	Randomly oriented brine needles in pure ice
First-year columnar ice	Vertically oriented brine needles in pure ice
Multi-year hummock	Any shape of air inclusions in pure ice
Pond ice	Vertically oriented brine needles and spherical air bubbles

assumptions are related to the sea ice formation processes described in Chapter 2. Typical values for the permittivity of FY ice are between 3 and 5.

Multi-year (MY) ice contains much less salt, so the permittivity of MY ice is also a function of the volume of air included in the ice. Typical values for the permittivity of MY ice are between 2.5 and 3.

3.1.3.5 Penetration Depth – Snow, Ice, Water

At microwave wavelengths, radiation can penetrate snow and to some extent also sea ice. The penetration depth D_p indicates approximately from which depth range the emitted or backscattered radiation contains information. In a non-scattering dielectric material, penetration depth is a function of the dielectric losses and thus the permittivity,

$$D_p = \frac{c\sqrt{\varepsilon'}}{2f\pi\varepsilon''}, \tag{3.7}$$

where ε' and ε'' are the real and imaginary parts of the permittivity, respectively, c is speed of light in vacuum and f is the frequency. It is seen that the penetration depth is inversely proportional to the imaginary part of the permittivity (the attenuation or loss factor), which is itself proportional to conductivity. Figure 3.4 shows the penetration depth of microwaves into ice and snow.

The much larger loss factor in water means that the penetration of microwaves into seawater as well as freshwater is in the order of millimetres. Penetration in FY ice (salty ice) is limited by the conductivity of the brine included in this ice type, and penetration into MY ice is limited by scattering in the inhomogeneous material. Penetration in dry snow can be up to several metres, whereas penetration into wet snow is only a few centimetres or less and strongly decreases with snow wetness (volumetric water content) (see Figure 3.4). The penetration of microwaves into snow and sea ice enables derivation of snow properties, ice type and the thickness of thin ice from microwave measurements. However, penetration into materials with a very steep temperature gradient (in winter) provides challenges in analysis of thermal microwave data. The uncertainty in estimating sea ice free-board using radar altimetry is also related to variations in penetration depth.

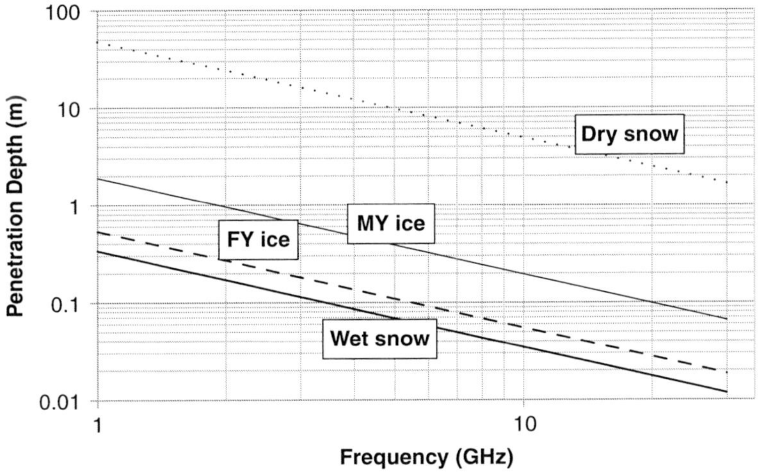

Figure 3.4: Typical values of penetration of microwaves into snow and ice (in metres) as a function of microwave frequency (in GHz). Actual values will vary with temperature, snow wetness etc.

3.2 Passive Remote Sensing

Passive remote sensing instruments are devices that rely on a natural source of radiation such as reflected and scattered sunlight or thermal emission.

3.2.1 Microwave Radiometry

A microwave radiometer is a very sensitive passive microwave (PM) receiver that measures microwave radiation (the antenna temperature T_A) incident on an antenna pointing towards the ground.

Space-borne microwave radiometers have been in orbit since the 1970s. One of the earliest was the Electrically Scanning Microwave Radiometer (ESMR) on board Nimbus 5. It measured at a single microwave frequency (19 GHz) horizontal linear polarization using an electronic across track scan method. Most radiometers following ESMR have either been across track scanning instruments or conically scanning instruments using mechanical scanning mechanisms (rotating antennas). The across track scanning instruments have mostly been used for atmospheric sounding and meteorological applications, while the conically scanning instruments including Scanning Multi-channel Microwave Radiometer (SMMR), the Special Sensor Microwave/Imager (SSM/I), the Advanced Microwave Scanning Radiometer (AMSR-E and AMSR2) and the Special Sensor Microwave Imager Sounder (SSMIS) have been used for surface applications and in particular for monitoring sea ice. Microwave radiometer data for sea ice concentration monitoring is one of the most successful and long-lasting applications of satellite remote sensing. The large difference between thermal microwave emission from water and from ice due to the large difference in permittivity enables clear distinction between these two surface types. This has led to microwave radiometer data being used in strategic planning, operational applications and in climate monitoring of sea ice for many years.

An advantage of conically scanning instruments is that they measure at a constant incidence angle and with a constant plane of polarization. This facilitates implementation of simple algorithms to retrieve, for example, sea ice concentration or sea ice drift. A more complicated imaging geometry requires more sophisticated scattering or emission models to process and compare neighbouring pixels and derive snow and ice physical properties. Therefore, it is easier to build applications using conically scanning instruments. However, the use of scattering and emission models when deriving the physical snow and ice state has several other advantages, as we will discuss further in Section 3.9.

3.2.1.1 Antenna Beam Width – Resolution

The spatial resolution of a satellite observation is given by the footprint size, which is the projection of the antenna beam of a microwave radiometer on the ground. It is in general inversely proportional to the size of the microwave antenna (typically a parabolic reflector) and proportional to the microwave wavelength. This means that to obtain a fine spatial resolution on the ground, a large antenna is needed, and with a given antenna size the finest resolution will be achieved at the highest frequencies (shortest wavelengths).

The actual footprint size also depends on satellite altitude and incidence angle. At vertical incidence footprints would be near circular, but at the typical incidence angles of 50–55 degrees, footprints become elongated. Typical antennas are designed to have maximum sensitivity (gain) in one direction and within a given beamwidth. However, since an infinitely sharp cut-off cannot be obtained, there will also be some sensitivity to radiation from areas/directions outside the footprint. This property is the origin of the so-called land spill-over effect, where radiometrically warm areas (land or ice) in the neighbourhood of radiometrically colder areas (ocean) still influence the measurements from the cold areas and vice versa.

Microwave radiometers such as SMMR, SSMI(S) and AMSR use only one parabolic reflector antenna at all wavelengths from a few millimetres to a few centimetres, and thus provide data where the resolution varies by up to a factor of 10 between the shortest and the longest wavelength. The SMMR and SSM/I featured antenna sizes of less than 1 m whereas the AMSR-E has a 1.6 m dish and the AMSR2 has a 2 m antenna. This explains the higher resolution of the AMSR instruments compared to SMMR and SSMI(S) (see Table 3.3).

In 2009 ESA launched the MIRAS (Microwave Imaging Radiometer with Aperture Synthesis) on board the Soil Moisture and Ocean Salinity (SMOS) satellite. It deploys a huge antenna structure and synthetic aperture processing in order to obtain a resolution of 35–50 km. NASA also operates an L-band radiometer on the Soil Moisture Active/Passive (SMAP) satellite. The SMAP satellite deploys a 6 m rotating antenna dish to obtain a resolution of 36 km. The penetration depth at the long L-band wavelength allows the use of these data to derive thin ice thickness (see Section 3.5).

A microwave radiometer antenna integrates the T_B distribution incident upon it by its antenna pattern weighting function. This integral is called the antenna temperature T_A. T_A is what a microwave radiometer actually measures; T_B is computed by correcting T_A for the effect of the antenna pattern. Consequently, T_B represents a linear sum of the emission from the materials in the footprint.

Since the interaction between microwave radiation and Earth surface materials depends on the polarization of the electromagnetic radiation (Section 3.1), a distinction between horizontal (H) and vertical (V) polarization is necessary. With multichannel measurements at different frequencies and polarizations it is common to refer to the individual channels of a microwave radiometer by an abbreviation of the frequency and polarization. An 18 GHz vertical polarization channel thus is referred to as 18V.

The measurement uncertainty, also called the *sensitivity*, of a microwave radiometer can be expressed in terms of its bandwidth and integration time as

$$\Delta T = \frac{T_n + T_A}{\sqrt{B\tau}}, \tag{3.8}$$

where T_n is the internal receiver noise from the electronic components of the radiometer, T_A is the antenna temperature (the quantity measured by the radiometer), B is the bandwidth of the receiver (frequency range) and τ is the integration time (dwell time at each footprint). It is clear that better sensitivity can be obtained by having less receiver noise, or by a longer integration time or more receiver bandwidth. A very sensitive (low noise) receiver is used to minimize T_n. Bandwidth is limited by the frequency band allocation (frequencies that are used by active transmitters on the ground cannot be used because these transmitters would interfere with the measurements) and in some cases by technology. Integration time has limitations on satellite instruments because fine resolution necessarily limits the dwell time per (small) antenna footprint during the scan of the antenna across the swath. The microwave radiometers listed in Table 3.3 typically have sensitivities between 0.3 and 1.5 K, corresponding to a dwell time of a few milliseconds, bandwidth of a few hundred MHz, and radiometer noise (T_n) in the order of a few hundred K.

Table 3.3: *Spatial resolution in terms of the footprint size in km of some common satellite microwave radiometers used for sea ice observations. SMMR did not have a radiometer in the 85–90 GHz range, and SSMI(S) do not have radiometers at the lower frequencies.*

Frequency (GHz)	SMMR	SSMI(S)	AMSR-E	AMSR2
6/7	95 × 148		43 × 74	35 × 62
10	59 × 41		30 × 51	24 × 42
18/19	41 × 55	45 × 70	16 × 27	14 × 22
21/22/23	30 × 46	40 × 60	18 × 31	15 × 26
36/37	18 × 27	30 × 38	8 × 14	7 × 12
85/89/91		14 × 16	4 × 6	3 × 5

3.2.2 Visible and Infrared Radiometry

The first successful Earth Observation satellite mission carrying a visible and infrared radiometer was the Television InfraRed Observation Satellite (TIROS-1), launched in 1960. Since then, there has been a wealth of visible and infrared sensors in orbit because these instruments have many important applications including providing information on ocean, sea ice and snow surface temperature, aside from their applications for weather prediction. Clear-sky thermal infrared emission measured by satellite radiometers of snow-covered sea ice in the atmospheric window around 8–13 micrometres is primarily a function of the snow surface temperature. The snow emissivity at these wavelengths is very close to 1 for dry as well as for wet snow. Some smaller emissivity variability in the infrared occurs at large incidence angles. Examples of visible (almost true colour) MODIS image can be seen in Figure 3.5, and examples of surface temperatures derived from Infrared measurements can be seen in Figure 3.6.

Examples of wide swath visible and infrared radiometers useful for operational monitoring of polar regions are the Advanced Very High Resolution Radiometer (AVHRR) on the NOAA and MetOp satellites, the MODerate-resolution Imaging Spectroradiometer (MODIS), the Advanced Along-Track Scanning Radiometer (AATSR), the Sea and Land Surface Temperature Radiometer (SLSTR) and the Visible Infrared Imaging Radiometer Suite (VIIRS) (Table 3.4).

In addition, ocean colour missions such as the Coastal Zone Color Scanner (CZCS) on Nimbus-7, the MEdium Resolution Imaging Spectrometer (MERIS) on Envisat and the

Table 3.4: *Visible and infrared instruments used for ice observations. VIS, NIR and THIR refer to the visible, near infrared and thermal infrared part of the electromagnetic spectrum. Resolution is for nadir looking. For many instruments, resolution deteriorates towards the edges of swath.*

Satellite(s)	Instrument	Resolution (m)	Channels
NOAA (1979–)	AVHRR	1100	4–5 (VIS, NIR, THIR)
MetOp (2006–)	AVHRR/3	1100	6 (VIS, NIR, THIR)
AQUA + TERRA (1999–)	MODIS	250–1000	36 (VIS, NIR, THIR)
Envisat (2002–2012)	AATSR	1000	7 (VIS, NIR, THIR)
Envisat (2002–2012)	MERIS	300–1000	15 (VIS, NIR)
NPP (2011–)	VIIRS	375–750	22 (VIS, NIR, THIR)
Nimbus-7 (1978–1987)	CZCS	825	6 (VIS)
Sentinel-3 (2016–)	SLSTR	500–1000	9 (VIS, NIR, THIR)
Sentinel-3 (2016–)	OLCI	300–1200	21 (VIS, NIR)
Landsat (1972–)	MSS+TM+ETM+OLI	15–80	4–11 (VIS and NIR, later also THIR)
Sentinel-2 (2015–)	MSI	10–60	13 (VIS, NIR)

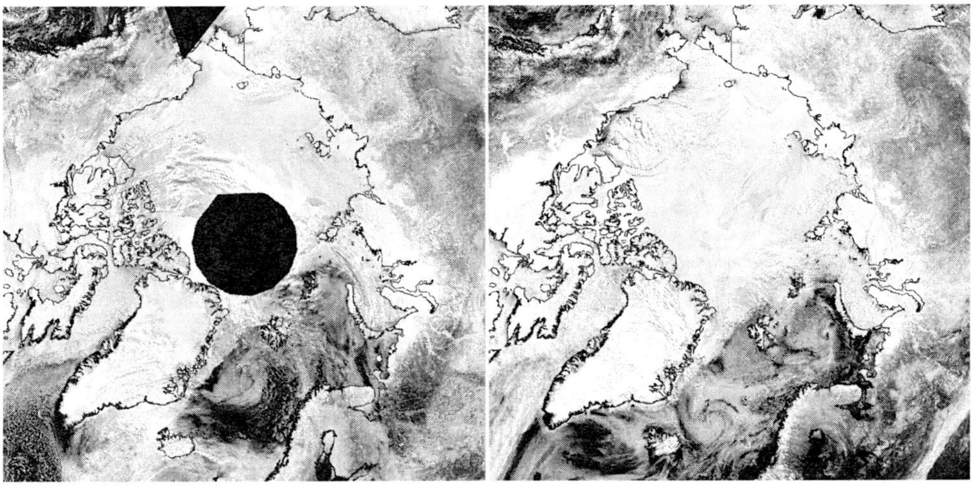

Figure 3.5: MODIS Arctic Mosaic 13 March 2016 (left) and 11 April 2016 (right). Note the black area around the North Pole on the left due to lack of sunlight. Source: https://lance.modaps.eosdis.nasa.gov/imagery/subsets/?mosaic=Arctic (A black and white version of this figure will appear in some formats. For the colour version, please refer to the plate section.)

Ocean and Land Colour Instrument (OLCI) on Sentinel-3 provide multispectral visible imagery useful for ice monitoring such as ice charting, spectral albedo and melt pond detection. Sentinel-2 and Landsat deliver even higher resolution (10–30 m) visible images. Landsat 7 and 8 also include thermal infrared radiometers. However, the Landsat satellites are mainly operated over land areas and obtain only limited coverage of sea ice.

Visible data are limited by the availability of sunlight as an illumination source. During the polar night there will be no visible imagery (see example of MODIS images from March and April in Figure 3.5). NPP VIIRS includes a night-time channel that functions using moonlight and has shown some potential during the polar night.

3.3 Active Remote Sensing

An active remote sensing instrument is an instrument that provides its own source of illumination and detects the returned radiation after it has interacted with the surroundings, typically the Earth's surface. Examples of active instruments are radars (microwave) and Lidars (laser-based instruments). Active instruments are used to measure not only the magnitude of reflected and scattered radiation from the Earth, but also to measure the distance between a satellite and the Earth (altimeters). Calibrated radars can be used to quantitatively measure the radar cross section which is a property of the Earth's surface primarily determined by the surface and volume scattering of the material.

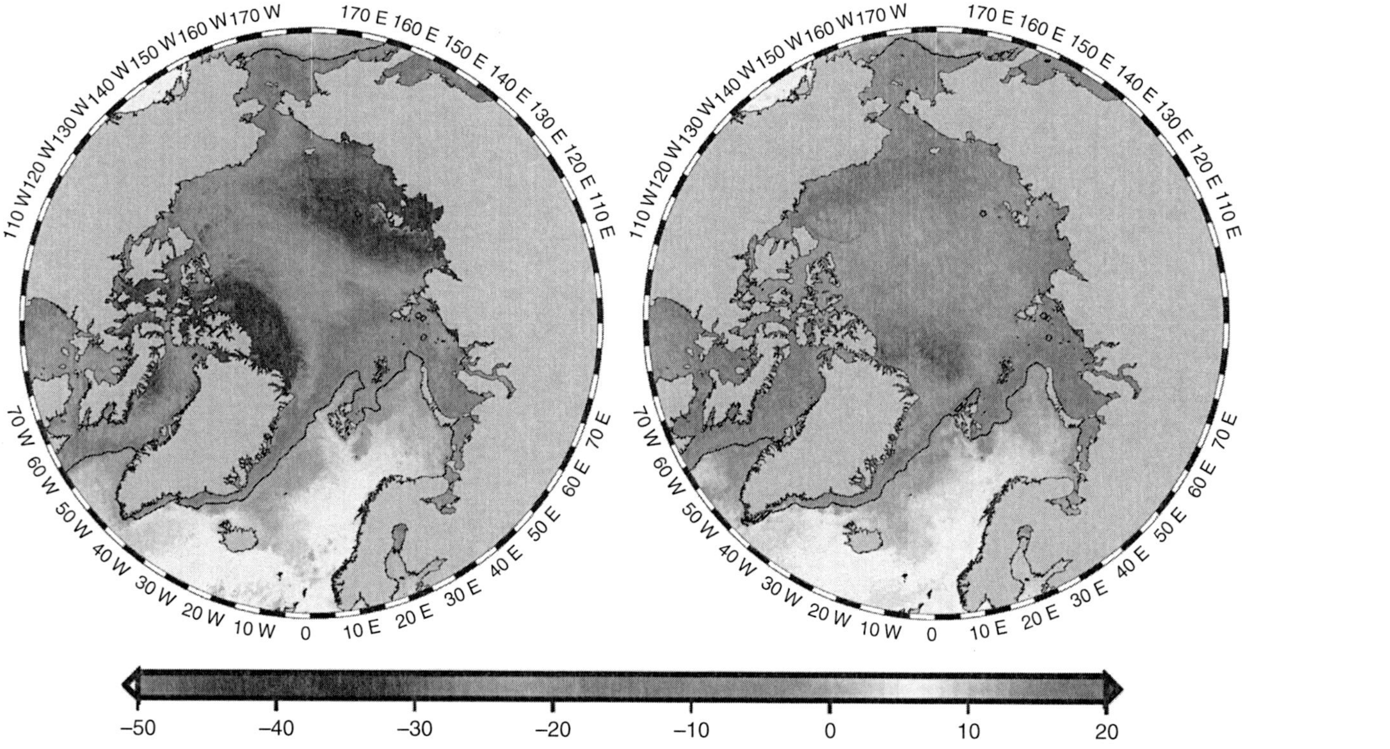

Figure 3.6: Surface Temperature computed from thermal infrared measurements by the MetOp AVHRR instrument on March 13 2016 (left) and April 11 2016 (right). http://ocean.dmi.dk/arctic/index.uk.php and the Copernicus Marine Environment Monitoring Service. (A black and white version of this figure will appear in some formats. For the colour version, please refer to the plate section.)

3.3.1 Scatterometry

A scatterometer is a radar designed to measure very accurately the magnitude of the backscattered radar signal from the surface of the Earth. Scatterometers are typically used to detect the wind-generated roughness of the ocean surface for wind speed and direction retrieval. Scatterometers typically operate at C-band (5–6 cm wavelength) or at Ku-band (2–3 cm wavelength). Since surface scattering and volume scattering in ice are substantial at these wavelengths, scatterometer data contain information on ice and snow (Barber *et al.*, 1998) and are often used to supplement other satellite data from Polar Regions.

The ERS-1 and ERS-2 satellites of the European Space Agency (ESA) carried C-band scatterometers and EUMETSAT's[1] MetOp satellites (2006–) also carry the C-band Advanced Scatterometer (ASCAT). NASA's SCATterometer (NSCAT) (launched 1996) was the first of a series of Ku-band scatterometers, followed by the SeaWinds scatterometer on the QuikSCAT satellite (1999–2009) and the Indian scanning scatterometer OSCAT on OceanSat-2 (2009–2014).

A scatterometer can be a calibrated version of a side-looking radar, such as ASCAT and NSCAT, which uses fixed antennas pointing in three directions and with incidence angles varying in the range direction or it can deploy a scanning antenna to obtain measurements at constant incidence angles, such as QuikSCAT and OSCAT.

Spatial resolution of data from space-borne scatterometers is typically in the order of 12–25 km. Dual polarization measurements (HH + VV, i.e. horizontal transmit–horizontal receive and vertical transmit–vertical receive) such as from QuikSCAT or triple incidence angle measurements such as ASCAT are used to resolve wind direction ambiguities in the ocean data but also provides information relevant for ice/ocean classification. ASCAT is a fan-bean scatterometer system with antennas pointing fore, mid and aft on both sides of the spacecraft. For every pixel along the swath there are measurements at two different incidence angles, and in three different directions. Both the backscatter as a function of incidence angle and the backscatter as a function of azimuth angle are different for ice and open water, and these qualities are often used to distinguish ocean surface signals from the isotropic backscatter from sea ice.

The C-band radar backscatter coefficient (σ^0) at oblique incidence angles is primarily a function of surface scattering processes from rough water or ice surfaces. The volume backscatter contribution, especially from the upper layer of MY ice, is however not negligible. In X- and Ku-band (2–3 cm wavelength) both surface and volume scattering processes are important for the total backscatter coefficient. Radars thus offer data that allow distinction between ice types, either due to their surface roughness differences or due to differences in volume scattering. It is worth noting that the backscatter coefficient from MY ice (typically −6 dB at Ku-band) is an order of magnitude higher than the backscatter coefficient from FY ice (typically −16 dB at Ku-band). Therefore, the backscatter coefficient is normally given on a logarithmic scale in decibels. The conversion between linear (*lin*) and logarithmic (*dB*) backscatter coefficients σ^0 is:

$$\sigma^0(dB) = 10x \log_{10}(\sigma^0(lin)) \tag{3.9}$$

[1] European Organization for the Exploitation of Meteorological Satellites

Because of the large differences between the backscatter coefficient of different surface types or surface features it is not possible to derive sub-pixel distributions of surface types as it is with a microwave radiometer. It is also important to use the linear backscatter coefficients when computing the statistical properties of a dataset such as mean and standard deviation or when computing ratios of backscatter at different angles or polarizations.

Surface scattering in radar applications is a function of interface roughness and permittivity differences and can be approximated using, for example, the Integral Equation Method, geometric optics etc. For a more in depth treatment of these topics see Ulaby (2013). Volume scattering inside materials is a function of the size and distribution of the scatterers (snow grains, air bubbles or brine inclusions), the wavelength, the volume fraction of scatterers, the layer depth and of the permittivity of the background and the scatterer materials. A mathematical description of the volume scattering coefficient such as the Improved Born Approximation is appropriate for snow and sea ice where the scatterers are closely packed so that the scattering from individual particles influences the scattering of its neighbours (Tsang and Kong, 2001 and Fung, 1994).

Figure 3.7 illustrates how the difference between VV and HH backscatter in QuikSCAT data can be used to distinguish ice from water. Even areas of rather small ice concentrations are detected as ice.

3.3.2 Synthetic Aperture Radar

A Side-Looking Airborne Radar (SLAR) is a radar similar to a ship radar but with an antenna that does not rotate, but is mounted on the side of an aircraft or a satellite. The antenna illuminates an area to the side of the aircraft from which a line in an image is constructed by recording the returned power of the radar-pulse as a function of time. Strong targets (large backscatter) appear bright in the image line and weaker targets appear darker. An image is constructed by appending return signals from consecutive radar pulses in consecutive image lines. The direction along the recorded lines is referred to as the range direction and the direction along the flight direction as the azimuth direction. Incidence angle varies with range.

A major limitation of a SLAR is the azimuth resolution. The azimuth (along-track) resolution X_a is determined by the antenna beam width, which is inversely proportional to the horizontal size of the antenna and proportional to the distance (range) between the target and the antenna.

$$X_{a,SLAR} = \lambda R/l, \tag{3.10}$$

where λ is the wavelength, R is the distance from the radar to the target and l is the length of the antenna. The range resolution X_r is a function of the pulse length of the radar and in order to generate a very short pulse (for achieving high range resolution) a large microwave bandwidth (frequency band) is needed

$$X_r = c\tau/2\sin\theta, \tag{3.11}$$

where τ is pulse length, θ is incidence angle. A pulse length of 10 ns at the speed of light corresponds to a distance (in the air) of about 30 m. Due to the two-way propagation of

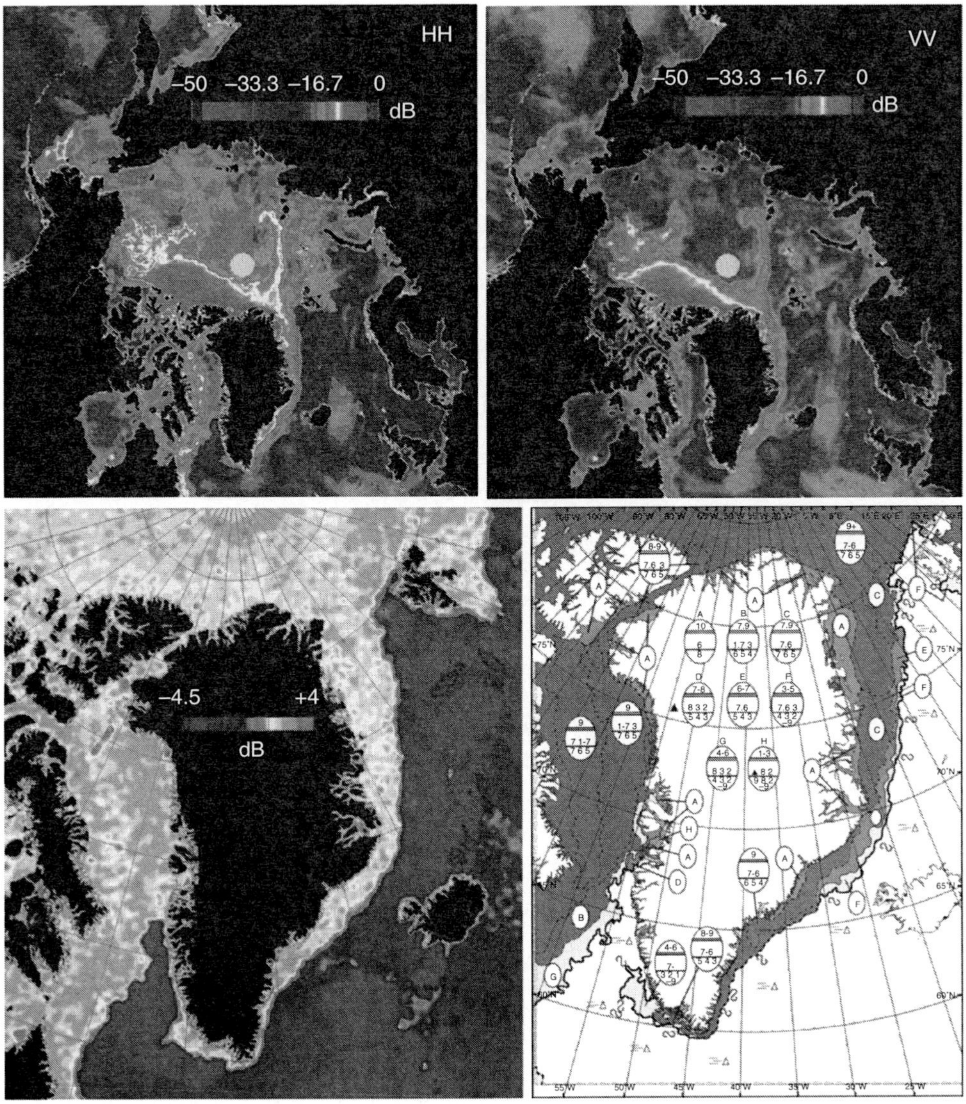

Figure 3.7: QuikSCAT scatterometer data from 11 April 2008. HH-polarization (upper left) and VV-polarization (upper right), ratio of HH and VV (lower left). An ice chart from the Danish Meteorological Institute from 13 April 2008 (lower right). Source: www.seaice.dk (A black and white version of this figure will appear in some formats. For the colour version, please refer to the plate section.)

a radar signal, a 30 m pulse length allows two targets to be distinguished, spaced >15 m apart. This means that the spatial resolution in range in this case is about 15 m.

A Synthetic Aperture Radar (SAR) is a side-looking radar system where signal-processing techniques are applied to synthesize a very long antenna and thus obtain

Table 3.5: *Some satellite SAR missions used for sea ice monitoring. The swath width, resolution and polarizations given are for the commonly used combinations for sea ice monitoring. Most SAR instruments can be operated in many different modes with different resolutions, polarizations and swath width. Typically, more polarization options are available only with a narrower swath.*

Satellite(s)	Instrument	Swath (km) / Resolution (m)	Polarizations	Band
Seasat (1978)	SAR	100 / 25	HH	L
ERS-1 & 2 (1991–2011)	SAR	100 / 30	VV	C
ALMAZ (1991–1992)	SAR	40 /15	HH	S
RADARSAT (1995–2013)	SAR	500 / 100	HH	C
RADARSAT 2 (2007–)	SAR	500 / 100	HH+HV or VV+VH	C
Envisat (2002–2012)	ASAR	400 / 120	HH/VV	C
COSMO–SkyMed (2007–)	SAR	200/100	HH or VV	X
TerraSAR-X (2007–)	SAR	30 / 20 and more	HH, VV etc	X
ALOS (2006–11)	PALSAR	250–350 / 100	HH or VV	L
ALOS-2 (2014–)	PALSAR-2	350–490 / 60–100	HH+HV or VV+VH	L
Sentinel-1 (2014–)	SAR	400 / 90	HH+HV or VV+VH	C

a very high resolution along the flight track (azimuth), independent of the range to the target.

$$X_{a,SAR} = l/2, \tag{3.12}$$

where l is still the physical length of the antenna. The azimuth resolution of a SAR is thus independent of range to the target. The synthetic antenna length can be several kilometres. Typical SAR resolutions are a few metres, but in order to reduce noise in the data, spatial averaging (so-called multi-looking) is often applied, which reduces resolution to 10–100 m.

SAR observations are currently the primary satellite data sources for ice charting. The images of ice are characterized by their relatively high resolution in addition to their all-weather capability to see through clouds and without solar illumination. Differences in surface roughness between sea ice and the ocean and between different types of sea ice are the primary image characteristics of interest to National Ice Information Services (NIIS). Ice analysts use the magnitude of the radar backscatter, as well as the spatial pattern (texture and shape), to distinguish ice from ocean, to identify ice types, and to derive floe size. Examples of SAR images of sea ice can be seen in Figure 3.8.

Since both ocean and sea ice backscatter are a function of incidence angle, an incidence angle effect is a distinct feature of most SAR images. Images are generally brighter in the near range (steep incidence angle) and darker in the far range. The exact incidence angle variation for the ocean depends on wind speed. For ice the variation is,

Figure 3.8: Left: Sentinel-1 SAR image of the West coast of Greenland on 2 April 2016. The image shows typical areas of FY ice in Baffin Bay to the left and areas of fast ice in the fjords and along the coast. Right: Sentinel-1 SAR image of the East coast of Greenland on 12 April 2016. The image shows typical rounded MY ice floes at the top and meandering patterns of broken ice in the marginal ice zone. Source: www.seaice.dk

in general, much smaller than for the wind-roughened ocean. Cases where ice is darker than the ocean in near range and brighter than the ocean in far range are thus often seen, and constitute one of the ambiguities that obstruct the automatic retrieval of ice concentration from SAR images. Since the incidence angle effect over the ocean depends on wind speed, automatic correction procedures are difficult to apply. In cases where such procedures are applied, they often just 'correct' for the effect over sea ice, and a residual ocean effect remains.

Satellite radars require transmitting an electromagnetic signal towards the ground, and this requires power on board the satellite. Since satellite power is limited by the size of the solar panels charging the batteries, typical satellite SAR missions can only operate the radar between 10 and 25 percent of the time. This fact, together with the limited swath of max 400–500 km, means that one satellite such as RADARSAT or Envisat cannot deliver daily SAR images of all sea ice covered regions of the Earth. To compensate, the European Sentinel-1 mission is a constellation of two satellites, and Canada's RADARSAT Constellation Mission (RCM) will consist of three satellites. The Italian COSMO–SkyMed SAR constellation consists of four satellites.

3.3.3 Radar Altimetry

A radar altimeter measures the distance from the satellite to the ground, or changes in this distance, along the satellite flight track. The instrument measures the travel time of an electromagnetic signal sent towards the Earth directly from above (nadir) and reflected back to the sensor. This travel time is converted to a distance using the speed of light in air. By precisely knowing the position of the satellite in its orbit relative to the Earth's ellipsoid,

Table 3.6: *Some examples of space-borne altimeters used for sea ice thickness monitoring.*

Satellit	Instrument	Type	Frequency/band
ERS-1 & 2 (1991–2011)	RA-1	Radar Altimeter	Ku-band (13.8 GHz)
Envisat (2002–2012)	RA-2	Dual band Radar Altimeter	Ku-band (13.575 GHz) and S-band (3.2 GHz)
Cryosat (2010–)	SIRAL (SAR Interferometer Radar Altimeter)	SAR Interferometer Radar Altimeter	Ku-band (13.575 GHz)
SARAL/AltiKa (2013–)	AltiKa	Radar Altimeter	Ka-band (35.75 GHz)
Sentinel-3 (2016–)	SRAL (SAR Radar Altimeter)	Dual band SAR Altimeter	C-band (5.41 GHz) and Ku-band (13.575 GHz)
ICESat (2003–2009)	GLAS (Geoscience Laser Altimeter System)	Laser Altimeter	1064 nm (IR) and 532 nm (visible green)
ICESat-2 (2018–)	ATLAS (Advanced Topographic Laser Altimeter System)	Laser Altimeter (photon counting)	532 nm (visible green)

changes in the distance to water and floes is used to infer the elevation of sea ice floating on the ocean, relative to the sea surface. The elevation of the ice above the sea surface is called the freeboard. The sea ice freeboard is converted to sea ice thickness using the Archimedes principle and a priori knowledge of snow depth and density along with ice and water density (see Section 3.5.1).

Radar altimeters typically operate at frequencies in the Ku-Band of the electromagnetic spectrum. The radar signal penetrates dry snow and is thus assumed to be reflected at the ice–snow interface under cold and dry conditions. A radar altimeter therefore is used to determine the sea ice freeboard directly.

Radar altimeters, such as on ERS-1 and -2, Envisat, Cryosat-2 and Sentinel-3, are profiling instruments with a typical spatial footprint size of a few kilometres. The azimuth resolution is in the order of 7 km for all these altimeters. In order to determine the reference sea level along the flight track, it is important that a number of the distance measurements are from open water inside the icepack. The SAR (Synthetic Aperture Radar) altimeter technique was developed to reduce the footprint of the altimeter to allow more measurements from the often narrow leads during winter as well as from smaller floes and areas of homogenous ice thickness. Instruments on Cryosat-2 and Sentinel-3 employ the SAR-altimeter technique which improves their along track resolution to about 250 m.

The nadir-looking Ku-band radar backscatter is dominated by surface scattering processes, i.e. backscattering from smooth snow, ice and lead surfaces or interfaces. Volume scattering

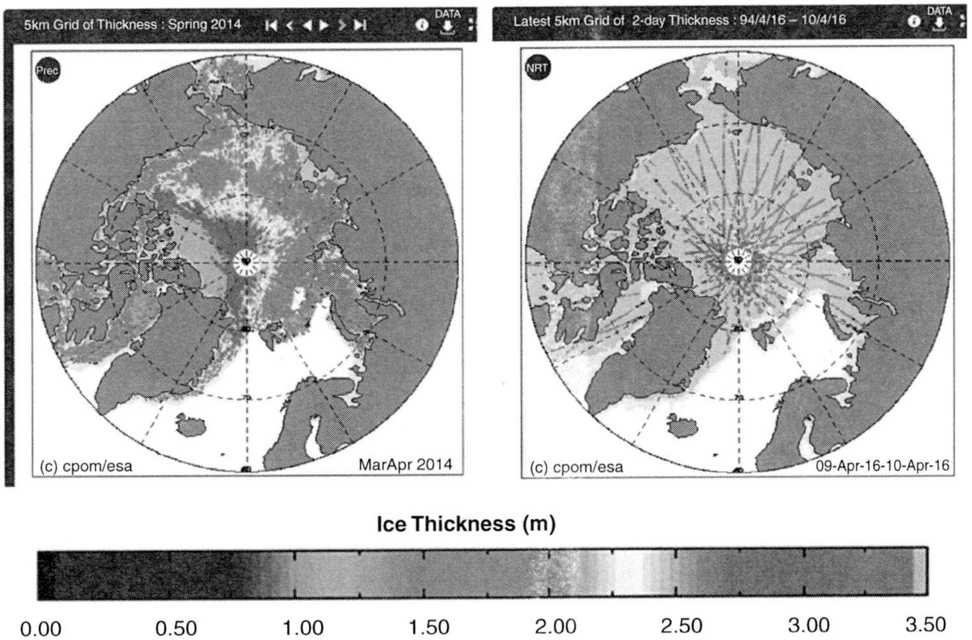

Figure 3.9: Cryosat-2 radar altimeter data showing retrieved sea ice thickness during March–April 2014 (left) and an example of two days' profiles from April 2016 (right). Source: www.cpom.ucl.ac.uk/csopr/seaice.html (A black and white version of this figure will appear in some formats. For the colour version, please refer to the plate section.)

may contribute to attenuation of the signal within the snow cover. However, volume scattering from snow grains or air bubbles in the ice is negligible as a source of backscatter. It is normally assumed that the effective scattering horizon is synonymous with the snow–ice interface during winter when the snow is dry and considered transparent. However, the snow does affect the backscatter and the range estimation (Tonboe *et al.*, 2010). Microwave absorption is high if the snow is moist or saline and in such cases the effective scattering surface is not necessarily the snow–ice interface. For these reasons and due to the presence of melt ponds which dominates the backscatter during summer, radar altimetry is mainly used during the colder part of the year when the snow is dry and considered transparent. However, even in winter horizontal crust layers in the snow may still influence the effective depth and apparent scattering surface. The influence of snow properties on the effective depth will be stronger at higher microwave frequencies such as Ka-band where the effective scattering horizon cannot be considered to be the snow – ice interface.

3.3.4 Laser Altimetry (Lidar)

A laser altimeter accurately measures the distance from the instrument (aeroplane or satellite) to the surface. Laser altimeters were first flown on aeroplanes to make detailed maps of (ice) surface topography. Airborne laser altimeters are often scanning, enabling them to produce 3-D maps of ice surface topography.

A laser altimeter operates in the visible or near infrared frequency range (Table 3.6) where penetration into snow/ice is very limited and the laser signal is thus mostly reflected at the snow surface or, if the ice is bare, at the ice surface. A laser altimeter therefore is used to determine the so-called total (sea ice plus snow) freeboard. Like radar altimeters, laser altimeters require regular measurements of the reference sea level in order to convert the distance measurement to freeboard. The laser does not penetrate clouds and therefore space-borne systems can only be used for sea ice applications when there are no or only thin clouds.

In 2003, NASA launched the ICESat satellite with a profiling (green) laser on board for measuring cloud top height and ground topography. ICESat provided the first large-scale laser-based maps of sea ice freeboard and derived thickness from the Arctic Ocean. Due to laser reliability problems ICESat was only operated for 2–3 periods of approximately 1 month duration per year from 2003 to 2009. Current NASA plans are to launch an ICESat-2 in 2018. The laser footprint size of ICESat was 70 m on the ground and footprints were located 170 m apart along the flight track.

Since a laser altimeter measures the snow plus ice freeboard and a radar altimeter more closely measures the ice freeboard (in winter) a combination of radar and laser potentially could be used to estimate snow thickness.

3.4 Observation of Ice Concentration and Ice Type

Ice concentration is defined as the fraction of a specific ocean area covered by ice of any type and any thickness. Ice type concentration (or partial concentration) refers to the fraction of an ocean area covered by ice of a particular type (e.g. MY ice, FY ice and new ice). Ice concentration is normally given as a percentage, but may also be specified as a fraction (a number between 0–1) or as tenths (0–10), the latter being used in ice charts (Chapter 1 and Section 3.8).

3.4.1 Microwave Radiometer Observations

Microwave radiometer measurements at 19 and 37 GHz can distinguish at least two different ice types, i.e. MY ice and FY ice, which are the most abundant in Polar Regions. These ice types are distinct because of different surface permittivity and, in particular, scattering magnitude and snow cover as described earlier in this chapter. The physically distinct ice types have different microwave signatures (brightness temperatures or derived quantities like their ratios or differences) at 19 and 37 GHz (Tucker *et al.*, 1991). At higher frequencies (>50 GHz) where the radiation penetrates only into the snow, ice types are less distinct and snow surface roughness, density and grain size dominate the emissivity variability.

Most algorithms for retrieving sea ice concentration from microwave radiometer data are based on the large difference in passive microwave signatures of sea ice and open water. This difference allows estimating the percentage of sea ice within the satellite footprint. For example, the difference between vertically and horizontally polarized brightness temperatures at 19 or 37 GHz is substantially smaller for ice than for water, and this difference is a key element in many algorithms, such as NASA Team (Cavalieri *et al.*, 1984), for retrieval of sea ice concentration. In addition, the increased volume scattering of MY ice relative to FY ice reduces the brightness temperature at higher frequencies and this can be used to distinguish the two ice types using differences between brightness temperatures at 19 and 37 GHz. A modified version of the NASA Team algorithm (NT2 – Markus and Cavalieri, 2000) includes, additionally, the 85 GHz channels allegedly to decrease the surface effects on the emissivity and also uses a lookup table of modelled atmospheres to potentially reduce weather effects. Apparent performance improvements over the original NASA Team algorithm were mainly due to overestimation of high concentrations and underestimation of low concentrations combined with the capping of the concentrations to 0–100 per cent (Ivanova *et al.*, 2015).

The Bootstrap algorithm (Comiso, 1986) obtains sea ice concentration from cluster analysis in the brightness temperature space utilizing channels 19V, 37V and 37H, and has two respective modes – polarization mode (channels 37V and 37H) and frequency mode (channels 19V and 37V). Frequency mode is used at low concentration since it is less sensitive to atmospheric noise and wind roughening of the ocean surface, whereas the polarization mode is used at higher concentration where the air is dryer and where the use of 37 GHz only, improves the spatial resolution without too much atmospheric noise.

The Bristol algorithm (Smith, 1996) was developed to overcome the discontinuity in the Bootstrap concentrations when switching from frequency to polarization mode. Similar to the Bootstrap algorithm, cluster analysis is used, (here in three dimensions using 19V, 37H and 37V channels simultaneously). In the Bristol algorithm the polarization and frequency modes are combined into one by introducing a coordinate transformation from the three dimensions mentioned above into two dimensions of ice concentration and ice type using principal component analysis.

More recent ice concentration algorithm developments are from EUMETSAT Ocean and Sea Ice Satellite Application Facility (OSISAF) (Tonboe *et al.*, 2016) and ESAs Sea Ice Climate Change Initiative (SICCI) (Ivanova *et al.*, 2015). These are hybrid algorithms combining the Bristol and the Bootstrap frequency mode for consolidated ice and open water, respectively, thus improving the accuracy of the concentration retrievals. In addition, these algorithms use a priori knowledge of atmospheric variables to reduce weather contamination. An example of a sea ice concentration retrieval obtained by the SICCI/OSISAF algorithm with per-pixel uncertainty is shown in Figure 3.10. The total uncertainty shown in the figure is composed of the retrieval algorithm uncertainty and the smearing uncertainty. The latter causes relatively large uncertainties in the marginal ice zone, where the footprint mismatch effect (combining data at different microwave frequencies that have different footprint sizes) plays an important role when producing ice concentrations on a finer grid spacing than the footprint size.

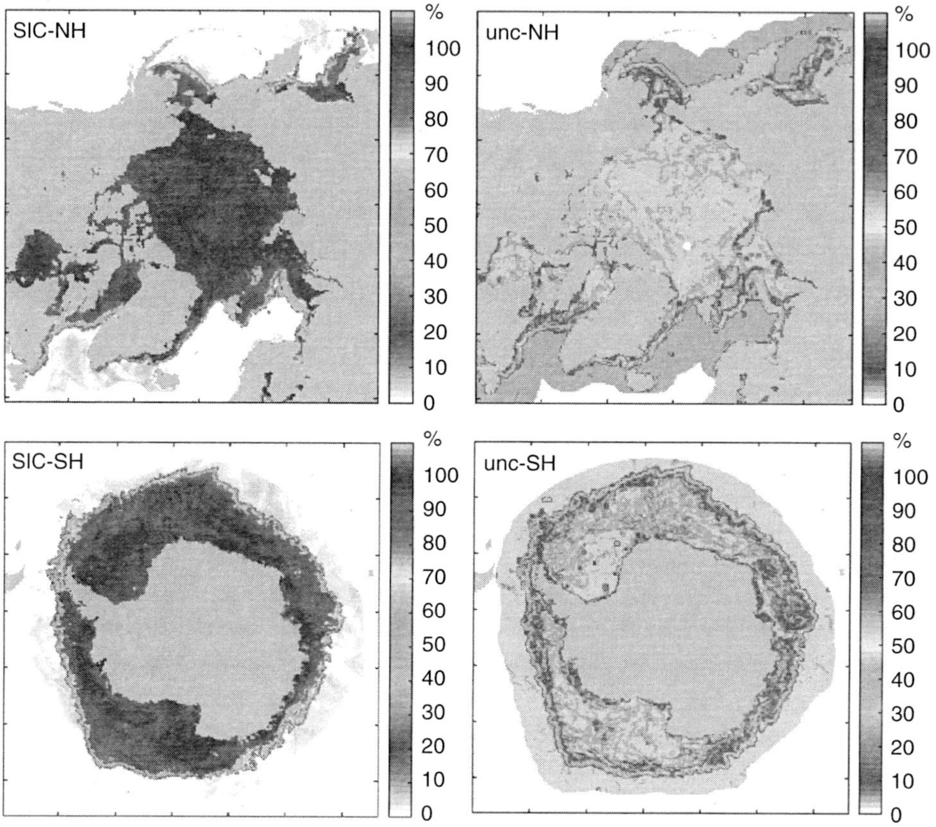

Figure 3.10: Sea ice concentration (SIC) (left) and uncertainty (right) retrieved by the SICCI algorithm from AMSR-E brightness temperatures. Northern Hemisphere (NH): 15 March 2011, Southern Hemisphere (SH): 15 September 2011 (A black and white version of this figure will appear in some formats. For the colour version, please refer to the plate section.)

Spurious ice, caused by ambiguities between signatures of lower sea ice concentrations and atmospheric emission from water vapour and clouds and from the wind roughened ocean surface, is present in most sea ice concentration data. The effect can be reduced by applying atmospheric corrections to the input brightness temperatures using data from numerical weather prediction such as are done in the OSISAF/SICCI algorithms. Alternative methods to remove such artefacts, are the so-called weather filters (i.e. ratios of brightness temperatures in microwave channels with different sensitivity to atmospheric features), sea surface temperature filters that removes ice when SST from ocean analysis is above some threshold (e.g. 5 degrees), or sea ice climatology filters that removes spurious ice in areas where no ice is supposed to appear. The drawback of using weather filters is that they are designed to also remove most sea ice up to 15 per cent concentrations and have been shown to sometime remove sea ice up to 25 per cent (Ivanova *et al.*, 2015). Thus, ice concentration data products where weather filters have been applied do not contain lower ice concentrations.

In order to obtain higher spatial resolution, algorithms such as the ASI (Artist Sea Ice) algorithm (Kaleschke *et al.*, 2001; Spreen *et al.*, 2008) use only higher frequency channels (85/89/91 GHz). However, such algorithms are more sensitive to atmospheric influence since the atmosphere is less transparent at these higher frequencies. Such influence results in spurious sea ice retrieval in ice-free areas. Therefore, the ASI and other near 90 GHz algorithms normally employ weather filters.

The OSISAF/SICCI algorithms do not apply weather filters but rather corrects the measured brightness temperatures using data from a numerical weather prediction model (in order to preserve low concentrations in the product). As can be seen in Figure 3.10, some spurious ice remains around southern Greenland (in the example) due to imperfections in the atmospheric data.

There is normally snow on FY ice, and variability in the snow properties dominates the microwave signature variability of ice (Barber *et al.*, 1998). The formation environment, deformation and snow and ice metamorphosis processes affect FY ice surface roughness. Snow melt processes change the microwave signatures during summer where melt water also drains most of the brine from the ice. After the melt period, the remaining (now MY) ice appears with an undulating landscape with elevated, rounded hummocks (which were ridges and deformation features before the summer), and with old snow, porous desalinated ice, and refrozen melt ponds with smooth surfaces even after refreezing. Ice signature variability (except for the FY/MY distinction) is normally not taken into account in ice concentration algorithms and is the main source of uncertainty at higher ice concentrations.

In spite of the issues outlined above, sea ice concentrations derived from the 40 years of microwave radiometer measurements are used for strategic planning of ship routes and as input (boundary conditions) for operational modelling of the ocean and atmosphere. In addition, they are used to derive climate records of sea ice extent (Figure 3.11).

The microwave signatures of thin ice differ slightly from those of thicker FY ice even at higher microwave frequencies, and the majority of sea ice concentration algorithms tend to underestimate the sea ice concentration when thin ice dominates. Studies (e.g. Heygster *et al.*, 2014) have shown that this underestimation is strongest at lower frequencies (6–37 GHz). Ice thicknesses up to 20–30 cm lead to underestimation of concentration. Figure 3.12 shows apparent concentration of thin ice for a number of sea ice concentration algorithms. The physical reasons for this effect is the generally smoother surface and higher salinity of thin ice and the lack of snow cover resulting in higher reflectivity and hence lower emissivity.

Figure 3.13 illustrates the performance of two widely used ice concentration algorithms during summer. Passive microwave algorithms cannot distinguish open water (ocean) from melt ponds due to the lack of penetration into water at microwave frequencies. Some algorithms, however, overestimate the ice surface fraction during summer melt due to the high brightness temperatures of wet snow. Such an overestimation may partly compensate for the fact that the melt ponds are seen as open water and thus not included in the ice concentration. However, the algorithms cannot distinguish, for example, 100 per cent ice with 20 per cent melt ponds from 80 per cent ice and 20 per cent open water! The actual

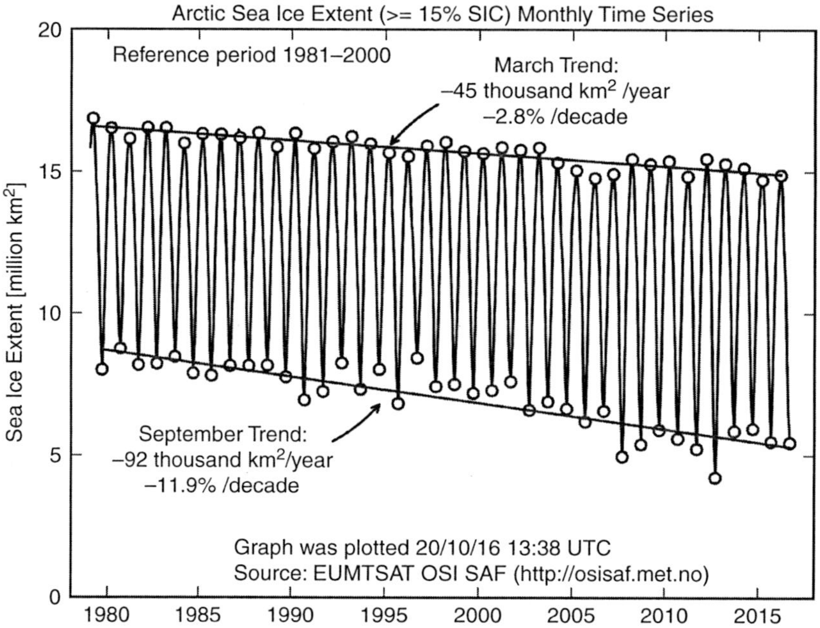

Figure 3.11: Arctic monthly sea ice extent from 1979 to the present time based on sea ice concentration computed from SMMR, SSM/I and SSMIS microwave radiometers using the OSISAF sea ice concentration algorithm. Source: http://osisaf.met.no.

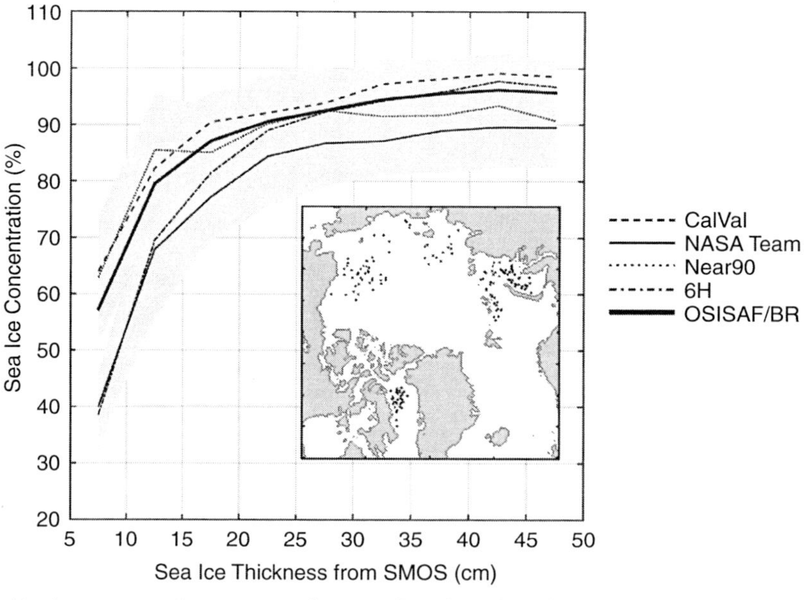

Figure 3.12: Apparent sea ice concentration as a function of sea ice thickness for some of the most common sea ice concentration algorithms. The CalVal algorithm is similar to the Bootstrap frequency mode algorithm. The locations of near 100 per cent thin ice reference data identified in Envisat SAR image during the fall of 2010 shown in insert map. Sea ice thickness in these locations was subsequently estimated from SMOS measurements The algorithms are CalVal (Ramseier, 1991), NASA Team, Near90 (Svendsen *et al.*, 1987), 6H (Pedersen, 1994) and OSISAF (Tonboe *et al.*, 2016).

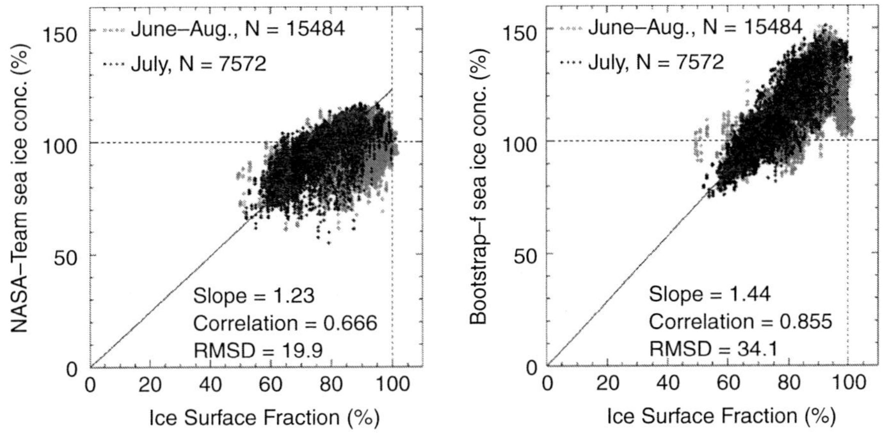

Figure 3.13: Retrieved sea ice concentration from microwave radiometry vs. ice surface fraction determined from cloud free MODIS images. Note the underestimation of sea ice concentration during melt pond period. All data have near 100 per cent ice concentration.

effect of melt ponds and summer ice signatures on a particular algorithm should be considered when using these data in summer periods (Ivanova *et al.*, 2015).

Identifying the ice edge from coarse resolution satellite data is not straightforward. Even when the edge is sharp at the metre scale, it may be difficult to delineate in coarse resolution data from microwave radiometers seeing both ice and water within its 10–50 km footprint. In sea ice concentration data from microwave radiometers, the edge is often defined as the 15 per cent contour. This generally coincides fairly well with the actual edge, and the reason for selecting a higher ice concentration threshold than 0 for defining the edge is to avoid the smearing of the data due to the large footprint as well as the influence of measurement/ atmospheric noise over open water and due to the fact that algorithms deploying weather filters on average have removed ice below 15 per cent. The extent of sea ice is defined as the total area of all grid points/pixels with an ice concentration above a certain value (e.g. 15 per cent). Including all areas with an ice concentration above 15 per cent in the ice extent also means including most melt-ponded ice during summer (see Figures 3.11 and 3.13), even though melt-ponds, when not frozen, are seen as open water by microwave radiometers. Ice extent is therefore considered as a more robust measure of summer sea ice than the area of ice, which does not include the melt ponds. Note also that, with this definition, ice extent will depend on the grid cell size used in the computation.

3.4.1.1 Evaluation and Quality Control

For evaluation of sea ice concentration retrievals from passive microwave measurements, independent data sources are used: ice charts, in situ and ship observations, and data from other satellites with high-resolution optical instruments (LANDSAT, Sentinel-2). Due to the relatively large footprint of satellite passive microwave observations, comparison to in situ and ship observations is very difficult (Ivanova *et al.*, 2015). An example of algorithm evaluation at sea ice concentrations of 0 per cent and 100 per cent is shown in Figure 3.14. Reference areas of 0 per cent ice were identified based on ice charts, and areas of 100 per cent ice were found as areas of convergent ice motion in subsequent daily winter SAR images (see Section 3.6). The figure shows the standard deviation of the difference between the reference sea ice concentrations and sea ice concentrations estimated using a large number of algorithms applied to brightness temperature measurements from the AMSR-E and AMSR2 microwave radiometers. All results are derived without applying weather filters. Typical standard deviations of the better sea ice concentration algorithms are 2–5 per cent. These results confirm earlier findings by Kwok (2002) and Andersen *et al.* (2007), who showed that passive microwave sea ice concentration algorithms do not reflect the ice concentration variability in the Arctic adequately when the concentration is near 100 per cent. Variability due to actual ice concentration changes in the order of less than 3 per cent is below the noise floor of the algorithms.

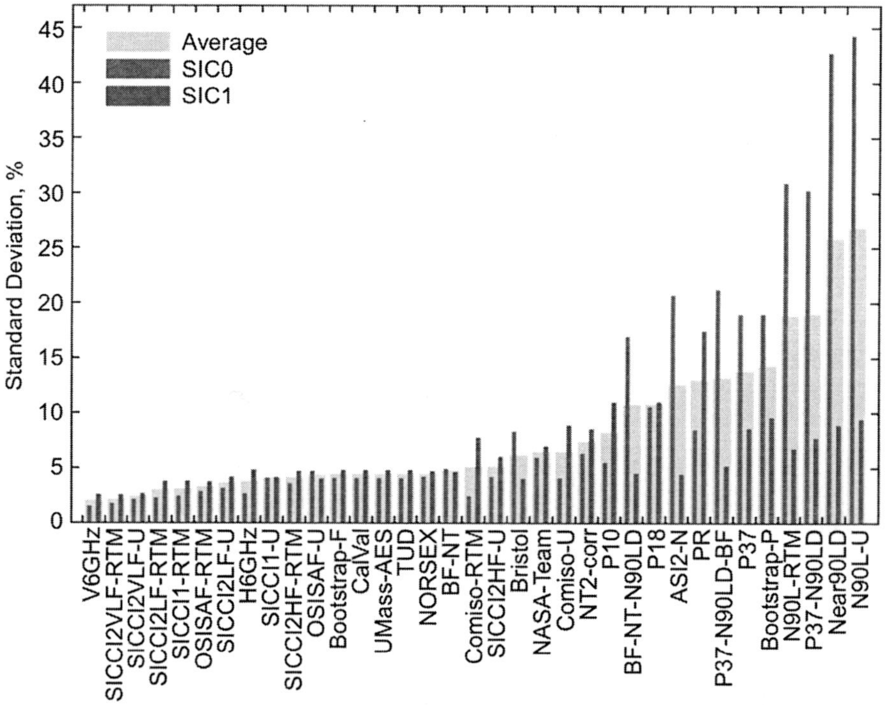

Figure 3.14: Standard deviation of difference between the reference sea ice concentration and that estimated the retrievals by various sea ice concentration algorithms for the instruments AMSR-E and AMSR2. Red bars are for high concentrations (100 per cent, SIC1) and blue bars are for low concentrations (0 per cent, SIC0). Grey bars are averages of low- and high-concentration results. Note that these results are obtained without weather filters and that all algorithms were tuned to AMSR-E and AMSR2 data separately to obtain near zero biases. (A black and white version of this figure will appear in some formats. For the colour version, please refer to the plate section.)

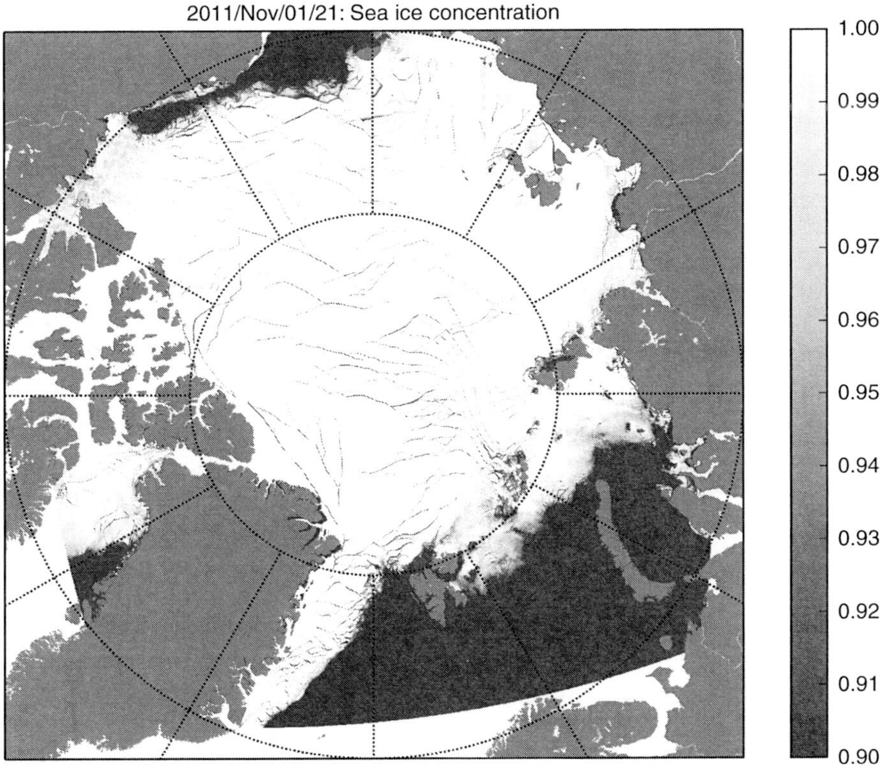

Figure 2.3: Sea ice concentration on 21 November 2011 as simulated by the MITgcm. The spatial resolution is 1 km.

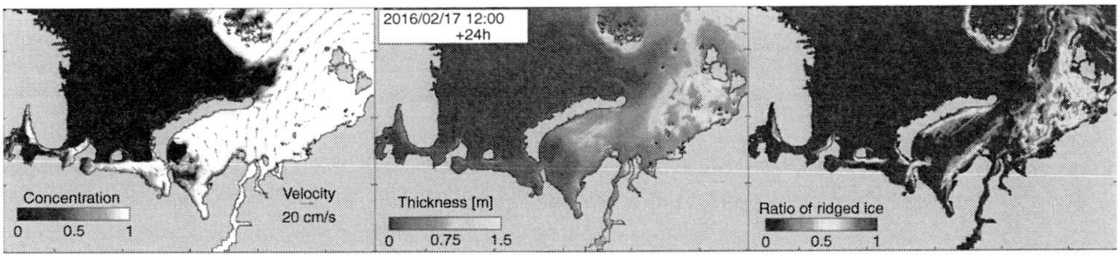

Figure 2.13: 24-hour forecast of sea ice concentration (left), drift (left), mean thickness (middle) with neXtSIM-F at 3 km resolution. The right panel displays the volume fraction of ridged ice in each grid cell of the model mesh.

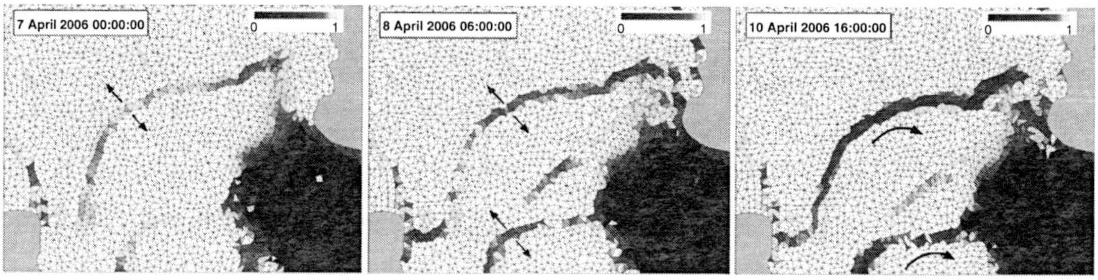

Figure 2.14: Example of a simulated sea ice concentration field and the underlying moving mesh obtained using a 10 km pan-Arctic neXtSIM configuration. The panels show a zoom on the Fram Strait, between the Northern East tip of Greenland (lower left corner) and the Spitzbergen Island (upper right corner). neXtSIM is a continuum-based sea ice model defined in a Lagrangian framework. As it uses an unstructured grid, it is discretized with the finite-element method instead of finite differences for structured grids. The panels illustrate how localized divergence generates discontinuities in the ice cover and how these discontinuities are preserved over time by the Lagrangian advection scheme.

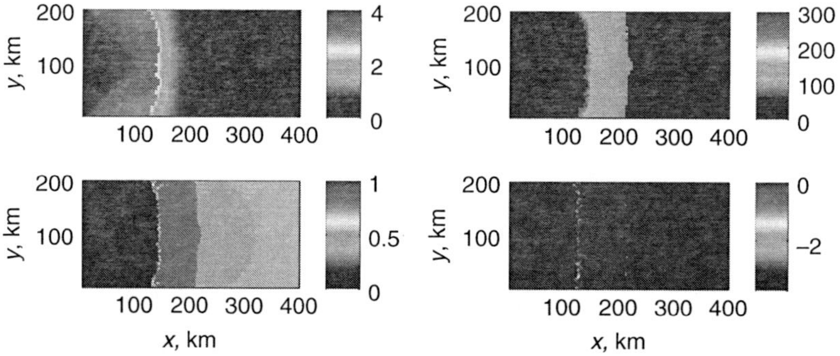

Figure 2.15: Waves breaking ice in an idealized experiment (the right-hand, upper, lower lines of grid cells correspond to land). The wave model, based on Williams *et al.* (2013a,b), is coupled to the neXtSIM sea ice model. The figure shows results after 48 h of steady pushing by a wave spectrum with significant wave height of 4 m and a peak period of 12 s, that is arriving from the left. Top left: significant wave height (m); top right: maximum floe size (m); bottom left: ice concentration; bottom right: $\log_{10}(1 - d)$, where d is the damage. Note the movement of and compression at the ice edge due to wave radiation stress as the waves continue to arrive at the ice. The ice has initial constant concentration of 0.7, thickness of 1 m, maximum floe size of 300 m, and $d = 0$.

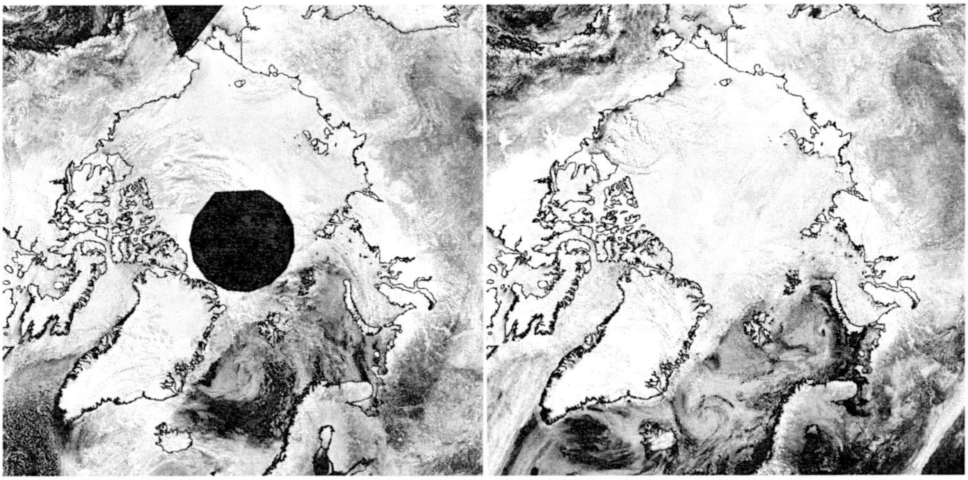

Figure 3.5: MODIS Arctic Mosaic 13 March 2016 (left) and 11 April 2016 (right). Note the black area around the North Pole on the left due to lack of sunlight. Source: https://lance.modaps.eosdis.nasa.gov/imagery/subsets/?mosaic=Arctic

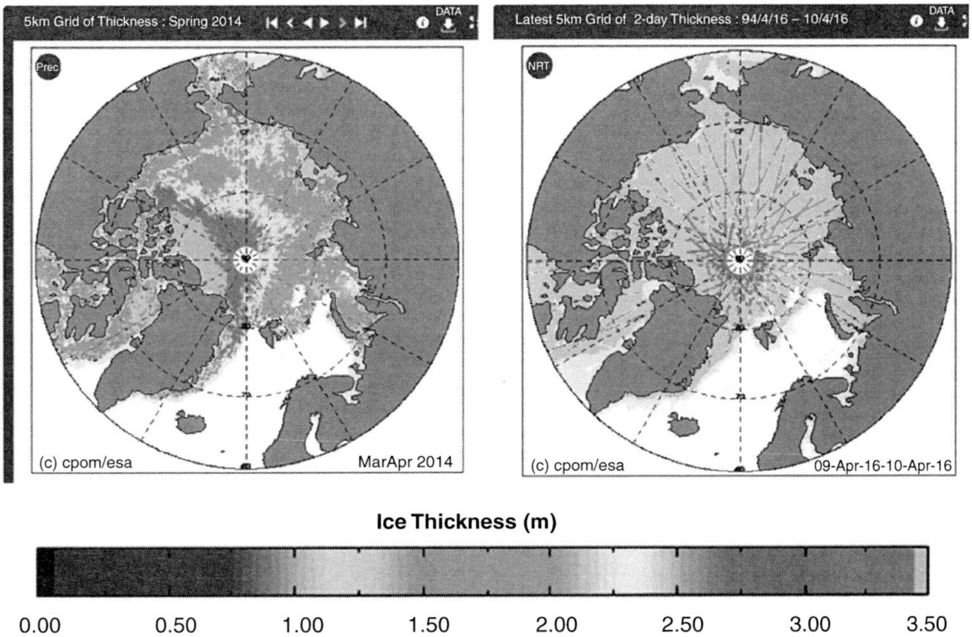

Figure 3.9: Cryosat-2 radar altimeter data showing retrieved sea ice thickness during March–April 2014 (left) and an example of two days' profiles from April 2016 (right). Source: www.cpom.ucl.ac.uk/csopr/seaice.html

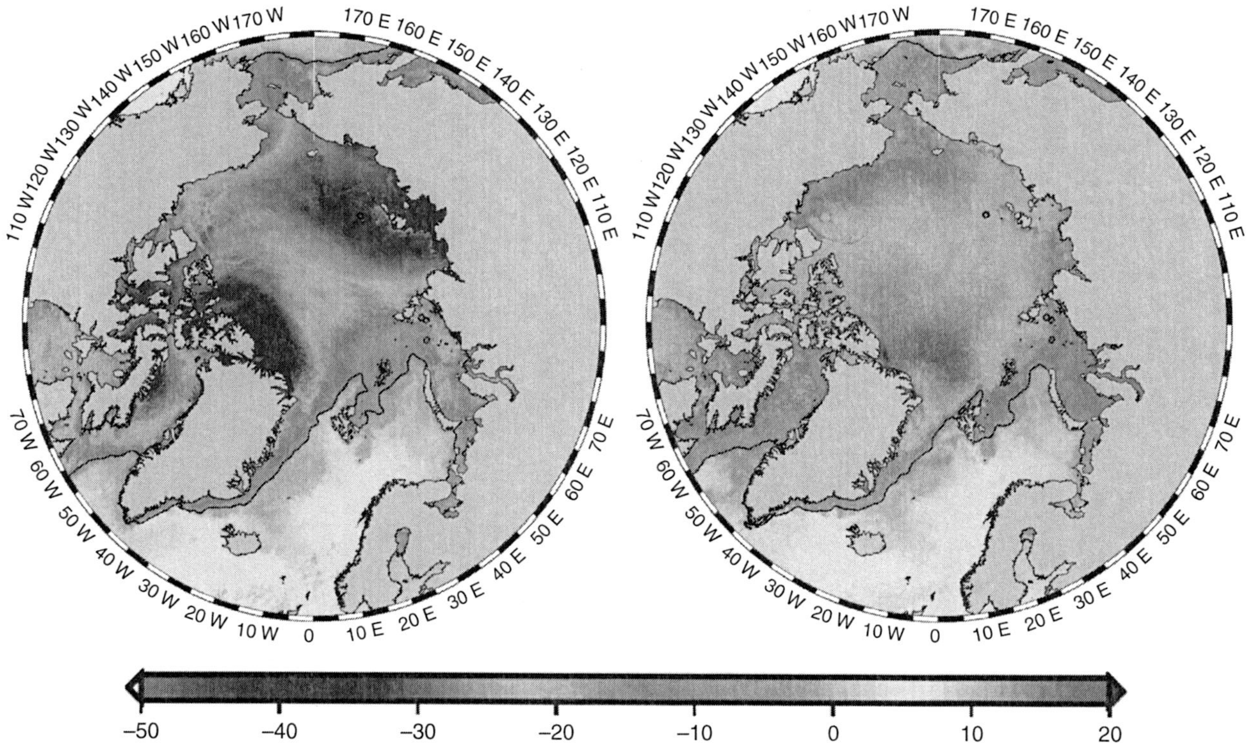

Figure 3.6: Surface Temperature computed from thermal infrared measurements by the MetOp AVHRR instrument on March 13 2016 (left) and April 11 2016 (right). http://ocean.dmi.dk/arctic/index.uk.php and the Copernicus Marine Environment Monitoring Service.

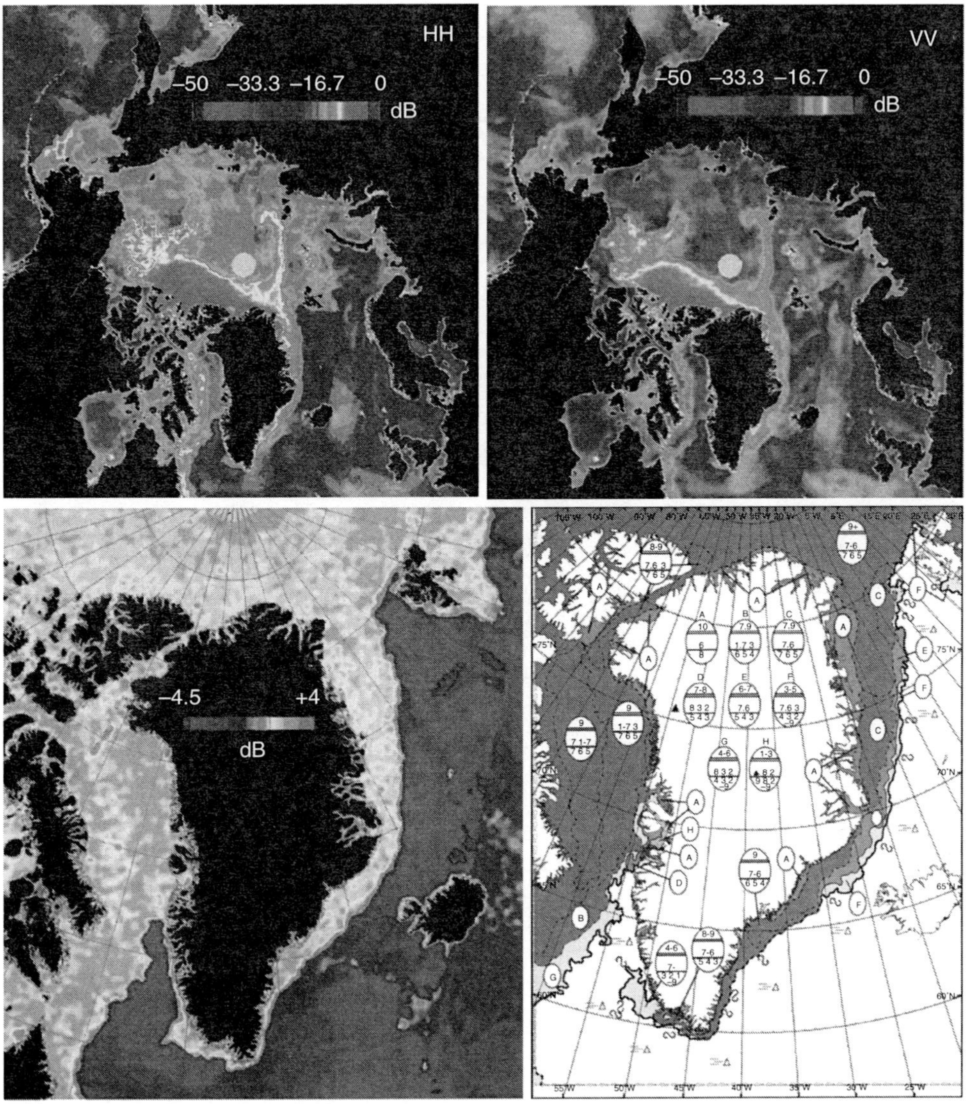

Figure 3.7: QuikSCAT scatterometer data from 11 April 2008. HH-polarization (upper left) and VV-polarization (upper right), ratio of HH and VV (lower left). An Ice chart from the Danish Meteorological Institute from 13 April 2008 (lower right). Source: www.seaice.dk

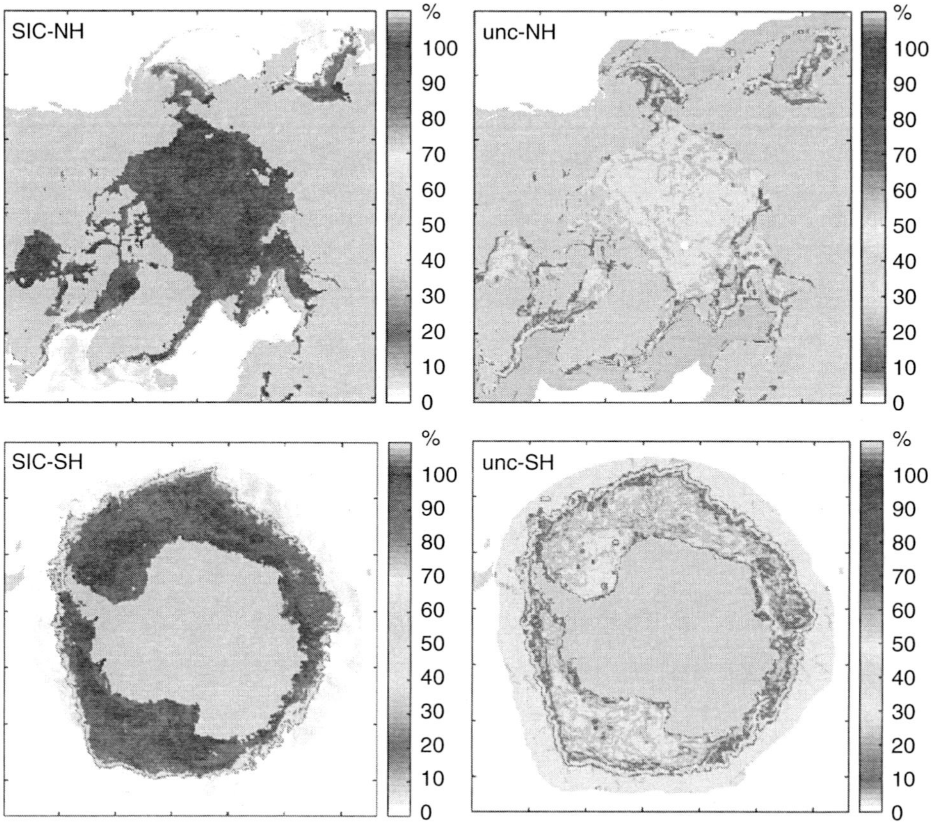

Figure 3.10: Sea ice concentration (SIC) (left) and uncertainty (right) retrieved by the SICCI algorithm from AMSR-E brightness temperatures. Northern Hemisphere (NH): 15 March 2011, Southern Hemisphere (SH): 15 September 2011

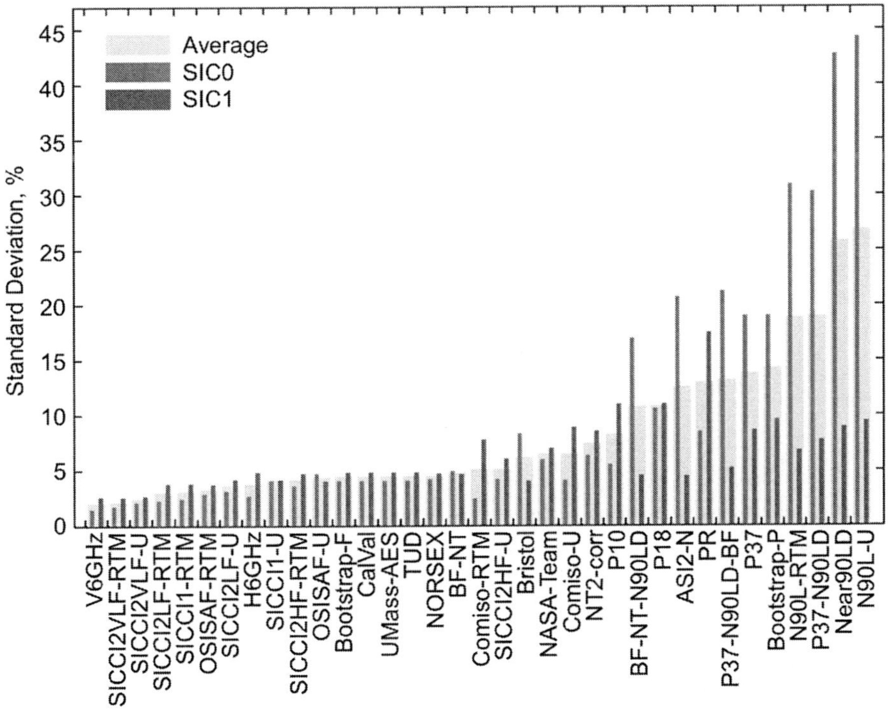

Figure 3.14: Standard deviation of difference between the reference sea ice concentration and that estimated the retrievals by various sea ice concentration algorithms for the instruments AMSR-E and AMSR2. Red bars are for high concentrations (100 per cent, SIC1) and blue bars are for low concentrations (0 per cent, SIC0). Grey bars are averages of low- and high-concentration results. Note that these results are obtained without weather filters and that all algorithms were tuned to AMSR-E and AMSR2 data separately to obtain near zero biases.

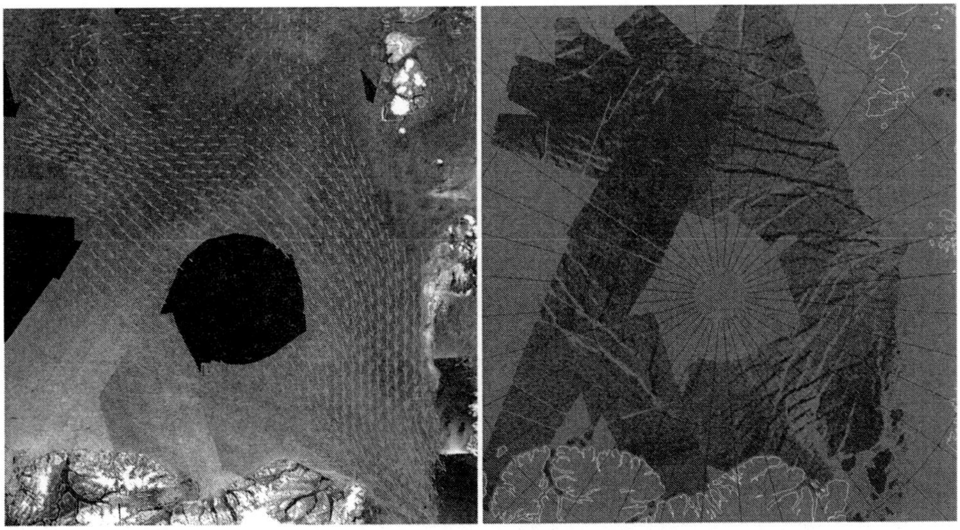

Figure 3.19: High-resolution ice drift field derived from Sentinel-1 images on 27 and 28 February 2016. Only every 25th original drift vector shown. Mosaic of 26–28 February Sentinel-1 coverage as background (left). Example of divergence deformation field derived from the Sentinel-1 ice drift (right). Blue colours correspond to divergence (openings) and red colours to convergence (closings/potential pressure ridge formation). Source: www.seaice.dk

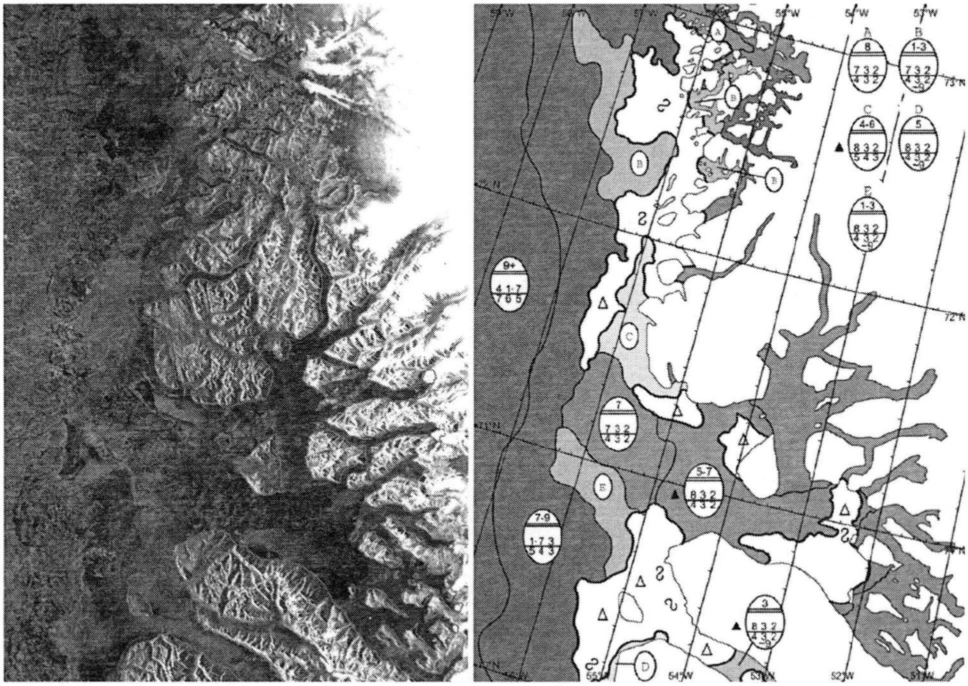

Figure 3.15: Sentinel-1 SAR image from 29 March 2016 and the corresponding ice chart from DMI. For explanation of the WMO egg code, see Chapter 1 and Section 3.8. In the ice chart, colours correspond to ice concentrations as follows: Green: 1–3/10, yellow: 4–6/10, orange: 7–8/10 and Red: 9–10/10. Grey indicate land-fast ice.

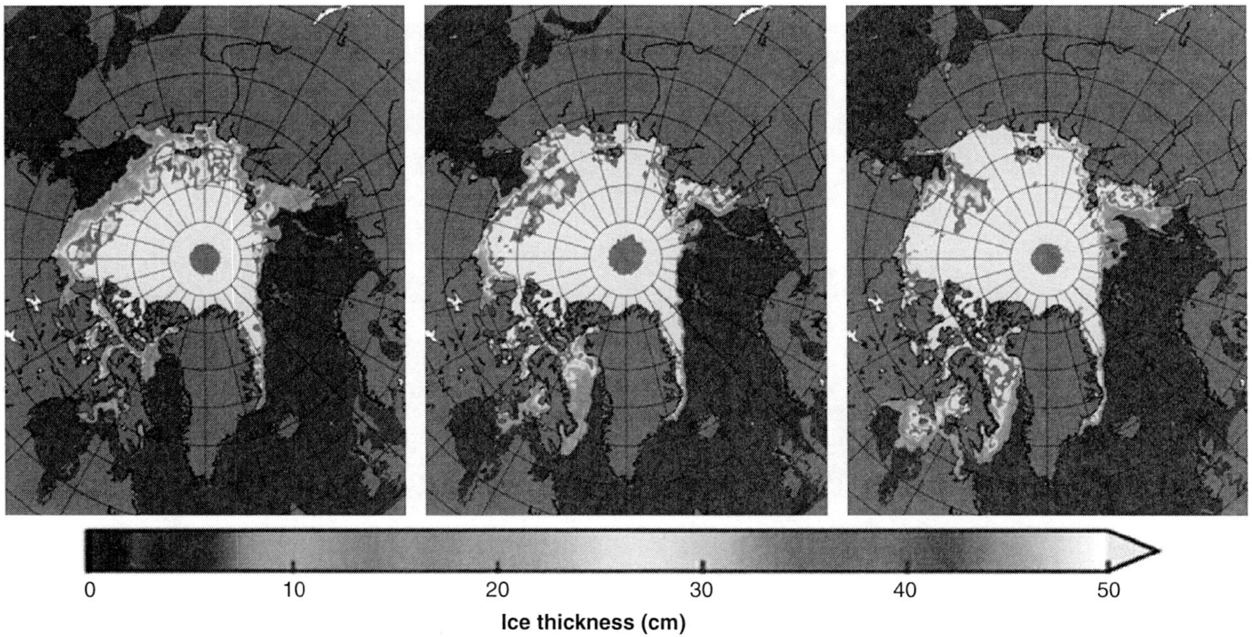

Figure 3.16: Thin ice thickness computed from SMOS data, 1 November 2015 (left), 16 November 2015 (middle) and 1 December 2015 (right). Source: www.iup.uni

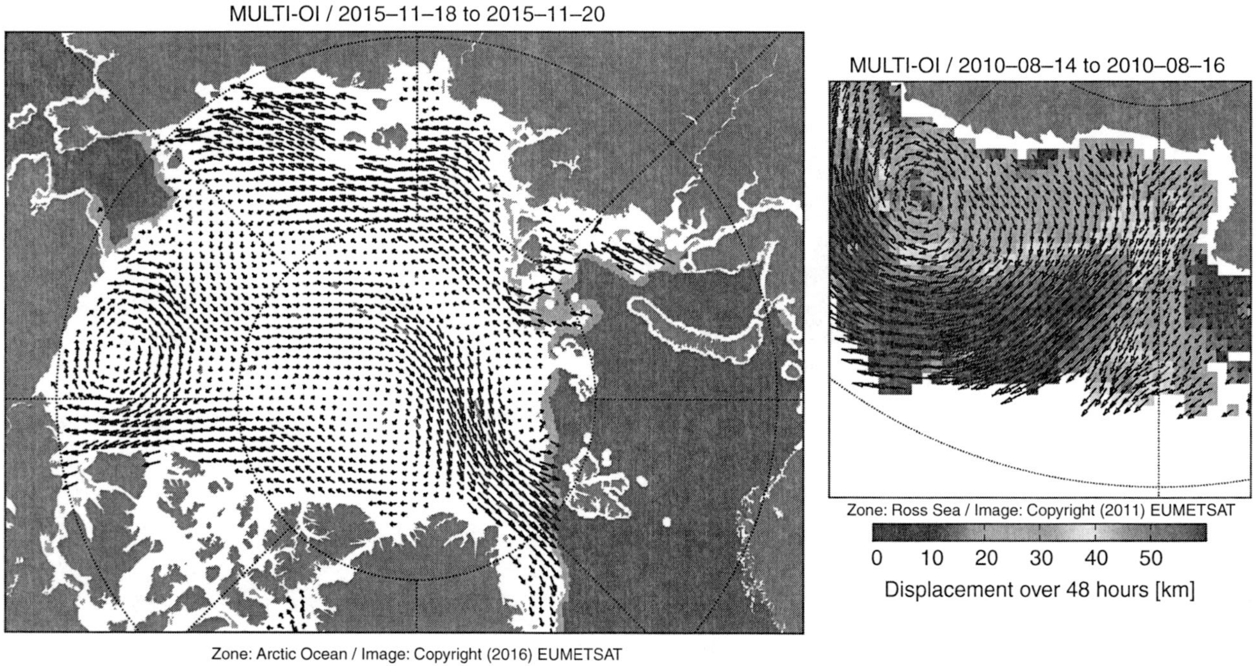

Figure 3.18: Example maps of sea ice drift as distributed from the EUMETSAT OSISAF. Left panel shows an example of the drift in the Arctic Ocean from 18 to 20 November 2015. The superimposed red vectors are the displacements measured by in situ buoy trajectories. The right panel shows the motion field over the Ross Sea from 14 to 16 August 2010 with colours indicating the magnitude of the vectors.

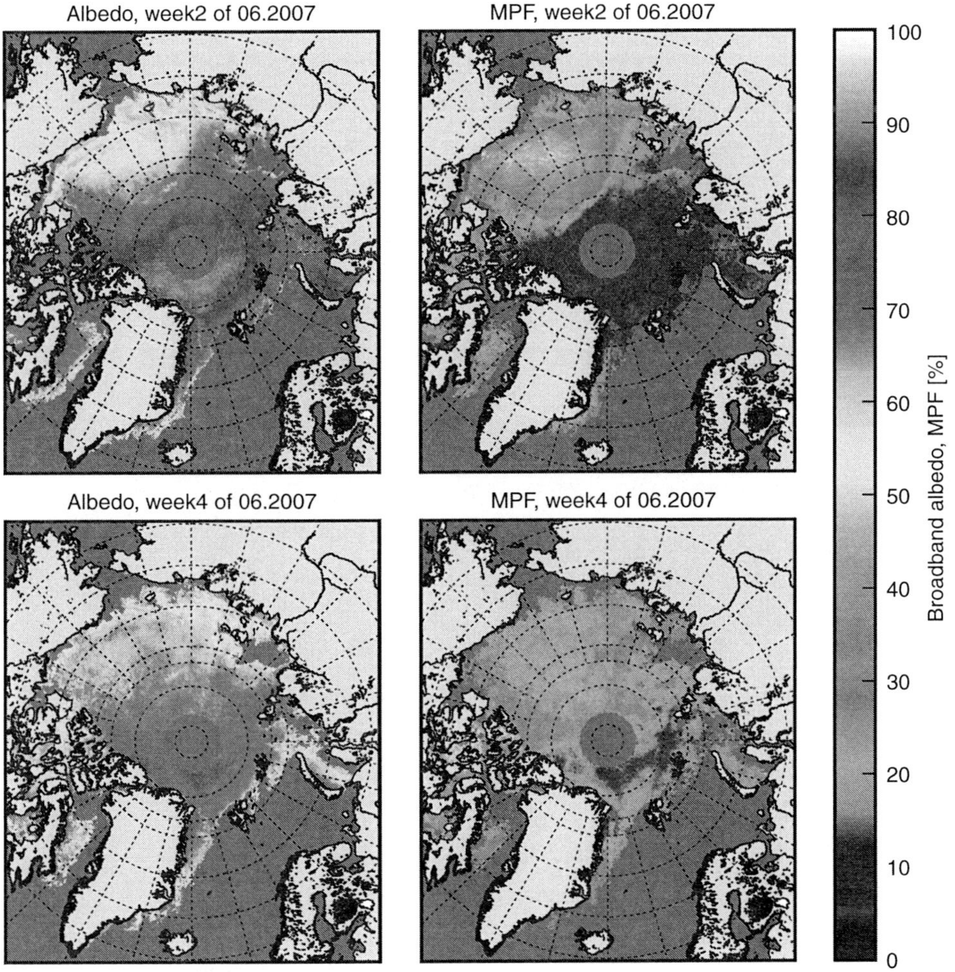

Figure 3.20: Eight-day composites of Arctic broadband albedo (left) and melt pond fraction (right). The left-to-right comparisons show that melt pond fraction and albedo are statistically but not strictly correlated. The top-to-bottom comparisons illustrate the fast development of both quantities during the melting period (Istomina et al., 2015a).

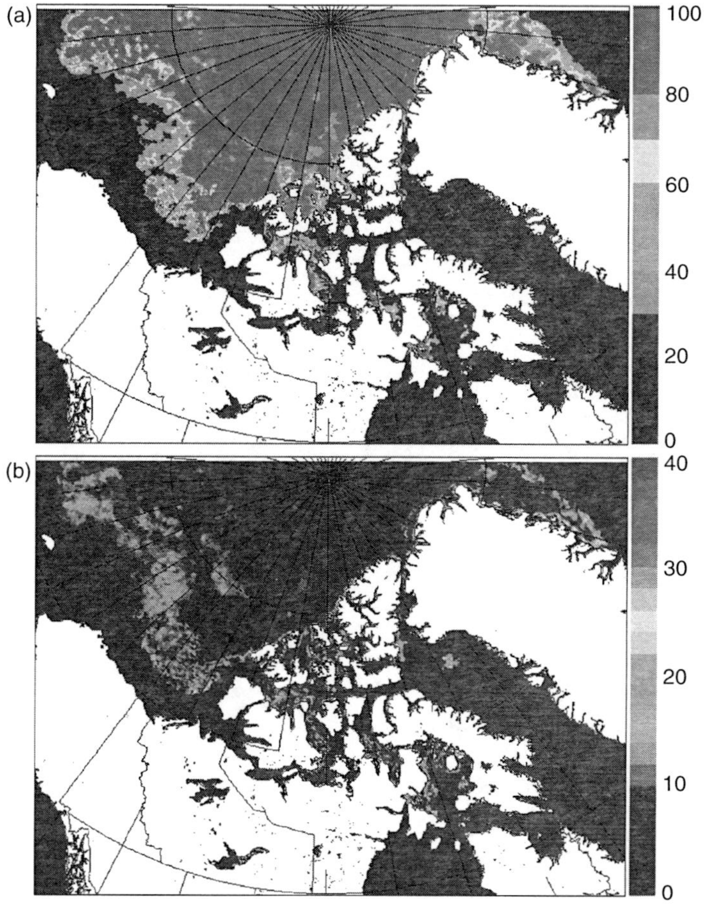

Figure 4.5: The background (a) mean sea ice concentration and (b) standard deviation of sea ice concentration computed from a 60-member ensemble. The ensemble was constructed using six-member ensembles of six-hour forecasts from 10 consecutive assimilation times, separated by six hours (from 15 August 2011 at 1200 UTC to 17 August 2011 at 1800 UTC). Note that the units are in per cent (similar to Figure 13 from Buehner and Shlyaeva, 2015).

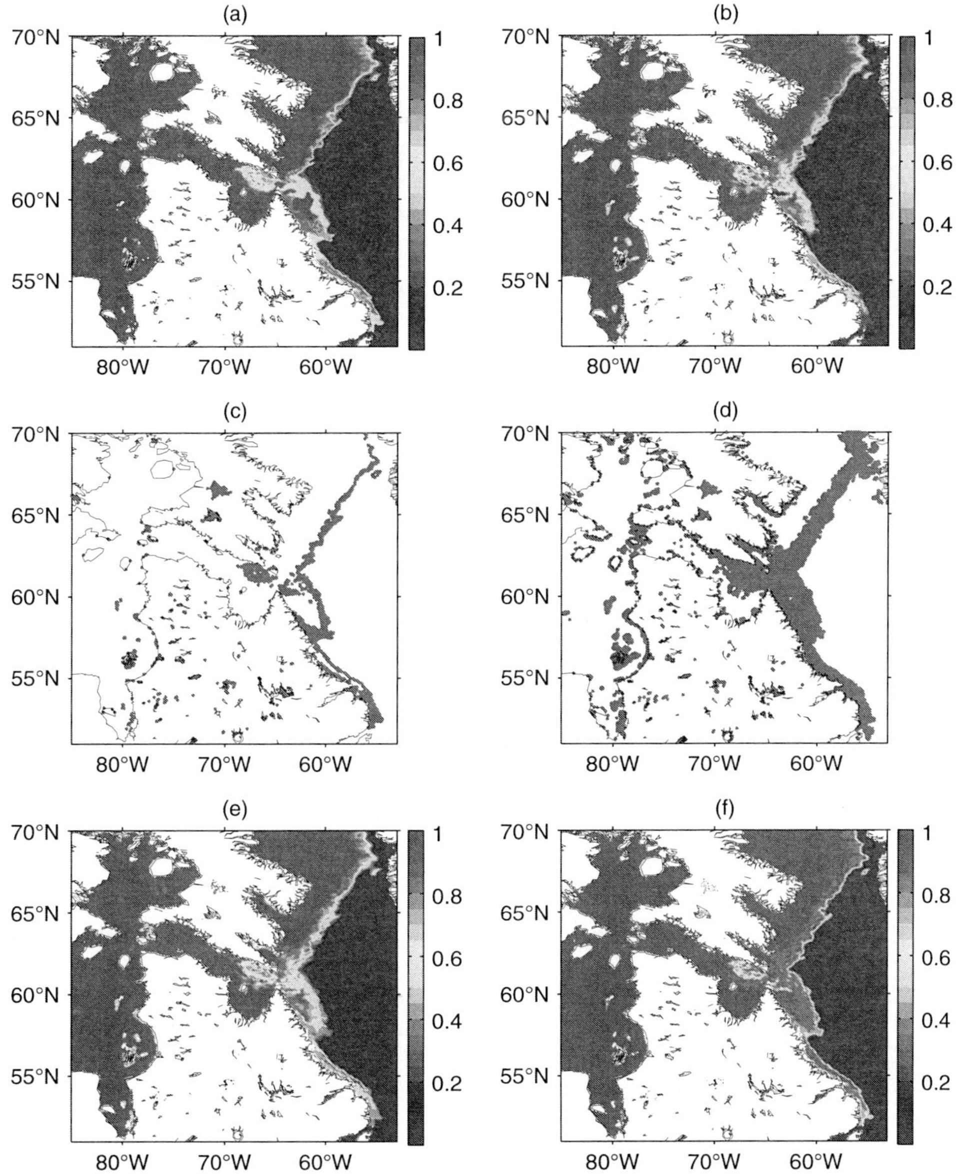

Figure 4.7: The ice concentration fields used for idealized numerical experiments to illustrate different approaches for assimilating categorical observations: (a) the 'true' ice concentration, (b) the background ice concentration to be corrected by assimilating observations. Locations of the ice and open water observations assimilated using two different sets of concentration threshold values: (c) 50 per cent for both ice and water, (d) 90 per cent for ice and 10 per cent for water. The case of threshold values 100 per cent for ice and 0 per cent for water are not shown, since the observations cover the entire domain. The resulting analysed ice concentration after assimilating high-resolution ice and open water observations using different concentration threshold values: (e) 50 per cent for both ice and water, (f) 90 per cent for ice and 10 per cent for water.

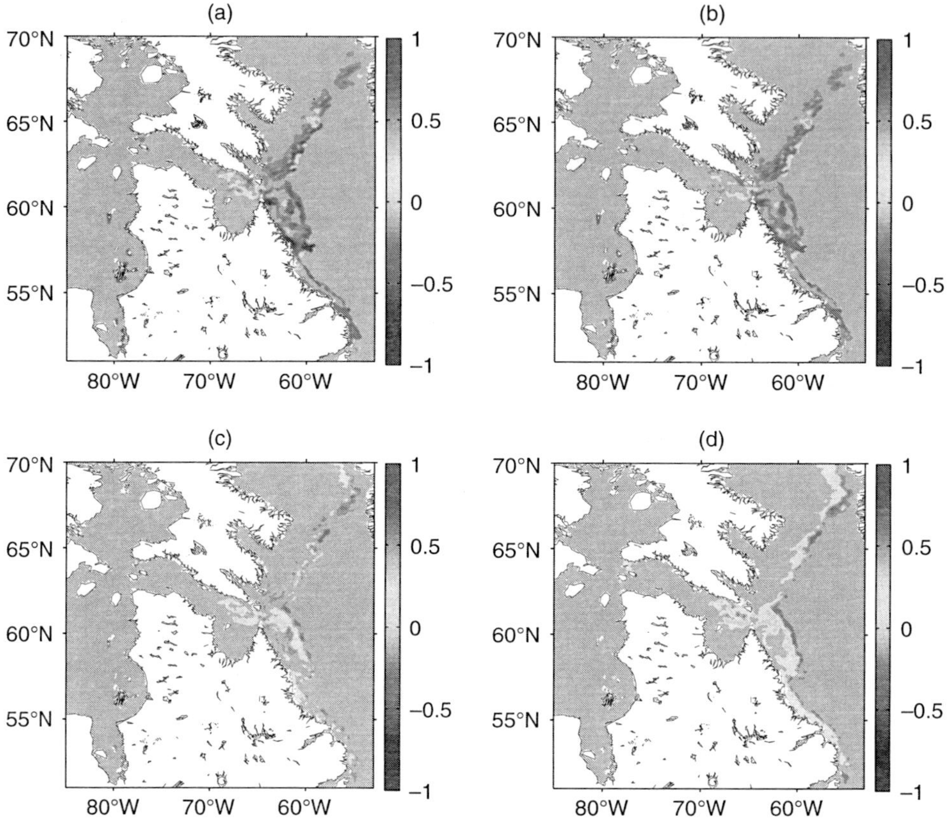

Figure 4.8: The ice concentration error for (a) the background state (RMSE of 15.5 per cent) and the analysis state after assimilating high resolution ice and open water observations using various concentration threshold values: (b) 50 per cent for both ice and open water (RMSE of 14.1 per cent), (c) 90 per cent for ice and 10 per cent for open water (RMSE of 12.8 per cent), (d) 100 per cent for ice and 0 per cent for open water (RMSE of 12.9 per cent).

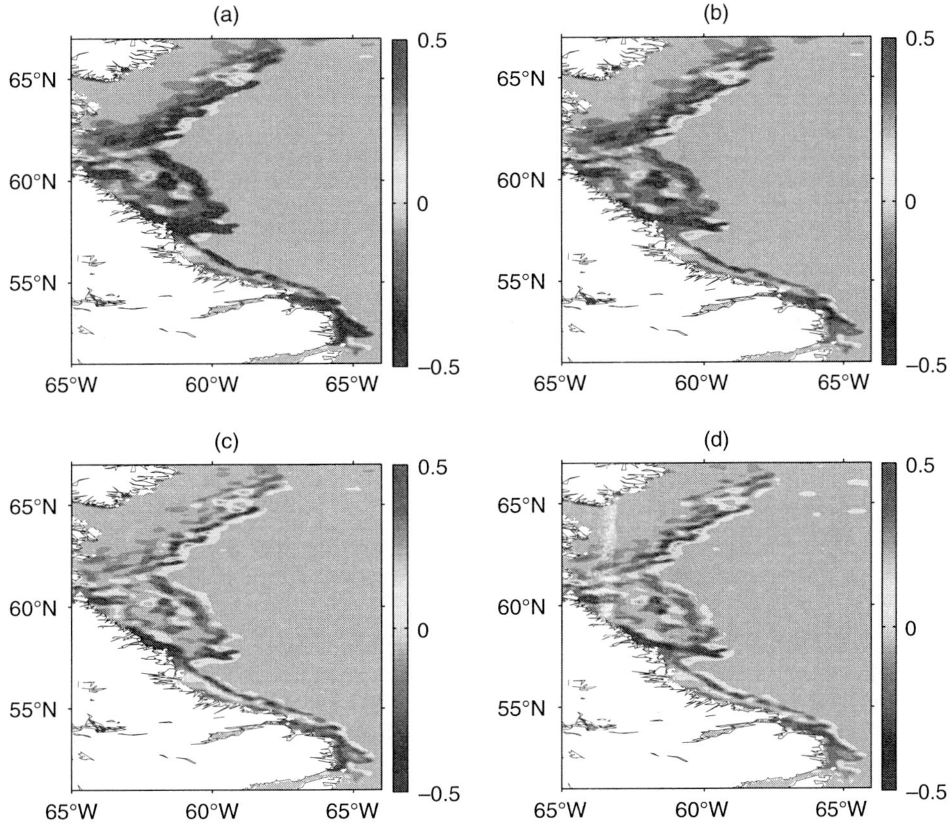

Figure 4.9: Idealized example showing the impact of assimilating low-resolution retrievals of ice concentration from passive microwave data in combination with high-resolution binary retrieval of ice and open water. The ice concentration error in (a) the background state (RMSE of 14 per cent) is shown in addition to the error in the analysis states after assimilating (b) only high-resolution binary observations with concentration threshold values of 50 per cent for both ice and open water (RMSE of 11 per cent), or (c) only low-resolution ice concentration observations (RMSE of 9.8 per cent), or (d) both low-resolution ice concentration observations and high-resolution binary observations (RMSE of 8.7 per cent).

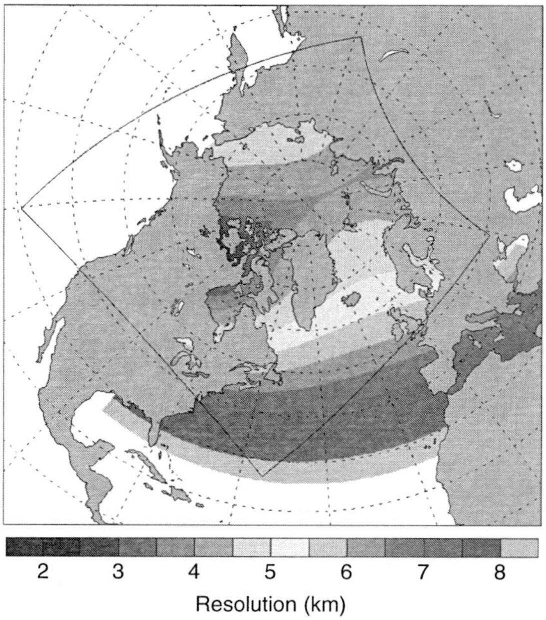

Figure 5.1: The RIPS analysis domain is a rotated latitude–longitude grid with grid points evenly spaced every 0.045° (area enclosed by black line). The colour contours indicate the average grid spacing of the RIPS forecast model grid which is a tri-polar grid with 12 grid points evenly spaced every degree. Note that most of the larger inland lakes are included in the analysis but not in the forecast system.

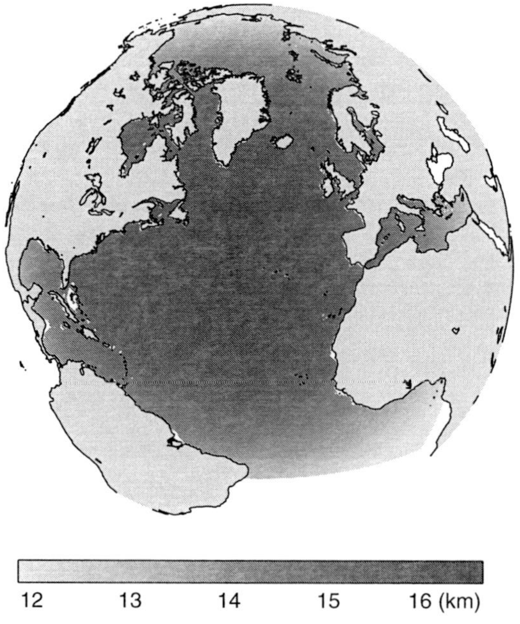

Figure 5.3: Map of the TOPAZ4 domain with grid resolution in km.

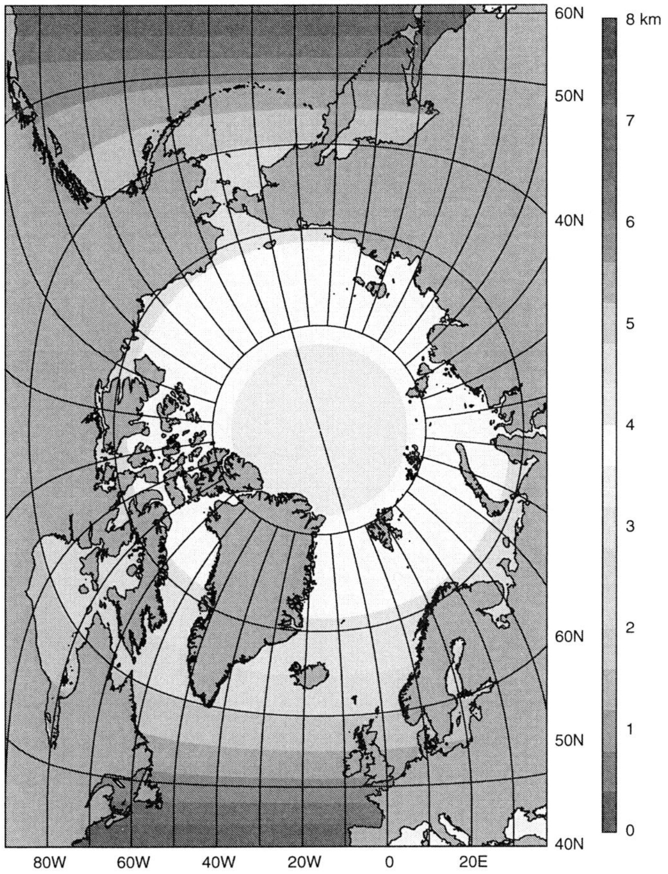

Figure 5.4: The Arctic Cap Nowcast/Forecast System (ACNFS) and Global Ocean Forecast System (GOFS 3.1) grid resolution (km).

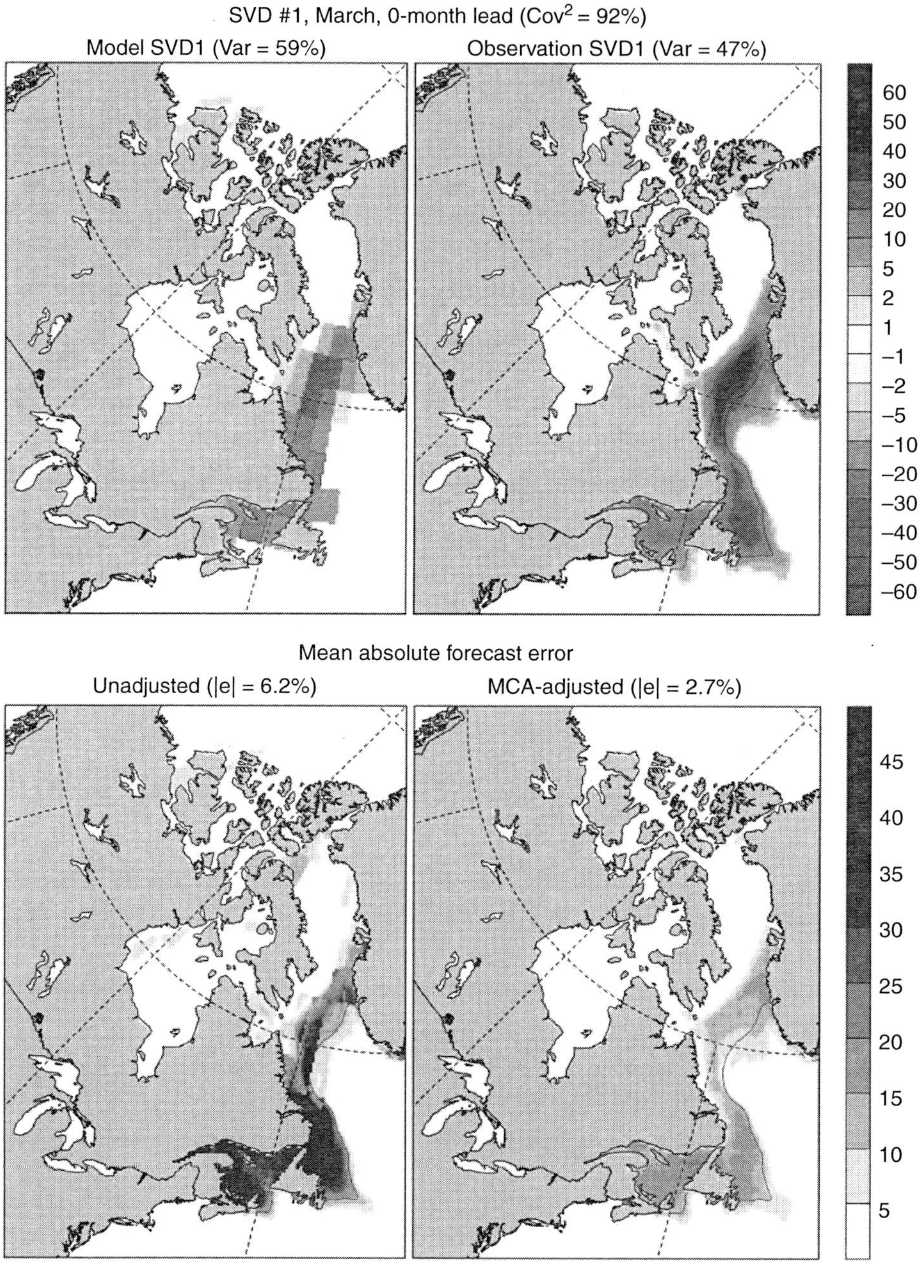

Figure 5.7: The first pair of SVD patterns (top) obtained for monthly mean CanSIPS predictions of March sea ice concentrations initialized on 1 March (lead = 0) in 1979–2012 (left) and the corresponding observed (NSIDC) field (right). This pair of patterns explains more than 90 per cent of the squared covariability in the two fields. The mean absolute error (bottom) of 0-lead CANSIPS predictions of March sea ice concentration for the original unadjusted (left) and MCA-adjusted (right) forecasts, with domain-averaged error indicated in the panel titles.

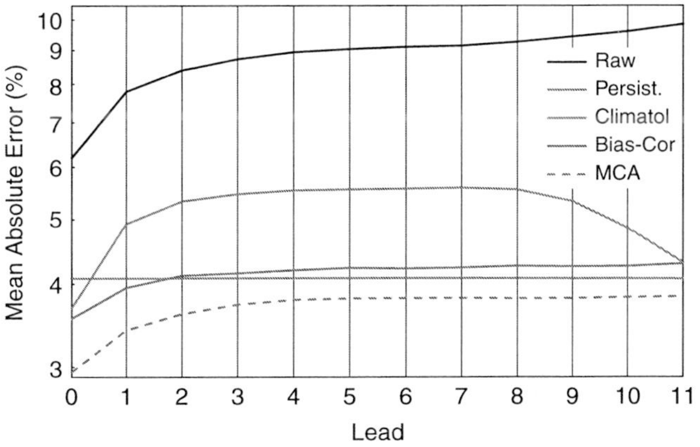

Figure 5.8: The domain-averaged mean absolute error of CanSIPS hindcasts averaged over all target months as a function of lead time (in months), for different forecasting methodologies (details in text).

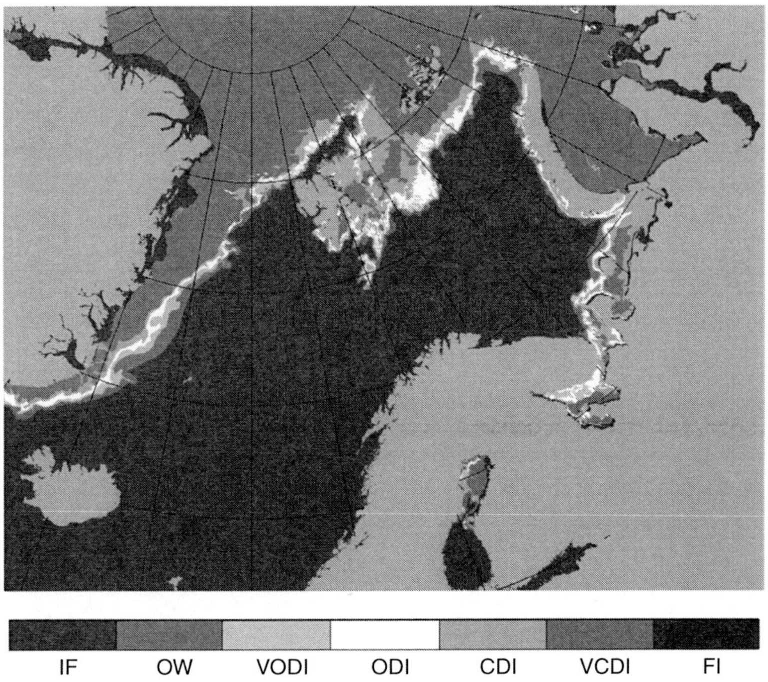

Figure 6.4: A sample map of the distribution of sea ice classes in the observational product. The entire area covered by the ice charts is displayed. Note that results for the Baltic Seas are discarded prior to evaluation. See the text for details on the definition of the various classes.

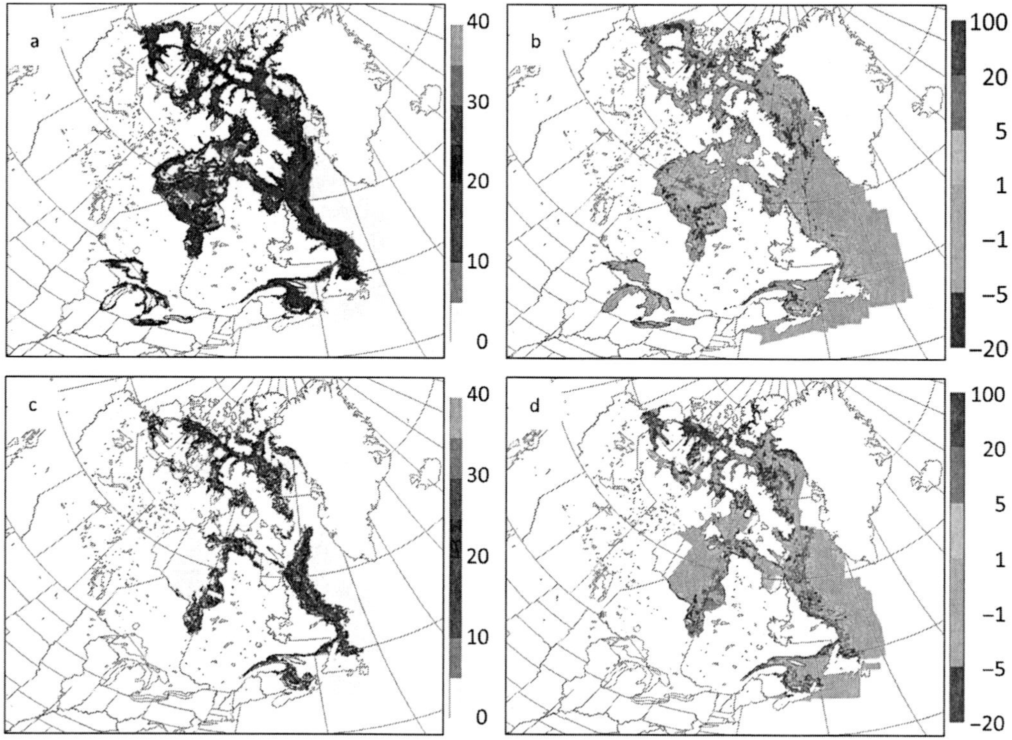

Figure 6.2: Evaluation of the RIPS ice concentration using CIS RADARSAT image analysis charts for all of 2013, (a, b) *ESD* and bias of 06 UTC analysis compared with charts valid from 09 to 15 UTC, (c, d) *ESD* and bias of 36-hour forecast valid at 12 UTC compared with charts valid from 09 to 15 UTC.

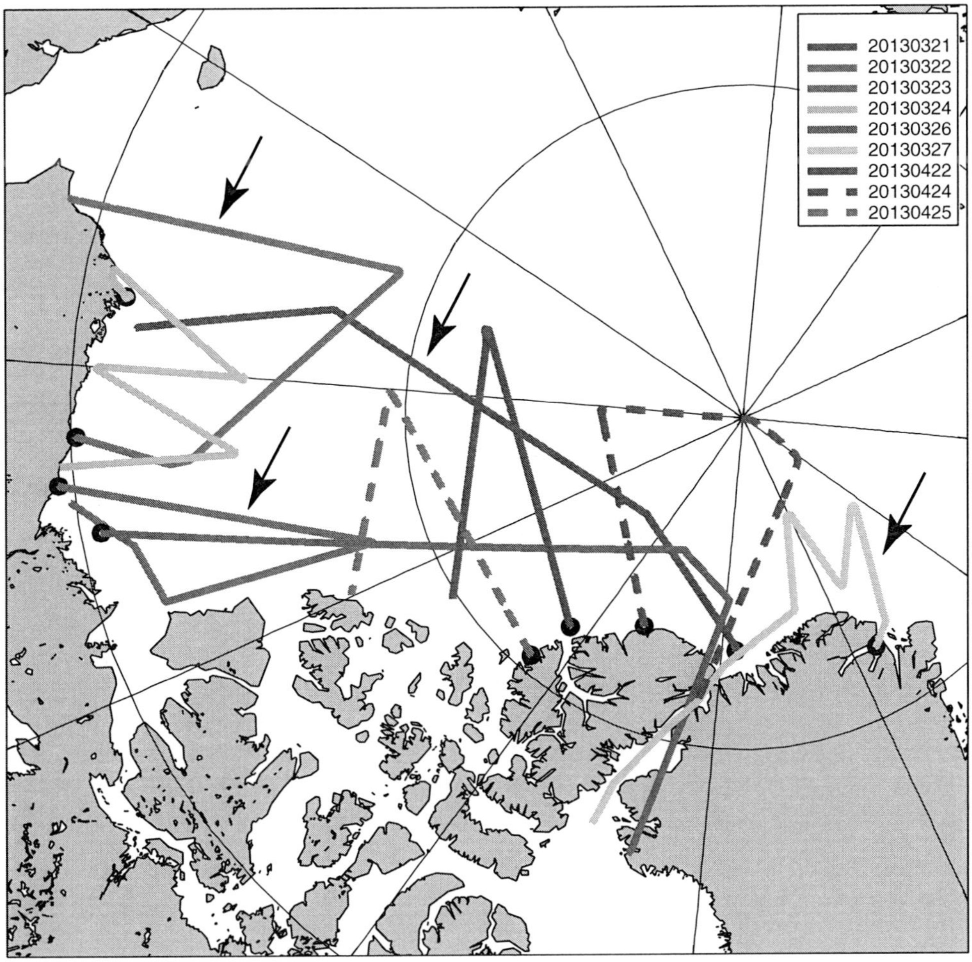

Figure 6.6: NASA Operation Ice Bridge aircraft flight paths over the Arctic Ocean in Spring 2013. The black circle indicates the starting point of each flight and the flight numbers correspond to those used in Table 6.4.

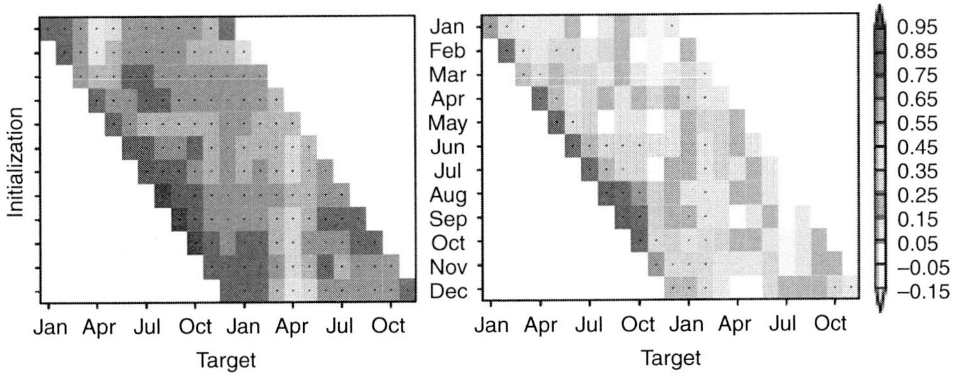

Figure 6.7: CANSIPS forecast skill (quantified by the anomaly correlation coefficient) for sea ice area as a function of initialization and target (predicted) month, evaluated against observations from the National Snow and Ice Data Center. Black dots represent statistical significance at the 95 per cent confidence level as determined by bootstrapping. The left panel is for full anomalies, and the right panel is for linearly de-trended anomalies (adapted from Sigmond *et al.*, 2013).

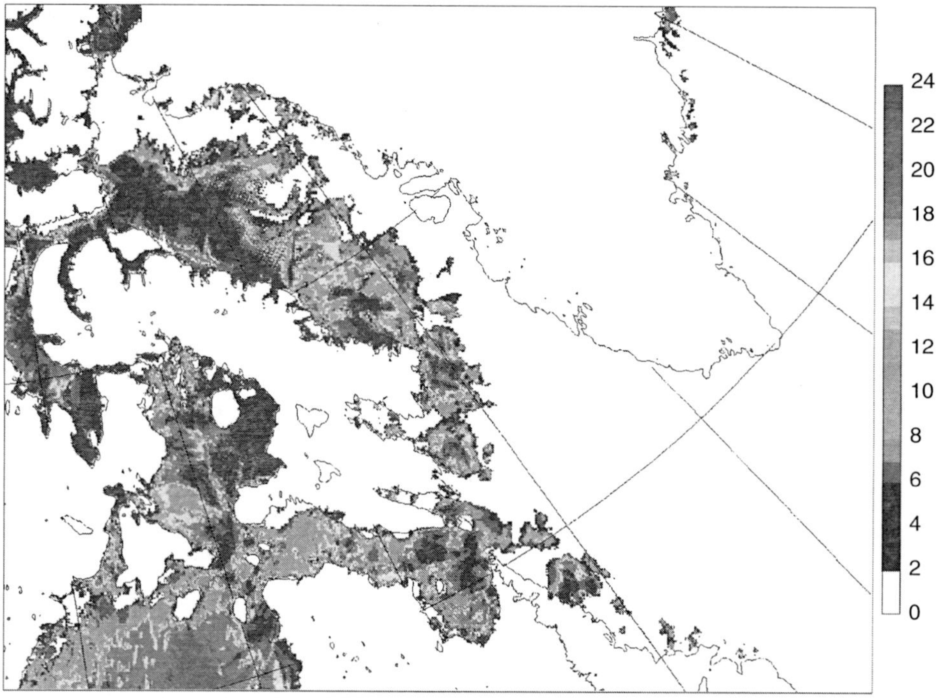

Figure 7.5: 24-hour forecast of the duration in hours of ice pressure greater than 2.5×10^3 Nm^{-1} from RIPS, valid 1200 UTC 11 March 2016.

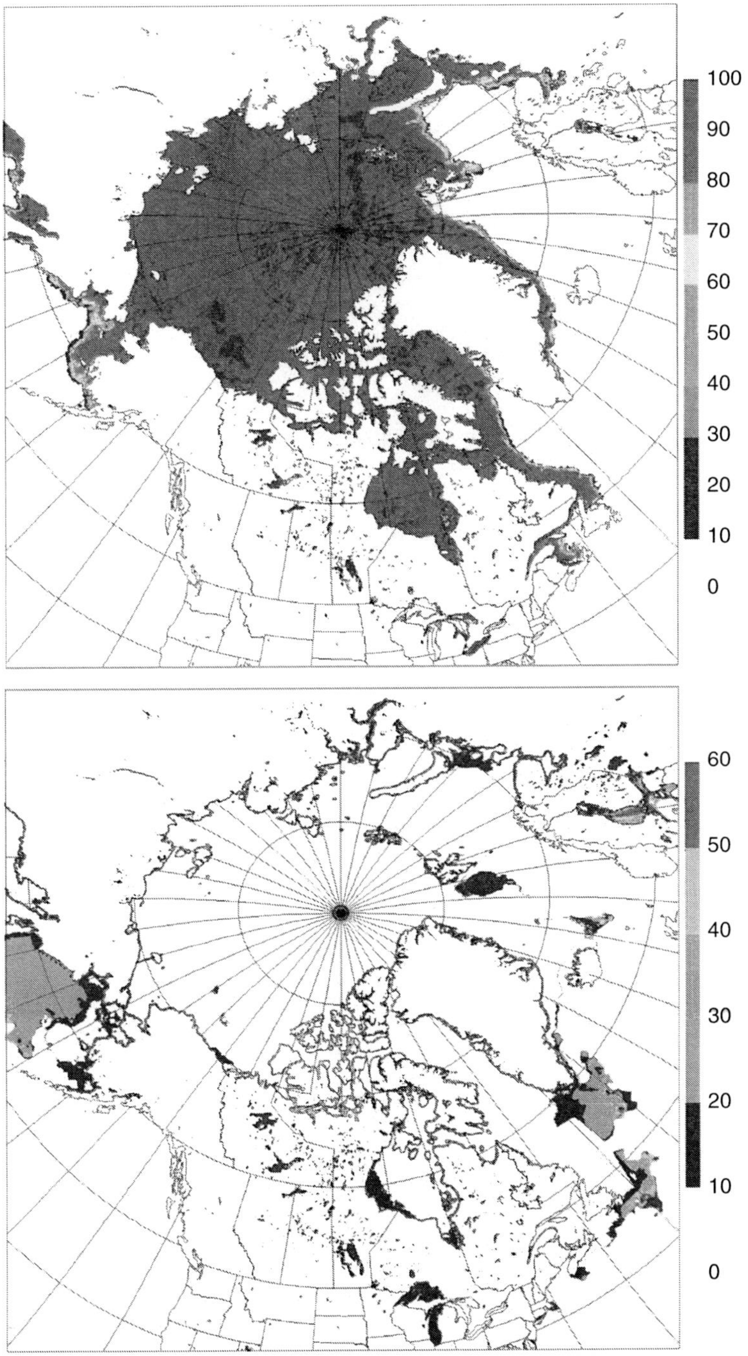

Figure 7.6: Ice concentration (top) and analysis uncertainty (bottom) from RIPS, valid 1800 UTC 31 January 2015. Units as percentages.

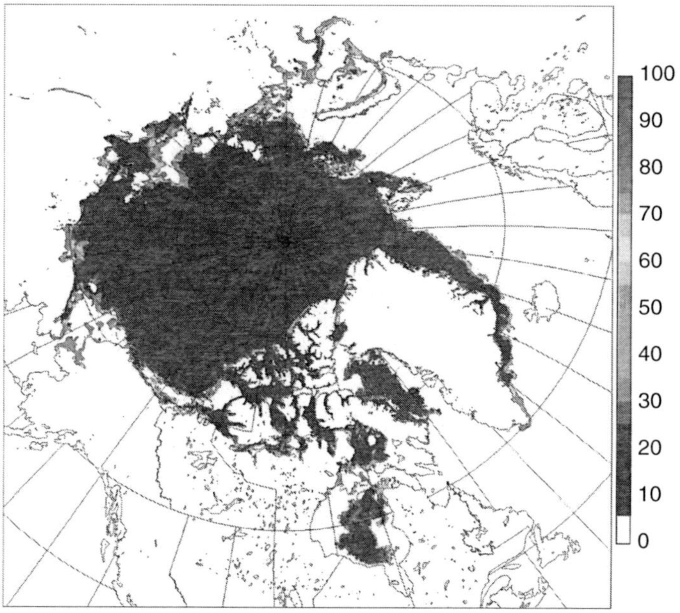

Figure 7.7: Forecast of the probability of encountering ice with concentration greater than 10 per cent, valid 0600 UTC 10 July 2011. This forecast is based on a 21-member ensemble system (Shlyaeva *et al.*, 2016).

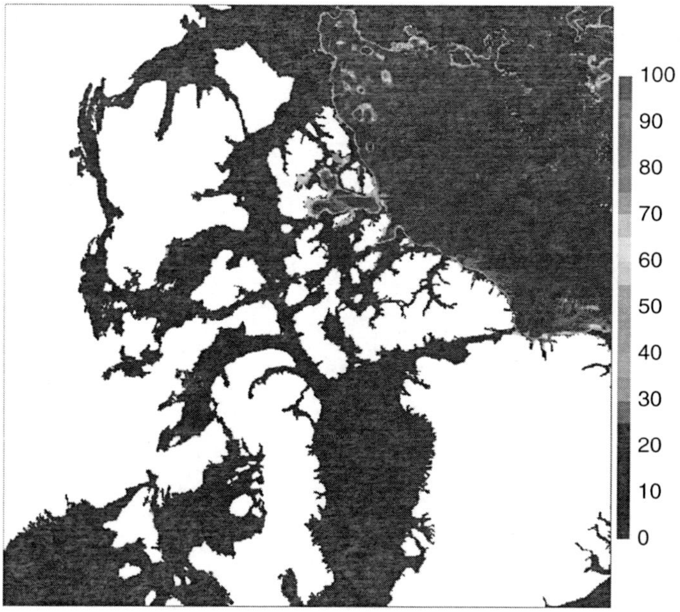

Figure 7.8: Forecast of the probability of encountering ice with concentration greater than 90 per cent with mean ice thickness in a grid cell greater than 1 m, valid 0600 UTC 31 August 2011. This forecast is based on a 21-member ensemble system (Shlyaeva *et al.*, 2016).

3.4.2 SAR Observations

Synthetic aperture radar (SAR) allows identification of different ice types, such as FY ice, MY ice and various forms of thin ice; and distinguishing these types from smooth or roughened open water. To quantify this into ice concentration automatically, different methods have been used, for example, image auto-correlation and neural networks. Common intermediate steps in SAR image processing are calculation of texture features, principal component analysis and cluster analysis. Ice/water discrimination from SAR is also used to retrieve regional high-resolution sea ice edge and ice concentration maps for operational purposes.

Due to the large number of ambiguities in SAR images of sea ice and the potentially large variability in backscatter of both water and ice, robust automated methods to derive sea ice concentration to this day have found only very limited use, and NIIS still rely on manual interpretation of SAR images when producing ice charts. Figure 3.15 shows an example of a Sentinel-1 SAR image of central west Greenland and a corresponding ice chart produced at the Danish Meteorological Institute (DMI).

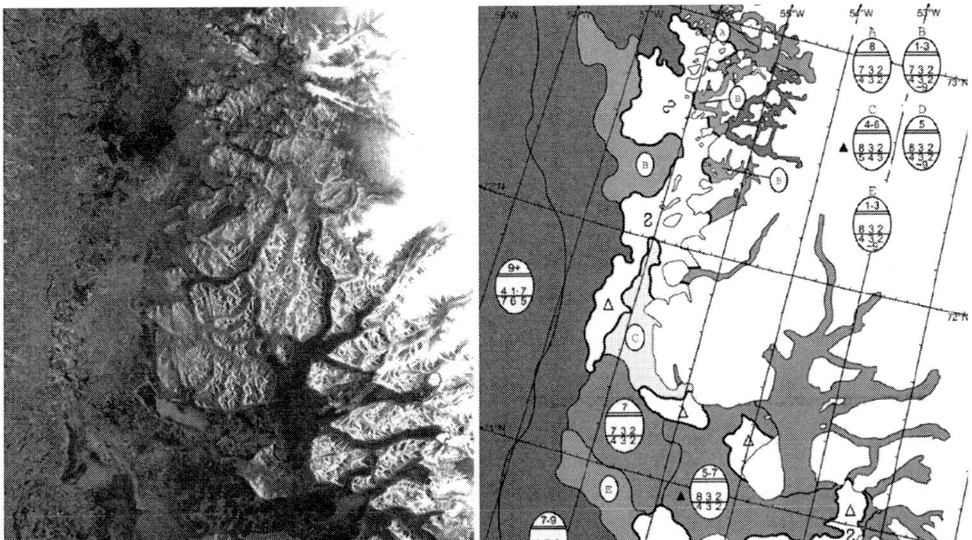

Figure 3.15: Sentinel-1 SAR image from 29 March 2016 and the corresponding ice chart from DMI. For explanation of the WMO egg code, see Chapter 1 and Section 3.8. In the ice chart, colours correspond to ice concentrations as follows: Green: 1–3/10, yellow: 4–6/10, orange: 7–8/10 and Red: 9–10/10. Grey indicate land-fast ice. (A black and white version of this figure will appear in some formats. For the colour version, please refer to the plate section.)

3.4.3 Other Sources

Ship observations of sea ice concentration are difficult to compare directly with satellite observations due to the completely different geometry of the observations (incidence angle, field of view). However, they have proven useful when intercomparing different satellite-based sea ice concentration retrievals by their correlation with ship-based observations (Spreen *et al.*, 2008).

Scatterometer data from polar orbiting satellites such as SeaWinds QuikSCAT (Ku-band, 1999–2009) and MetOp ASCAT (2006–) have been used to distinguish ice from water. The QuikSCAT scatterometer used the polarization ratio to distinguish ice from water and was therefore useful for delineating the ice edge (see also Section 3.4.1 and Figure 3.7). However, the ability to quantify the concentration of ice inside the ice edge is limited using scatterometer data alone due to the non-linear relationship between backscatter and ice concentration.

Medium resolution visible and infrared data (0.1 to 1 km resolution) such as AVHRR, MODIS and Sentinel-3 are excellent supplements to the microwave data when cloud-free images can be acquired. Most NIIS have near-real-time access to this type of data.

High-resolution optical data from satellites such as Landsat and Sentinel-2 are being used to derive very detailed spatial information on ice concentration during the daylight season, but so far operational applications have been limited by the typically 12–24 hour delay from satellite acquisition to availability of the data. High-resolution optical data are very useful for validation of other ice concentration products.

3.4.4 Some Examples of Operational Products

EUMETSAT's Ocean and Sea Ice Satellite Application Facility (OSISAF) produces operational ice concentration, ice type and ice edge data using the latest version of the OSISAF algorithms (Eastwood, 2012). The data are available from the OSISAF high latitude centre at http://osisaf.met.no/p/ice/.

Other sources of operational ice concentration data are the MASIE-AMSR2 (MASAM2) daily 4 km sea ice concentration (Fetterer *et al.*, 2015), which is a prototype concentration product that is a blend of two other daily sea ice data products: ice coverage from the Multisensor Analyzed Sea Ice Extent (MASIE) product at a 4 km grid cell size and ice concentration from the Advanced Microwave Scanning Radiometer 2 (AMSR2) at a 10 km grid cell size. These products are available from https://nsidc.org/data/docs/noaa/g10005-masam2/. The University of Bremen produce daily AMSR2 and SSMIS sea ice maps (Spreen *et al.*, 2008) using the ASI sea ice concentration algorithm. The data are available from https://seaice.uni-bremen.de/amsr2/index.html.

3.4.5 New Directions and Challenges To Be Addressed

The development of consistent observation operators that relate sea ice and snow variables to satellite measurements of T_B and backscatter can potentially lead to new inversion algorithms that integrate data from several satellite sensors such as microwave and infrared radiometers, scatterometers and SAR. Such integrated retrievals already exist for retrieving ocean and atmospheric variables over the ice-free ocean, and they could potentially allow retrieval of combined ice and snow variables (see Section 3.9).

The advent of constellations of C-band SAR satellites such as Sentinel-1 A+B and RADARSAT Constellation calls for the development of automatic sea ice classification methods from SAR data. Recent work using convolutional neural networks (Wang *et al.*, 2016a+b) has shown significant potential, but needs to be generalized and extended to more areas and seasons. The vast amount of details about the ice cover found in SAR images has a huge potential for being useful for assimilation in high-resolution sea ice models, but more general and robust observation operators needs to be developed to exploit the full potential. Initial work by Scott *et al.* (2015) has shown some potential approaches, but more work is needed.

3.5 Observations of Ice Thickness

3.5.1 Remote Sensing

The most commonly used method to obtain estimates of large-scale sea ice thickness is by means of satellite radar or laser altimetry.

The common way to convert sea ice freeboard into sea ice thickness is to assume that the sea ice is in isostatic balance and to compute the sea ice thickness following Archimedes' principle. This requires a priori knowledge of water, snow and sea ice densities as well as snow depth:

$$h_i = \frac{\rho_w h_{fb} + \rho_s h_s}{\rho_w - \rho_i}, \tag{3.13}$$

where h refers to thicknesses, ρ refers to densities and subscripts w, i, s and fb refers to water, ice, snow and freeboard respectively. Since a laser altimeter measures the total (ice plus snow) freeboard, Equation 3.13 is only valid for radar altimeters and a modified version should be used for laser altimeters.

The accuracy of the obtained sea ice thickness is determined by the freeboard accuracy and the uncertainty in the estimate of the a priori knowledge. Biases in the freeboard can be substantial because the estimation of freeboard from the altimeter measurements of elevation requires detection of areas of open water or thin ice – usually in leads – to estimate the reference sea surface height. Another important limitation is the small footprint of altimeters (~50 m (laser) to between a few hundreds of metres and a few kilometres (radar)) which in combination with its nadir viewing geometry results in freeboard and thickness being obtained only along the sub-satellite track on the Earth's surface. Consequently, pan-Arctic or – Antarctic sea ice thickness maps are only available at coarse temporal (14–30 days) and spatial (25 km to 100 km) grid resolution and often require a substantial amount of interpolation (see also Figure 3.9).

The radar backscatter at nadir from smooth surfaces (such as water in leads or thin undeformed ice) is much stronger than the backscatter from thicker/deformed ice and hence will dominate the return signal. This may lead to a bias towards thinner ice in radar altimeter measurements, also from SAR altimeters. This bias is referred to as preferential sampling and occurs if mixtures of thicker and thinner (level) ice exists within the satellite footprint (Tonboe *et al.*, 2010).

For close to 100 per cent sea ice concentration under cold conditions, sea ice thickness can also be estimated by imaging satellite microwave radiometers operating at L-Band frequencies such as SMOS or SMAP. Such sensors exploit the much deeper penetration of microwave radiation into sea ice at L-Band than at higher frequencies such as those used for sea ice concentration retrieval (see above). The largest limitation of this kind of thickness retrieval is the temperature- and salinity-dependency of the penetration depth (typically a few tens of centimetres in common saline seasonal sea ice and up to 1.5 to 2 m in less saline sea ice such as MY ice or sea ice of brackish waters such as in the Baltic Sea). Consequently, the maximum retrievable sea ice thickness is of the order of 0.5 m under cold conditions. The maximum thickness to be retrieved can be increased somewhat for saline sea ice if a priori information about the sea ice thickness distribution and salinity is added to the retrieval as a boundary condition (Kaleschke *et al.*, 2016). The method is also sensitive to deviations of the sea ice concentration from 100 per cent and another error source is the contribution of sea ice roughness. Despite these limitations SMOS based sea ice thickness maps are currently the only source of daily weather independent large scale sea ice thickness information from thin ice in the Arctic Ocean during the sea ice growth phase in fall and early winter. The swath width of SMOS is approximately 1000 km and the spatial resolution in the daily maps is approximately 35 km. Figure 3.16 shows an example of the development of the Arctic sea ice thickness during the months of November and December 2015.

Because of the lack of more direct methods of inferring large-scale sea ice thickness, methods to obtain ice thickness proxy information have been devised. Ice types can either be discriminated with passive microwave sensors using different tie points for thin ice, FY ice and MY ice or by means of their different radar backscatter coefficient in scatterometer or SAR imagery. Both approaches work only during winter conditions when the contrast in emissivity or backscatter of the different ice types is sufficiently large. In case of the FY versus MY ice discrimination the key physical difference is the lower salinity and larger porosity of the MY ice compared to the FY ice. This allows deeper penetration and thus volume scattering in the ice, increasing radar backscatter and decreasing microwave emissivity for MY ice. Once the overlying snow or ice surface is wet or flooded the abovementioned contrast is lost.

Another proxy for ice thickness could be the ice age, which has been derived by tracking sea ice using a sequence of sea ice drift maps (see Section 3.6).

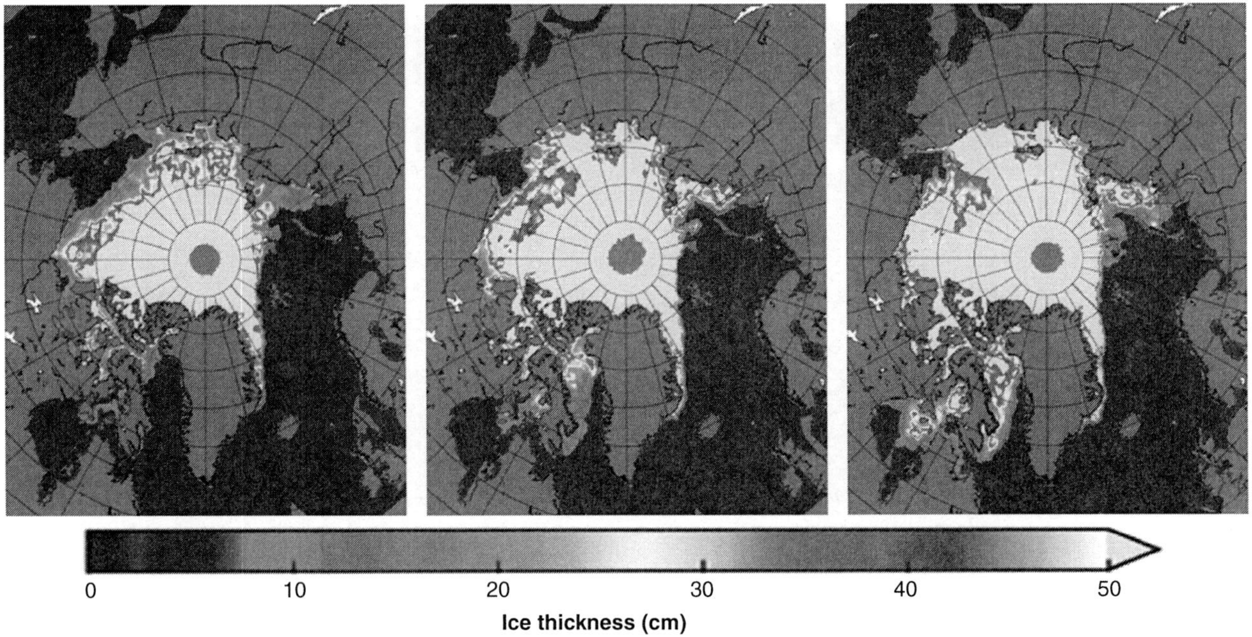

Figure 3.16: Thin ice thickness computed from SMOS data 1 November 2015 (left), 16 November 2015 (middle) and 1 December 2015 (right). Source: www.seaice.uni-bremen.de/thin-ice-thickness/ (A black and white version of this figure will appear in some formats. For the colour version, please refer to the plate section.)

3.5.1.1 Evaluation and Quality Control

Evaluation and quality control of sea ice thickness from altimeters is a difficult task. Indeed, as described above, such sea ice thickness products are usually available at a grid resolution of a few tens to hundreds of kilometres and at monthly temporal resolution or in near-real time only along the sub-satellite ground track. Co-location with reference data requires exact coordination of coincident airborne, surface-based or submarine activities for a direct evaluation of the sea ice thickness information obtained from satellite. In situ reference data require a vast number of measurements over large areas and over an extended time period in order to provide meaningful statistics. Difficulties arise since sea ice is moving and a set of co-located observations of sea ice thickness, snow depth, sea ice freeboard, degree of deformation and alignment of deformation is needed. Many of these can vary substantially on scales of less than a metre. Changes in location of reference data due to ice drift can thus be substantial. One needs to keep in mind that a satellite overpass occurs within seconds, airborne measurements are carried out within minutes to hours, and in situ measurements take hours or days. Despite these obstacles, and in lack of alternatives, direct comparison between in situ and satellite observations is still the preferred method for validating sea ice thickness from satellite altimeter measurements.

Another challenge is preferential sampling of a certain ice type by satellite altimetry or in situ measurements. For altimetry, the stronger reflection of the electromagnetic pulses from smoother surfaces means they will dominate the radar signal (see Section 3.3.3). For in situ measurements, limited accessibility of thin sea ice, limitations in drilling thick or deformed ice with reasonable effort, and, in general, the smaller number of research cruises and expeditions carried out during the winter season are the main reasons. Finally, worth mentioning is the problem of aliasing. The real sea ice thickness distribution is smoothed by most remote sensing sensors; the act of measurement can be regarded as a smoothing operator which causes the thin and particularly the thick parts of the natural sea ice thickness distribution to be folded towards thicker and thinner values, respectively (Geiger *et al.*, 2015).

Despite all these difficulties, reasonable attempts to evaluate currently available sea ice thickness products with the data mentioned in the previous section have been carried out (e.g. Kaleschke *et al.*, 2016; Ricker *et al.*, 2015; Kern *et al.*, 2016; Zygmuntowska *et al.*, 2014; Connor *et al.*, 2009; Kwok and Cunningham, 2008). Based on what is detailed in the section about potential biases and limitations and on what is written in the previous paragraphs it is recommended that the results of evaluation studies be regarded with a critical eye. Available data used for evaluation can be biased low, can be subject to an ice topography or ice drift bias, can suffer from limitations of the sensors and methodologies used and therefore, the results might not be globally applicable.

3.5.2 Other Sources (In Situ, Submarine and Airborne)

In situ measurements of the sea ice thickness (drilling holes through the ice and measuring the thickness directly) provide the most accurate information of this sea ice variable.

However, the vastness of the sea ice coverage, over several million square kilometres, paired with the nature of the sea ice, being rafted and ridged and hence of blocky structure, particularly underneath, makes in situ measurements primarily useful for evaluation purposes. This is particularly true if these measurements are coordinated with other measurements that allow appropriate scaling to satellite observations.

Another way to get sea ice thickness information is by visual observations from a ship's bridge according to protocols like ASPeCt (Antarctic Sea Ice ProcEsses and ClimaTe) and the Arctic counterpart called IceWatch. While a ship is cruising through sea ice, hourly observations are undertaken, providing, among several other variables, the estimated thickness of the first three thickest ice types encountered. Thickness is estimated by referencing the vertical extent of ice floes tilted by the ship's hull to a ruler stick or pole mounted perpendicular to the side of the ship. The accuracy of ASPeCt/IceWatch sea ice thickness estimates is on average 0.1 m. ASPeCt/IceWatch sea ice thickness estimates are known to be biased low because parts of the ice broken by the ship's hull might slip underneath the ship or adjacent ice and because ships tend to avoid thick ice. Despite this, ASPeCt/IceWatch are often the only thickness information in regions otherwise void of observations.

Another important source of in situ sea ice thickness information is manned drift stations such as the Russian North Pole stations. At such stations snow depth and sea ice thickness measurements are carried out at regular time intervals along several transects a few hundred metres long while the station drifts through the Arctic Ocean.

So-called ice mass balance buoys (IMBs) have a thermistor string frozen into the sea ice and snow cover and provide measurements of the temperature profile through the snow–ice system from which information about the sea ice thickness can be inferred. Several of such buoys have provided information about the seasonal sea ice thickness evolution over the last 15 years. One disadvantage of these types of in situ measurements is that they are carried out mostly on MY ice as such buoys are typically deployed in late summer or early fall on the ice that survived the melt season. The second disadvantage is that these measurements are limited to one ice floe and might not represent the typical conditions around that floe.

The advantage of using sensors mounted on airborne vehicles such as helicopters and fixed-wing aeroplanes is the far larger spatial range covered. Airborne sensors of various kinds provide very detailed information about the ice thickness distribution and are currently an essential tool for the evaluation of sea ice thickness obtained from satellite remote sensing. Aircraft observations have the potential to bridge between the spatial scale of in situ measurements and satellite remote sensing.

For sea ice thickness the most important sensor combination is either an ElectroMagnetic (EM) sensor or a snow penetrating radar operating at S- to C-band frequencies used in conjunction with a laser altimeter. The EM sensor exploits the conductivity differences between the ice and underlying sea water to measure the distance from the sensor to the ice–water interface. The accuracy of this sensor is about 0.1 m or better over undeformed sea ice. Over ridged sea ice the accuracy degrades as a function of the ratio between the EM footprint and the across-ridge dimension, and the degree of blockiness of the ridge. EM sensors are most often used from airborne platforms. Other versions of this instrument are mounted on a sledge and towed over the sea ice, a method that is used in support of in situ measurements and airborne measurements. A ship-based version of this instrument also exists. As with satellite systems, the laser altimeter measures the distance between the sensor and the surface. This provides the topography of the snow surface or ice surface (when the sea ice is bare) and the leads between the floes. When combined with airborne EM measurement, one may obtain the total sea ice plus snow thickness. The snow penetrating radar locates the ice–snow interface with an accuracy between 0.05 m and 0.1 m, provided that the snow is dry and vertically homogenous. Using simultaneous and coincident measurements from these three systems provides an estimate of the snow and ice thickness.

A combination of snow-penetrating radar and laser altimetry is used for the NASA Operation Ice Bridge (OIB) flight campaigns to bridge the gap between ICESat-1 and ICESat-2. OIB measurements have a limited vertical resolution of about 0.05 m and are less accurate for ridged ice. OIB measurements can only be carried out during the cold season because snow wetness hampers the measurements or, at least, degrades the accuracy. Nevertheless, for the period 2010–17, a number of studies were conducted in which sea ice thickness, freeboard and snow depth were inter-compared using OIB measurements for evaluation and algorithm improvement. OIB measurement analysis is more mature for the Arctic than for Antarctic sea ice.

One caveat of satellite altimetry is the high sensitivity of the retrieved sea ice thickness to the accuracy of the observed freeboard. A relatively small bias in freeboard can translate into a large bias in the obtained sea ice thickness. This is different for measurements of the sea ice draft, i.e. the depth the floating sea ice penetrates into the seawater. Draft measurements have been carried out using upward looking sonar (ULS). These sensors measure the travel time of an acoustic pulse transmitted upwards and received after reflection from the ice underside. When properly taking into account signal runtime variations due to different water temperatures and salinity, draft measurements can be as accurate as 0.1 m. ULS sensors have been employed on submarines for decades and have been the sole source of large-scale sea ice thickness information in the Arctic Ocean before the satellite era. Our current knowledge about sea ice volume decrease in the Arctic Ocean is strongly based on submarine ULS measurements. In the 1990s moored ULS sensors were introduced in both the Arctic and the Southern Ocean. The seasonal development in sea ice draft measured by a moored ULS is a combination of local sea ice growth and ice advection due to ice drift. In the Arctic the longest time series (more than 20 years) from moored ULS exist from locations south of the Fram Strait and in the Beaufort Sea. In the Southern Ocean, the Weddell Sea and at the Greenwich Meridian are locations with the longest measurements series. Submarine ULS measurements are completely lacking in the Southern Ocean. Recently, first attempts using a multibeam ULS from Autonomous Underwater Vehicles (AUVs) were tested successfully at several locations in the Arctic and the Antarctic providing detailed sea ice draft distribution information.

3.5.3 Some Examples of Operational Products

Operational ice thickness products are produced from Cryosat data "Arctic Sea Ice Thickness Product from CryoSat" (see Figure 3.9) and are available from the Centre for Polar Observation and Modelling (CPOM) Data Portal at www.cpom.ucl.ac.uk/csopr/seaice.html. Operational thickness of ice up to 50 cm is produced from SMOS data by the University of Bremen (Figure 3.16). These are available from www.iup.uni-bremen.de:8084/smos/.

3.5.4 New Directions and Challenges to Be Addressed

A major limitation to the accuracy of sea ice thickness data derived from altimeter measurements is the lack of quantified snow data (thickness and density). Satellite observations of snow cover on sea ice are under investigation and the future should bring some progress. In addition, snow thickness modelling in operational ice models is improving. These developments should lead to a larger demand for distribution of the ice freeboard measurements from altimeters, so users can use their own choice of snow data for the freeboard to thickness conversion or for assimilating the freeboard measurements into their models.

3.6 Observations of Ice Drift and Deformation

Conversely to the sea ice variables covered above, remote sensing of sea ice drift does not require detailed understanding of physical or radiative properties of the sea ice media nor its upper and lower interfaces. The vast majority of satellite-based ice motion tracking methods work by analysing a pair or sequence of images and attempting to explain the change in intensity patterns between the images by two-dimensional shifts of small blocks of pixels. In that respect, motion tracking for sea ice is much like using images of clouds to retrieve atmospheric motion vectors (Schmetz *et al.*, 1993) or estimating ocean surface currents using thermal infrared images of sea surface temperature (SST) patterns (Emery *et al.*, 1986). Manually following distinctive ice floes is also sometimes used to detect ice drift.

3.6.1 Main Principles Behind Algorithms for Sea Ice Drift

The basic hypotheses behind the sea ice drift algorithms are that

(1) the intensities recorded by an imaging instrument over sea ice are the imprint of a property that is advected with the sea ice, and
(2) the signal-to-noise ratio of the imaging instrument allows the sea ice property to be recognized in a second image of the same general area, at a later time.

There is no limitation regarding the type of imager (active or passive), the type of wavelength (visible, microwave, etc.) or the actual property that is observed (snow, ice type, ocean in leads, melt pond patterns, etc.). This leads to the same algorithm being easily applied to several types of sea ice images, potentially with smaller or larger adaptations which are often driven by the need for better processing speed when using finer resolution images (typically SAR). Since the mid-1980s, the Maximum Cross-Correlation (MCC) algorithm (Ninnis *et al.*, 1986) has been successfully applied to sequences of sea ice images from AVHRR, microwave radiometers like AMSR2 and SSMIS, scatterometers such as ASCAT, and Envisat, Sentinel-1 and RADARSAT SAR.

In the MCC algorithm, an initial sub-window (e.g. 11 × 11 pixels) at a particular location in the first image is attempted to be matched with a set of equally sized target sub-windows from the second image. The set pertains to all the sub-windows in the neighbourhood of the initial sub-window. The size of the neighbourhood is often referred to as the search radius of the algorithm, and corresponds to a maximum expected displacement of sea ice between the acquisition times of the two images. The match between the initial and any of the target sub-windows is quantified by the value of the normalized correlation coefficient between them. This number can theoretically vary between −1 and +1, but in practice it ranges from small positive numbers (a poor match) to almost +1 (a good match). This correlation value is repeatedly computed between the initial sub-window and all target sub-windows in the neighbourhood and the maximum of these correlations is recorded. The target sub-window in the second image achieving highest correlation is viewed as an observation of the initial sub-window after a displacement (a shift) that occurred between the times of the two images. The displacement vector returned by the MCC is that joining the central locations of the initial sub-window and this target sub-window.

By construction, the vectors returned by the MCC are quantized. The vector components can only be an integer number of image pixels. For example, displacements that are less than half a pixel size are seen as no motion by the MCC. This quantization effect can seriously reduce the accuracy of the vector field when using the MCC with coarse resolution images (typically 10–20 km pixels) from microwave radiometers or scatterometers, and several strategies and algorithms were designed to achieve sub-pixel accuracies for the drift vectors (see Maslanik *et al.*, 1998 for a review). One such method is the Continuous Maximum Cross Correlation (CMCC) of Lavergne *et al.* (2010) that effectively removes the quantization effect (Figure 3.17).

Thanks to the much finer image resolution, the quantization effect of the MCC is not a limiting factor for SAR-based ice drift products using the MCC and algorithmic developments mainly target higher spatial resolution of the vector field and better computational efficiency. This can be obtained by processing the images in a pyramid of resolutions where the larger scale ice drift is found using a coarse resolution version of the images and subsequently at finer resolution the details of the ice drift pattern can be found using much smaller search radii around the coarser resolution drift vectors (e.g. Thomas *et al.*, 2008).

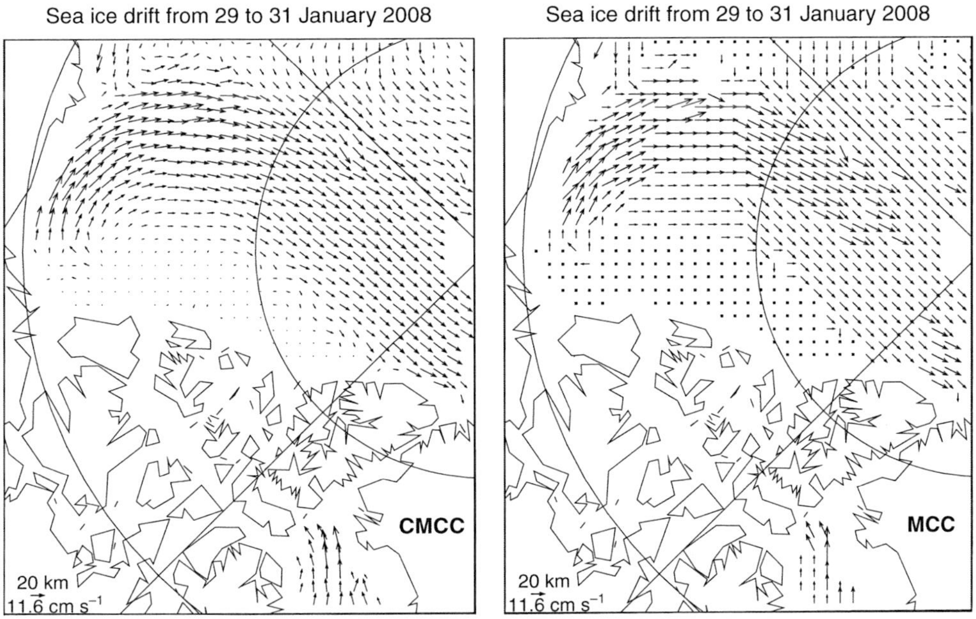

Figure 3.17: Example ice displacements from AMSR-E (37 GHz H and V channels) over the Beaufort Sea and Canadian Basin from 29 to 31 January 2008. The product was processed with the (left) Continuous Maximum Cross Correlation (CMCC) and (right) Maximum Cross Correlation (MCC) method from the same satellite images. On the MCC product, zero-length vectors are depicted with a small square symbol while tiny arrows are used for the CMCC product. The MCC product clearly exhibits a quantization noise. Such an artefact is not present in the CMCC drift data set. This figure and its caption are reproduced from Lavergne *et al.* (2010).

3.6.2 Sea Ice Drift Is Not Sea Ice Velocity

Since the motion vectors retrieved by sea ice drift algorithms are derived from a pair of satellite images separated in time, they do not hold any information about the path of sea ice *between* the image times. Consider the case where an ice drift algorithm is applied on two SAR images that are three days apart. The motion vectors measure the net Lagrangian displacement at the end of a three-day drift, and carry no information about the accelerations, idle time, loops, turnarounds, etc. of the observed sea ice during these three days. Consequently, *sea ice drift is not sea ice velocity*!

Due to the very nature of the motion algorithms that match a sub-window (e.g. 11 × 11 pixels) in one image to an equivalently sized sub-window in the second image, the retrieved drift vectors measure a spatial average of the motion field, inside the area of the sub-window. The drift vectors computed from matching pairs of satellite images are thus never point-like estimates but, rather should be used as area averages of the local motion field (similar to an APS net displacement forecast). An accurate use of an ice drift product requires knowledge of this sub-window size. Sea ice displacement obtained from matching similarities in pairs of satellite images are thus net Lagrangian, spatially averaged estimates of the motion field.

3.6.3 Resolution vs. Grid Spacing and Eulerian vs. Lagrangian View

The relation between the size of the sub-windows and the area averaging of the local motion was introduced just above. In many respects, the size of the sub-windows is the analogue to the size of the foot-prints of satellite sensors (their field-of-view projected on Earth surface) and can be viewed as the *spatial resolution* of the drift product. A second 'size' of importance is the *grid spacing* or distance between neighbouring vectors. This length is not imposed by the satellite images themselves, but rather chosen by the product developers. Ideally grid spacing should be twice the size of the sub-window. Shorter (longer) grid spacing will result in over- (under-) sampling of the motion field.

Thanks to the much finer resolution of the images, there is more freedom to choose a grid spacing that is more compatible with the sub-window size in a SAR-based ice drift product. Still, a note of caution is needed here: methodologies and algorithms that mix images at several spatial resolutions (e.g. all those using pyramids of resolutions) have a tendency to introduce dependencies between neighbouring vectors, especially if using coarse image resolutions to impose a priori knowledge of the motion field from finer image resolutions. SAR-based ice drift products are not necessarily free from oversampling of information and users of drift products should be aware of the difference there is between the spatial resolution and the grid spacing of the sea ice drift products they use.

Another design choice for sea ice drift products is whether to compute vectors on fixed equally spaced grids (Eulerian drift products) or at irregularly and varying locations (Lagrangian drift products). Eulerian drift products are the most common, and are appealing for comparison and data assimilation with sea ice models, since both estimate motion at a set of fixed locations on the Earth, at every time step. The initial location of the motion vectors are at the same locations at every time step and it is not the same ice parcel that is tracked over a longer period. Lagrangian drift products are designed to generate drift trajectories of sea ice parcels and follow them throughout a longer sequence of images. The initial locations of the motion vectors for a new image are at the final locations of the motion vectors from the previous image. The near-real-time SAR-based sea ice drift product from the Copernicus Marine Environment Monitoring Service (CMEMS[2]), processed at the Technical University of Denmark (DTU) is an Eulerian ice drift product. It will be introduced later. The most well-known Lagrangian ice drift product is that from the RADARSAT Geophysical Processing System (RGPS, Kwok *et al.*, 1990), but due to its latency it is not available for operational sea ice monitoring.

A third class of drift products exist, in which the initial locations are renewed for each matching image pair. The strategy is to seed drift vectors at locations that are expected to allow better retrieval accuracies. This is the basic principle of feature-tracking methods, as they locate vectors onto distinctive image features (typically edges, ridges, floes, etc.) and thus avoid attempting motion tracking in areas where image characteristics cannot guarantee success (see Komarov and Barber, 2014; Muckenhuber *et al.*, 2016).

[2] http://marine.copernicus.eu

As a final remark on these aspects, it is noteworthy that the spatial sampling strategy for the drift vectors (fixed Eulerian grid, Lagrangian grid, feature-driven grid) is a trait of the drift product chosen by the investigator or production centre. It is not a characteristic of the motion tracking algorithm, and the MCC can be (and is) used for all types of drift products.

3.6.4 In Situ Observations of Sea Ice Drift and Quality Control of Satellite Products

Autonomous drifting buoys deployed on ice are designed for measuring environmental variables above, inside, and below sea ice (e.g. wind speed, air temperature, snow depth, ice thickness and temperature or profiles of ocean temperature and salinity). A fortunate by-product is sea ice motion since all these buoys report their position and time as part of their data stream, irrespective of their main original purpose. In addition, technological developments and cheaper data transmission allow deployments of dispensable GPS beacons with the sole purpose of providing sea ice trajectories.

Evaluation of sea ice drift vectors from satellite products against trajectories requires slightly different collocation than when validating other sea ice satellite products or model fields, because the buoys report their positions at fixed times, while the ice drift vectors are net Lagrangian displacements over time. Collocation must be handled at both start and end times of the drift vectors. Like other satellite products, evaluation of ice drift vectors must take into account representativeness uncertainty which stems from the scale difference between point-like measurements from buoys, and the area-averaged measurements from satellites (with radii from tens to hundreds of kilometres). In situ positions are usually reported with GPS accuracies and often at hourly time resolution, which is sufficient for validating the basin-scale products based on microwave radiometers and scatterometers. Hourly (or sub-hourly) sampling is needed for useful evaluation of higher-resolution products based on SAR data.

It has become more common to report evaluation statistics on quantities that are close to what the ice drift satellite products actually measure: that is, the two vector components along the product grid as displacements between the times of the image pair used. This approach is generally preferred to reporting statistics in terms of velocities. To use velocities gives the false impression that satellite products can easily be compared to speeds.

The variety of choices from which components to validate, the regions covered, the quality of input data sources, the collocation methodologies and the time period covered make it impossible to directly compare evaluation results obtained using different strategies. Intercomparison of products must be carried out with common collocation methodologies, and over the same regions and time periods, as in Hwang (2013) or Sumata *et al.* (2015). When this is the case, the RMSE (root mean square error) of satellite products is typically reported to range from 400 m to 1 km for the components of SAR-based drift vectors, and 2 to 10 km for those of the vectors processed from passive microwave and scatterometer images. These numbers apply to the Arctic Ocean during winter.

The accuracy of Southern Hemisphere ice drift products is less well documented due to the relative lack of buoy data available.

Buoys are sometimes arranged in arrays, typically 5 to 30 buoys that aim at measuring local scales of ice deformation. These are, however, still mostly research infrastructures, not meant for operational monitoring.

3.6.5 Some Examples of Operational Products

This section briefly describes two operational products available at the time of writing. They are under continual development, and the characteristics given below might evolve.

3.6.5.1 The Low-Resolution Sea Ice Drift Product of the EUMETSAT OSISAF

OSISAF[3] has distributed an operational global sea ice drift product since late 2009 (Figure 3.18). It is currently processed from daily average maps of brightness temperatures of AMSR2 (36.4 and 18.7 GHz) and SSMIS (91 GHz), as well as C-band backscatter from ASCAT on board Metop satellites. It utilizes the CMCC method to deliver ice drift displacements with 62.5 km grid spacing (with about 125 × 125 km effective spatial resolution). The displacement vectors correspond to two full days of motion. Since summer 2016, the product has been extended to include the May–September period in the Northern Hemisphere. Maps of uncertainty document that the accuracy is reduced during the summer months (typically 5–6 km RMSE of the 48 hours x and y components) compared to winter (2–3 km RMSE). Figure 3.18 shows two examples of drift maps from this product. On the Arctic Ocean map (left panel), black vectors are from the product, while red vectors are the displacements reported by in situ drifters. There is generally very good agreement between the two datasets. The right panel shows an example in the Ross Sea, and the colours of the background field indicate the magnitude of the vectors. The product is able to capture the flow patterns driven by the frequent atmospheric low-pressure systems travelling over Southern Hemisphere sea ice.

A somewhat similar product is available from the IFREMER CERSAT, using the MCC for computing winter-only 2- and 3-day maps in the Arctic Ocean. The accuracy is generally less than that of the OSISAF (Hwang, 2013), but the archive holds maps back to 1992.

3.6.5.2 The SAR-Based Sea Ice Drift Product of CMEMS/DTU

The SAR-based sea ice drift product processed at DTU is distributed in near-real time through CMEMS. It is based on the MCC and a new map of ice motion is generated for each new pair of overlapping Sentinel-1 SAR images. Its grid spacing is 10 km (and about 10 × 10 km effective spatial resolution, thus no oversampling). The time span of the vectors is the time separation of the SAR images, which varies between 2 and 48 hours for the 'European' sector of the Arctic monitored by Sentinel-1. Most of the drift maps have time

[3] See http://osisaf.met.no

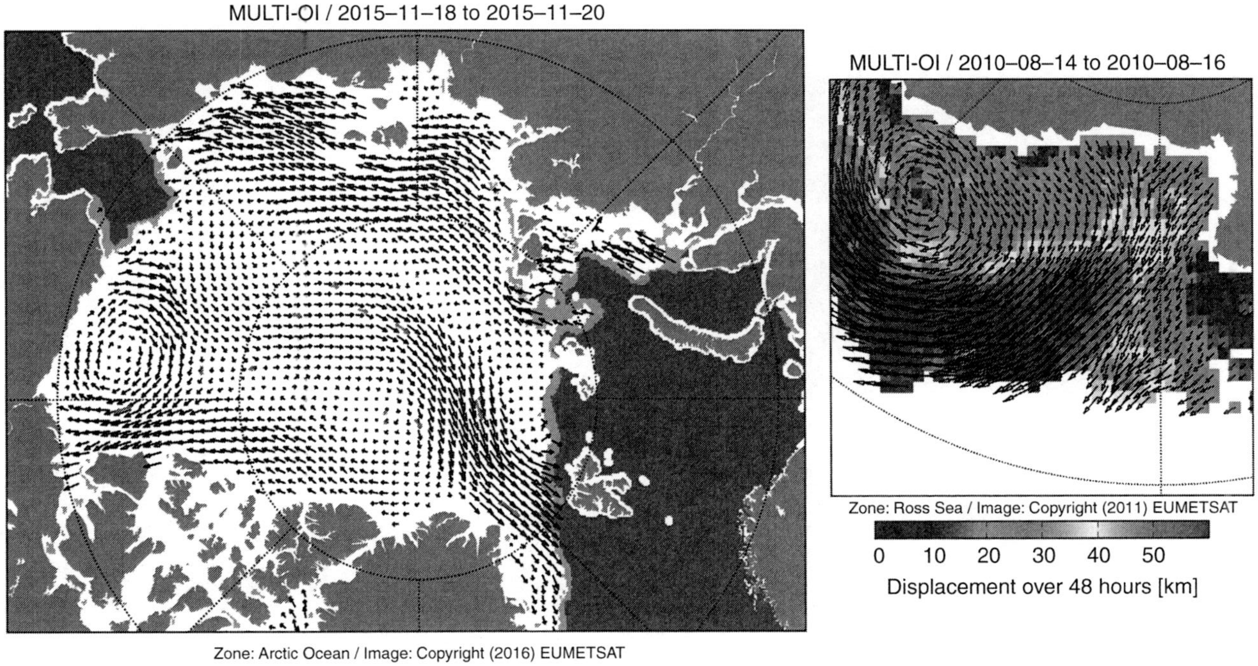

Figure 3.18: Example maps of sea ice drift as distributed from the EUMETSAT OSISAF. Left panel shows an example of the drift in the Arctic Ocean from 18 to 20 November 2015. The superimposed red vectors are the displacements measured by in situ buoy trajectories. The right panel shows the motion field over the Ross Sea from 14 to 16 August 2010 with colours indicating the magnitude of the vectors. (A black and white version of this figure will appear in some formats. For the colour version, please refer to the plate section.)

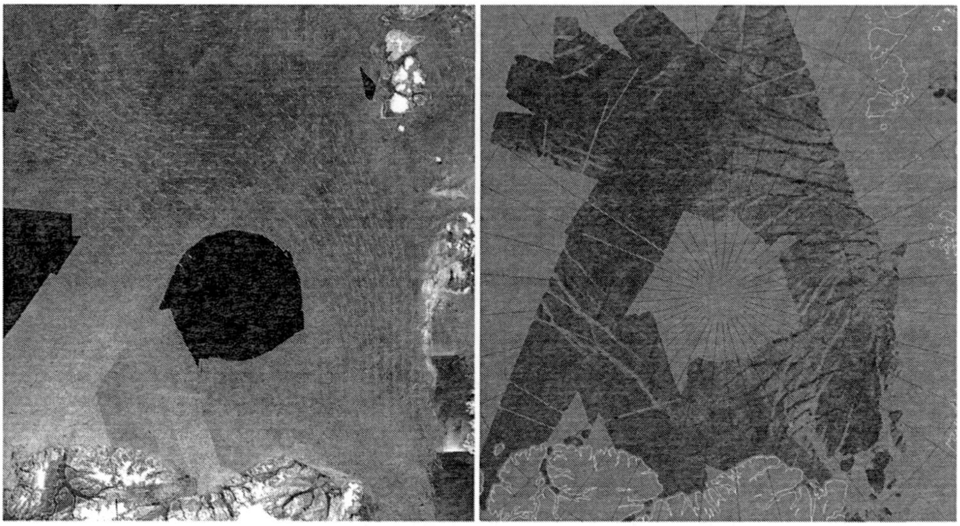

Figure 3.19: High-resolution ice drift field derived from Sentinel-1 images on 27 and 28 February 2016. Only every 25th original drift vector shown. Mosaic of 26–28 February Sentinel-1 coverage as background (left). Example of divergence deformation field derived from the Sentinel-1 ice drift (right). Blue colours correspond to divergence (openings) and red colours to convergence (closings/potential pressure ridge formation). Source: www.seaice.dk (A black and white version of this figure will appear in some formats. For the colour version, please refer to the plate section.)

spans close to 24 hours. Maps of sea ice deformations (divergence/convergence, shear, curl) are also computed from the maps of sea ice drift (see Figure 3.19).

Other SAR-based sea ice drift products are produced including at the Finnish Meteorological Institute (FMI) for the Baltic Sea and at the Canadian Ice Service (CIS) for Canadian waters.

3.6.6 New Directions and Challenges To Be Addressed

Most of the material and discussions above covered the retrieval of ice motion vectors from analysing pairs of images. This is by far the most widely used approach and the only one currently viable for operational applications. Improved satellite technologies, especially applied to SAR processing might provide new opportunities. One of these is the interpretation of Doppler shifts of the SAR signals into instantaneous, radial velocities of sea ice (e.g. Kræmer *et al.*, 2015). The estimates are only radial components of the vectors (line of sight direction), but are truly a velocity instead of a time-integrated displacement. Another recent development for tactical navigation in ice infested waters or near-shore break-up monitoring are marine radars, either on-shore such as at Point Barrow, Alaska, or onboard ships in the Baltic (Karvonen, 2016).

Research and development challenges also remain for algorithms and products based on pairs of images. One can mention the determination and distribution of maps of uncertainties along with the products, or the accurate merging of microwave radiometer-based products with SAR-based products, possibly including buoys, to ensure maximum spatial coverage and best accuracy where possible. Data assimilation provides the opportunity to optimally utilize observations from various sources with model forecasts. The ability to process drift products with spatial resolution on the order of a kilometre from SAR constellations (Sentinel-1s, RADARSAT Constellation) will require advanced methods to increase computation efficiency. At the same time, advanced Pan-Arctic or regional sea ice models with finer and finer resolutions will require accurate, higher resolution products of sea ice drift and deformation.

Higher temporal resolution observations of ice drift are sometimes available from the many VIS and IR instruments on weather satellites during cloud free conditions. Up to 20 acquisitions per day can be achieved using the suite of instruments and satellites from Table 3.4. At lower latitudes such as the Gulf of St Lawrence even geostationary satellites may be used to detect ice drift under cloud-free conditions.

3.7 Other Ice-Related Variables

3.7.1 Snow on Sea Ice

The key snow properties are its depth, density, albedo, water content and temperature profile. Other important properties are its grain size and vertical structure. While many of these quantities are frequently estimated from observations as single (average) values at each horizontal location, most of them vary significantly with depth. However, accurately estimating the complete vertical profiles of these variables from remote sensing data is often not possible. Therefore, it is more practical to represent the profiles as effective values, i.e. scalars which ideally lead to the same remote sensing observations when representing the snow pack as a single layer of constant (bulk) microphysical variables.

Most of these variables can be observed in situ, with similar limitations to the spatio-temporal coverage as mentioned for sea ice thickness observations. In situ observations made at manned drifting ice stations and aircraft landings in the Arctic Ocean have been compiled in a snow-depth climatology of the Arctic Ocean for the period 1954 to 1991 (Warren *et al.*, 1999). However, most of these observations were carried out on MY ice, which has been decreasing in areal coverage to less than half since the 1970s, i.e. from being the dominant ice type to now covering less than half of the Arctic Ocean. Consequently, intercomparison of the current snow depth on sea ice observed with airborne and satellite remote sensing reveals that, although still regularly used, this climatology is no longer valid today.

The previously mentioned ASPeCt and IceWatch data sets (Section 3.5.2) are the most comprehensive source for more recent ground based observations of snow depth on sea ice and have also been compiled into a climatology for the Antarctic (Worby and Allison, 1999; Worby *et al.*, 2008; Meiners *et al.*, 2012). Similar to the sea ice thickness, ASPeCt snow depths are likely underestimated.

Information about snow depth from remote sensing is provided by the snow radar flown on the Operation Ice Bridge (OIB) airborne campaigns (see Section 3.5.2). While these snow depth observations are only available for a limited period and region along the OIB flight tracks in the western Arctic Ocean, they are available over both FY and MY sea ice and are less limited for deformed sea ice than the ASPeCt/IceWatch data sets. The accuracy of OIB snow depth data is still under debate as thin snow depths seem to be underrepresented due to the limited range resolution of the radar and as many radar returns are missed over deformed sea ice leading also to an underrepresentation of the often thicker snow around ridges.

Ice-mass balance buoys (Perovich *et al.*, 2013) provide another (very local) source of snow depth on sea ice. They combine two acoustic sensors, one above the snow and one below the sea ice, with a thermistor chain to measure the temperature profile in the snow – ice system with near-centimetre vertical resolution.

Existing microwave radiometer based algorithms used to retrieve snow depth on sea ice all date back to the original Antarctic work by Markus and Cavalieri (1998). Their algorithm is based on brightness temperatures measured at vertical polarization at 19 and 37 GHz by the satellite microwave radiometers sensors SSM/I(S) and AMSR-E and exploits the differences in penetration depth and scattering in the snow layer at these two frequencies. The algorithm requires a correction of the brightness temperatures for the open water fraction within the sensor's footprint. The algorithm is known to:

(1) be only applicable over FY ice,
(2) underestimate snow depth over deformed sea ice, and
(3) be sensitive to snow wetness and grain size.

Despite these limitations it provides hemispheric maps of snow depth on seasonal FY sea ice.

Recent research demonstrated that snow depth can also be retrieved on thick Arctic sea ice, i.e. also over MY ice, using L-Band satellite radiometry (Maaß *et al.*, 2013), but hemispheric fields are not yet regularly available.

3.7.2 Melt Ponds and Albedo

Melt ponds are puddles of fresh melt water on top of sea ice forming during the melt season. Due to their low albedo, they play a fundamental role in the summertime Arctic Ocean net surface radiation balance and hence in the melt progress. They cover up to 50 per cent of the sea ice and are a serious source of bias for summertime sea ice concentration retrieved from satellite microwave radiometry (Section 3.4.1 and Figure 3.13). Moreover, melt ponds act as the light sources for the ocean below the sea ice, thus being crucial for sea ice bottom melt and for the microorganisms developing in and below the ice.

In addition to melt ponds, albedo information is also required in many ocean and atmosphere circulation models. Sea ice/snow albedo is high and nearly constant during winter. However, in spring, when the water content of snow starts increasing, the surface albedo of sea ice reduces well before melt ponds appear. Albedo retrieval is particularly important in summer because 96 per cent of the total annual solar heat input through sea ice occurs during the melt season from May to August (Arndt and Nicolaus, 2014). Sea ice leads, which may form during all seasons, also reduce the total surface albedo.

The small size of melt ponds, most commonly between 10 m^2 and 100 m^2, hampers their large-scale direct remote sensing with operational microwave or optical satellite sensors. High-resolution optical satellite sensors (e.g. Kwok, 2014) with a spatial resolution of about 1 m would suffice but frequent cloud cover, particularly during summer, along with the typically small swath, renders such sensors well suited only for case studies, but less so for regular monitoring of the melt pond coverage on Arctic sea ice. Therefore, more common approaches to retrieve melt pond coverage utilize optical sensors of coarser spatial resolution, but with a substantially larger spatial coverage such as MERIS, MODIS, VIIRS and OLCI. In order to obtain the melt pond fraction on sea ice from such sensors, one may first discriminate open water in leads from sea ice (with and without melt ponds) and then compute the melt pond fraction via the albedo of the sea ice. Alternatively, one uses a mixing approach for which it is assumed that the field-of-view of the sensor is covered by melt ponds and other surface types typically encountered during the melt season, i.e. wet snow, bare melting sea ice, and open leads between the floes. By employing typical spectral reflectance values of these surface types one can use a linear combination of the reflectances measured at different wavelengths to obtain the fraction of each surface type. The accuracy in the melt pond fraction obtained by such approaches is 5 to 11 per cent (Istomina *et al.*, 2015a + b). However, the accuracy strongly depends on:

(1) a proper elimination of cloud influence,
(2) the match of the number of different surface types and different channels used, and
(3) the stability of the signature of the surface types over time.

As the spectra of the various surface types show a considerable intra-class variability (Istomina *et al.*, 2013, 2016), a retrieval has been suggested based on physical forward models for the spectral reflectivity of the various surface types. A resulting daily data set covers the MERIS lifetime 2002–2012 (Istomina *et al.*, 2015a+b), and an eight-day composite data set based on fixed reflectances from MODIS covers the years 2000–2011 (Rösel *et al.*, 2012).

The retrieval of the broadband albedo is provided using the same procedure as that of melt pond fraction. Daily images are patchy because of cloud masking. For a visual impression, eight-day composites give a more complete picture; see Figure 3.20 for examples.

3.7.3 Surface Temperature

Ice surface temperature (IST) controls sea ice growth, air–sea heat exchange and snow metamorphosis and is thus an important quantity in large-scale modelling. Ice surface temperature can be retrieved from thermal infrared measurements in a similar way to

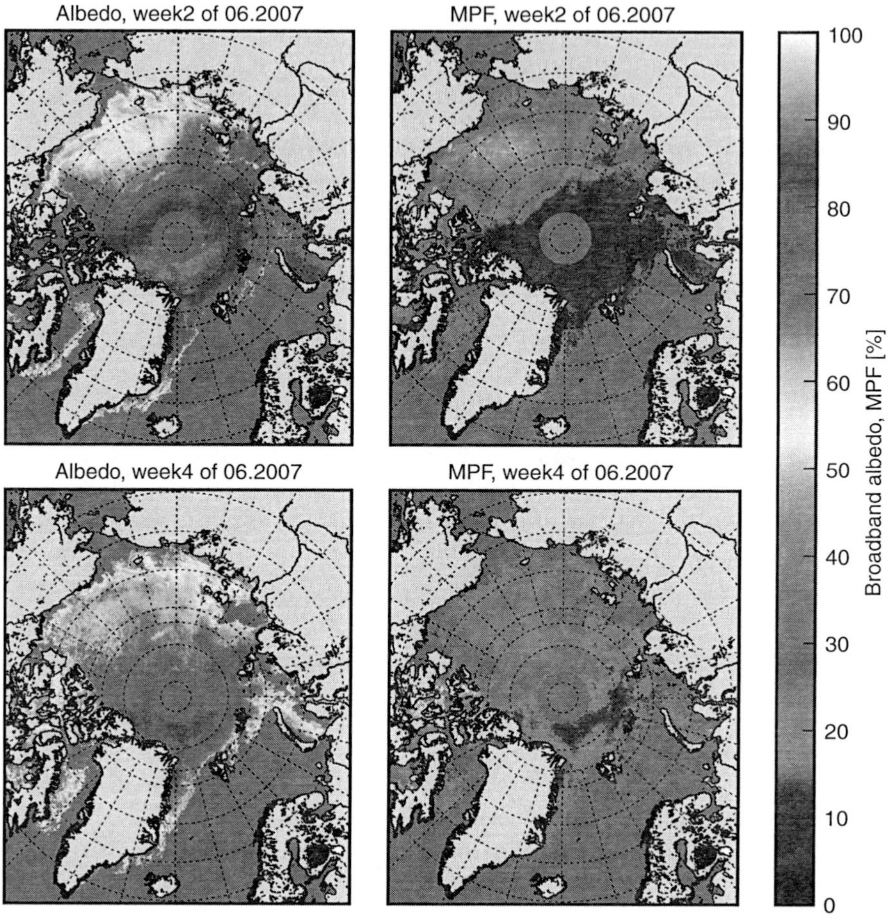

Figure 3.20: Eight-day composites of Arctic broadband albedo (left) and melt pond fraction (right). The left-to-right comparisons show that melt pond fraction and albedo are statistically but not strictly correlated. The top-to-bottom comparisons illustrate the fast development of both quantities during the melting period (Istomina *et al.,* 2015a). (A black and white version of this figure will appear in some formats. For the colour version, please refer to the plate section.)

retrieval of sea surface temperature. The emissivity at thermal infrared wavelengths is very close to 1 for ice, snow and water surfaces and the penetration depth is negligible. Hence the brightness temperature measured by an infrared radiometer can (after atmospheric correction) be interpreted as surface temperature.

The atmospheric correction step has two components, a screening for clouds and a correction for water vapour absorption based on measurements at two infrared wavelengths. Cloud screening may be quite difficult during the winter season where only thermal infrared data are available (Liu *et al.*, 2004), whereas, in summer, visible and especially near infrared channels of the same instrument facilitate the detection of areas of cloud cover (Ackerman *et al.*, 1998; Shi *et al.*, 2007; Istomina *et al.*, 2010; Liu *et al.*, 2010) which are subsequently discarded from processing. The two-channel approach to water vapour correction is often referred to as a split-window technique and is based on the assumption that, at the two similar wavelengths, the surface radiation is the same and that the difference in measurements can be ascribed to water vapour absorption which can then be quantified and used to correct both measurements. Examples of IST maps are shown in Figure 3.6. Attempts to use IST to infer ice thickness have been made, but the method depends on reliable a priori knowledge of snow properties including thickness.

The ocean brightness temperature at microwave frequencies is a function of surface temperature, wind speed and direction, and for frequencies less than about 10 GHz the sea surface salinity (Ulaby, 2013; Meissner and Wentz, 2012). Both the microwave and the infrared penetration depth into sea water are very shallow, resulting in the effective temperature being very close to the skin temperature or SST. Cloud masking schemes for measurement of SST using infrared channels are effective over open water which means that the global SST can be measured accurately by both polar orbiting and geostationary satellites (at lower latitudes) in cloud free conditions. The cloud penetrating microwave 6 GHz radiometers measuring SST from polar orbit provide complete global coverage at coarser spatial resolution (Table 3.3). The brightness temperature is almost proportional to SST near 6 GHz. Examples of SST maps are shown in Figure 3.6.

3.7.4 Floe Size Distribution

High-resolution images (Landsat, Sentinel-2 etc.) have been used to derive the ice floe size distribution (FSD) in the marginal ice zone for scientific applications (Toyota *et al.*, 2010). The FSD is also estimated from SAR or visible images by ice analysts and is part of the ice properties included in ice charts (Table 3.7). It has proven very difficult to automatically detect FSD in SAR images except for some relatively simple cases.

3.8 Ice Charts

Ice charts (see also Chapter 1) are often used as input to models or for model evaluation and also for evaluation of sea ice information retrieved automatically from satellite data. Ice charts are subjective analyses of information from a multitude of sources: satellites, weather and ice prediction systems, and in situ observations combined into a map of ice conditions at a given time (typically a fixed time of day (regional analysis) or at the time of a specific satellite image (image analysis)). In ice charts the ice concentration is typically given as 10ths of ice cover, and often fractional ice type concentrations (stage of development) are given as well. Figure 3.15 shows an example of a Sentinel-1 SAR image of central west Greenland and a corresponding ice chart produced at the Danish Meteorological Institute. Figure 1.1 shows an example of an ice chart from the Canadian Ice Service.

Ice charts contain information about sea ice type distribution in irregularly shaped polygons, the shape and size of which depict fairly uniform conditions.

Table 3.7: *Specification of stage of development and floe size in the WMO ice chart egg code.*

Egg code for Stage of Development for Sea Ice (thickness)	Code Figure S_x
New Ice-Frazil, Grease, Slush, Shuga (0–10 cm)	1
Nilas, Ice Rind (0–10 cm)	2
Young (10–30 cm)	3
Gray (10–15 cm)	4
Gray – White (15–30 cm)	5
First Year (30–200 cm)	6
First Year Thin (30–70 cm)	7
First Year Thin – First Stage (30–70 cm)	8
First Year Thin – Second Stage (30–70 cm)	9
Medium First Year (70–120 cm)	1
Thick First Year (>120 cm)	4
Old – Survived at least one season's melt (>2 m)	7
Second Year (>2 m)	8
Multi-Year (>2 m)	9
Ice of Land Origin	▲•
Egg code for Forms of Sea Ice (floe size)	**Code Figure F_x**
Belts and Strips symbol followed by ice concentration	~F
New Ice (0–10 cm)	X
Pancake Ice (30 cm – 3 m)	0
Brash Ice (< 2 m)	1
Ice Cake (3–20 m)	2
Small Ice Floe (20–100 m)	3
Medium Ice Floe (100–500 m)	4
Big Ice Floe (500 m–2 km)	5
Vast Ice Floe (2–10 km)	6
Giant Ice Floe (> 10 km)	7
Fast Ice	8
Ice of Land Origin	9
Undetermined or Unknown (Iceberg, Growlers, Bergy Bits)	/

Sea ice thickness in ice charts is only indirectly specified in terms of the *stage of development* (Table 3.7), which is interpreted based on information from multiple satellite sensors and a combination of proxy information about sea ice thickness such as ice type and sea ice age, information about meteorological conditions such as freezing degree days (FDD), and reports from ships and other observing and prediction systems, if available. While the ice thickness information obtained by remote sensing methods usually gives single sea ice thickness values per grid cell, albeit for quite a range of grid resolutions and temporal resolutions, stage of development information in ice charts is typically specified for several ice types within a polygon.

The primary remotely sensed information source for producing ice charts is SAR imagery, which provide high-resolution ice coverage and ice type information independent of cloud cover. Of secondary importance are medium resolution visible and infrared data from meteorological satellites. This information is blended by an Ice Analyst with information about the ice growth history, FDD, ship reports and other auxiliary and satellite information. Information about divergent and convergent ice motion may also be included in the analysis.

3.9 Observation Operators

Many of the assumptions that are inherent in the sea ice retrieval algorithms described above (especially for ice concentration and ice thickness) are due to the poor quality of auxiliary information that is often obtained from climatological estimates. Examples include the use of climatological snow cover in sea ice thickness algorithms or the use of climatological atmospheric variables in sea ice concentration retrieval. In many cases, however, automated prediction systems (APS) now or in the near future will possess much better estimates of these variables than climatology.

Applications where the retrieval algorithms directly use data from an APS are beginning to appear. The OSISAF sea ice concentration algorithm is one example where analysis fields or very short-term forecasts of wind speed and water vapour from a weather forecast model are used to correct the measured brightness temperatures before computing sea ice concentration. Results are generally good at low sea ice concentrations where the atmospheric noise is reduced by more than 50 per cent. At higher ice concentrations, the main factors driving the uncertainty is variability in snow cover and ice surface features and these are less well provided by current APS.

Satellite microwave radiometers measure brightness temperature at more wavelengths and polarization combinations than are used in standard ice concentration algorithms. Some of these measurements contain additional information about sea surface roughness, ice temperature and atmospheric variables that, potentially could be used instead of climatology, to reduce uncertainty in ice concentration retrievals. Also, other instruments on the same or other satellites measure quantities that contain such information.

Wind speed is just one of a number of the physical variables affecting the brightness temperature over the open ocean. Additionally, SST, atmospheric water vapour, cloud liquid water and air temperature can have a significant impact on the measurements. All of these can be retrieved simultaneously using multispectral microwave radiometer measurements and an emission model and, for example, using optimal estimation (Wentz *et al.*, 2000; 2007; Pedersen, 1994).

Exploiting all information in multispectral satellite data or even multi-instrument data requires models, referred to as *observation operators*, of the relationship between the

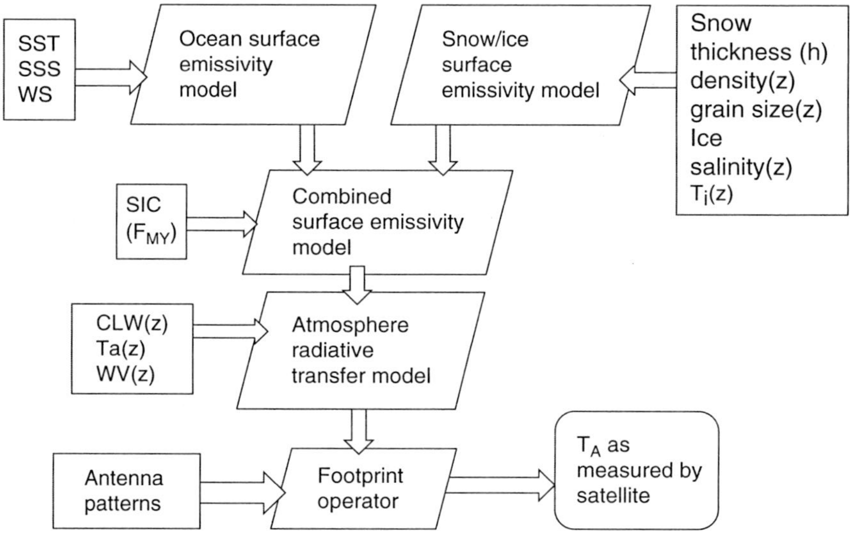

Figure 3.21: Example of an observation operator including emissivity models for the ocean and ice surfaces, a radiative transfer model for the atmosphere and a satellite footprint operator. Radiative Transfer for TOVS (RTTOV) is a radiative transfer model for the atmosphere. FASTEM is the surface emissivity model of RTTOV. The Wentz models are from Wentz & Meissner (2000), and MEMLS is a sea ice version of the Microwave emissivity model for layered snowpacks from Wiesmann & Mätzler (1999) and Tonboe *et al.* (2006).

physical properties of the snow–ice–water surface layer, the atmosphere and the electromagnetic radiation received by satellite instruments. A schematic of an observation operator for microwave radiometry is shown in Figure 3.21.

Observation operators are also necessary to directly use satellite measurements within a data assimilation system to compute corrections to short-term forecasts produced by ice–ocean–atmosphere models (see Chapter 4). They are also useful in blending several types of satellite data with a priori or background knowledge in the development of new sea ice retrieval algorithms.

Observation operators and model inversion techniques are integral parts of the data assimilation systems in NWP where microwave or infrared radiometer data are already used to provide atmospheric temperature and humidity information. Sea ice retrieval research is, in fact, working towards similar schemes for model inversion because it allows a more complete and consistent physical description of the snow and sea ice system emitting the radiation. Optimal estimation is one such blending technique that may use climatology or short-term forecasts as background state for blending with a variety of satellite measurements (Pedersen, 1994; Melsheimer *et al.*, 2009).

Observation operators already exist for the ocean surface and for the atmosphere, and they are under development for ice and snow. The sea ice emissivity, at a certain polarization, frequency and incidence angle is, as discussed earlier, a function of surface scattering, subsurface extinction and reflections between layers with different permittivity. Physical models exist that relate snow and ice properties such as density, temperature, snow crystal and brine inclusion size to microwave attenuation, scattering and reflectivity/emissivity. The Microwave Emissivity Model for Layered Snowpacks (MEMLS) (Mätzler, 1987, Wiesmann and Mätzler, 1999; Mätzler and Wiesmann, 1999) is an example. However, several of the relevant ice and snow emissivity model parameters are not currently prognostic variables in APS. Some of the input parameters and variables are even difficult to measure in the field. Some APS models do include a model of the snow dynamics and thermodynamics as well as a melt-pond parameterization or even a melt pond model. Use of data from such operational models in satellite data processing will eventually improve sea ice concentration retrieval also at higher concentration, and allow assimilation of the microwave measurements directly into the APS (Scott *et al.*, 2012).

3.10 Chapter Summary and Suggestions for Further Reading

This chapter introduced satellite remote sensing with an emphasis on observations of sea ice and polar oceans. Interaction between electromagnetic radiation and the surface of the Earth was introduced. The most commonly used satellite techniques and instruments were introduced and applications to observations of key sea ice variables such as concentration, thickness and drift were described. In situ or near surface techniques to collect evaluation data, such as using ship borne, helicopter borne or airborne instruments, were included where applicable.

More information about the physical and dielectric properties of sea ice can be found in Untersteiner (1986) and Wadhams (2000). Schanda (2012) gives an introduction to the physical fundamentals of remote sensing, and Ulaby (2013) provides an in depth description of remote sensing techniques in general, whereas Carsey (1992) and more recently Shokr (2015) provide thorough introductions to remote sensing of sea ice in particular. Detailed information about a long list of past, current and future Earth Observation satellite missions and their instruments can be found in the online Satellite Missions Database *eoPortal Directory* at https://directory.eoportal.org/web/eoportal/satellite-missions/.

References

Ackerman, S. A., Strabala, K. I., Menzel, W. P., Frey, R. A., Moeller, C. C. and Gumley, L. E. (1998). Discriminating clear sky from clouds with MODIS. *Journal of Geophysical Research*, **103D**, 32141–32157.

Andersen, S., Pedersen, L. T., Heygster, G., Tonboe, R., and Kaleschke, L. (2007). Intercomparison of passive microwave sea ice concentration retrievals over the high concentration Arctic sea ice. *Journal of Geophysical Research*, **112**, C08004. doi: 10.1029/2006JC003543.

Arndt, S. and Nicolaus, M. (2014). Seasonal cycle of long-term trend of solar energy fluxes through Arctic sea ice. *The Cryosphere*, **8**, 2219–2233. doi: 10.5149/tc-8–2219-2014.

Barber, D. G., Fung, A. K., Grenfell, T. C., Nghiem, S. V., Onstott, R. G., Lytle, V. I., Perovich, D. K. and Gow, A. J. (1998). The role of snow on microwave emission and scattering over first-year sea ice. *IEEE Transactions on Geoscience and Remote Sensing*, **36**(5), 1750–1763.

Carsey, F. D. (1992). *Microwave Remote Sensing of Sea Ice*. Published by the American Geophysical Union as part of the Geophysical Monograph Series, 68, 462 pages, ISBN: 9780875900339.

Cavalieri, D. J., Gloersen, P. and Campbell, W. J. (1984). Determination of sea ice parameters with the NIMBUS 7 SMMR. *Journal of Geophysical Research*, 89, D4, 5355–5369.

Comiso, J. C. (1986). Characteristics of arctic winter sea ice from satellite multispectral microwave observations. *Journal of Geophysical Research*, **91**, 975–994.

Connor, L. N., Laxon, S. W., Ridout, A. L., Krabill, W. B. and McAdoo, D. C. (2009). Comparison of Envisat radar and airborne laser altimeter measurements over Arctic sea ice. *Remote Sensing of Environment*, **113**, 563–570, 2009. doi: 10.1016/j.rse.2008.10.015.

Eastwood, S. (Ed.) (2012). *Ocean & Sea Ice SAF (OSISAF) Sea Ice Product Manual. Version 3.8.*, available at: http://osisaf.met.no, last access.

Emery, W. J., Thomas, A. C., Collins, M. J., Crawford, W. R. and Mackas, D. L. (1986). An objective method for computing advective surface velocities from sequential infrared satellite images. *Journal of Geophysical Research*, **91**, 12865–12878.

Fetterer, F., Stewart, J. S. and Meier, W. N., (2015). *MASAM2: Daily 4-Km Arctic Sea Ice Concentration*. Boulder, CO: National Snow and Ice Data Center. doi: http://dx.doi.org/10.7265/N5ZS2TFT.

Fung, A. K. (1994). *Microwave Scattering and Emission Models and Their Applications*. Norwood, MA: Artech House. ISBN-13: 978-0890065235.

Grenfell, T. C., Cavalieri, D. J., Comiso, J. C., Drinkwater, M. R., Onstott, R. G., Rubinstein, I., Hallikainen, M., Ulaby, F. T. and Abdelrazik, M. (1986). Dielectric properties of snow in the 3 to 37 GHz range. *IEEE Transactions on Antennas and Propagation* **34**(11), 1329–1340.

Geiger, C., Müller, H. R., Samluk, J. P., Bernstein, E. R. and Richter-Menge, J. (2015). Impact of spatial aliasing on sea-ice thickness measurements. *Annals of Glaciology*, **56**(69), 353–362. doi: 10.3189/2015AoG69A644.

Heygster, G., Huntemann, M., Ivanova, N., Saldo, R. and Pedersen, L. T. (2014). Response of passive microwave sea ice concentration algorithms to thin ice. *Proceedings Of the IEEE International Geoscience and Remote Sensing Symposium, IGARSS 2014*, IEEE, 2014, 3618–21. doi: 10.1109/IGARSS.2014.6947266.

Hwang, B. (2013). Inter-comparison of satellite sea ice motion with drifting buoy data. *International Journal of Remote Sensing*, 34, 24, 2013.

Istomina, L., Heygster, G., Huntemann, M., Schwarz, P., Birnbaum, G., Scharien, R., Polashenski, C., Perovich, D., Zege, E., Malinka, A., Prikhach, A. and Katsev, I. (2015a). Melt pond fraction and spectral sea ice albedo retrieval from MERIS data – 1: Validation against in situ, aerial, and ship cruise data. *The Cryosphere*, 9, 1551–1566. doi: 10.5194/tc-9-1551-2015.

Istomina, L., Heygster, G., Huntemann, M., Marks, H., Melsheimer, C., Zege, E., Malinka, A., Prikhach, A. and Katsev, I., (2015b). Melt pond fraction and spectral sea ice albedo retrieval from MERIS data – Part 2: Case studies and trends of sea ice albedo and melt ponds in the Arctic for years 2002–2011. *The Cryosphere*, 9:4, 1567–1578. doi: 10.5194/tc-9-1567-2015.

Istomina, L., Melsheimer, C., Huntemann, M., Nicolaus, M. and Heygster, G. (2016). Retrieval of sea ice thickness during melt season from in situ, airborne and satellite imagery. *IGARSS 2016 Proceedings*, Paper FR2.L12.3, July 10–15, 2016, Beijing, China.

Istomina, L. G., von Hoyningen-Huene, W., Kokhanovsky, A. A., and Burrows, J. P. (2010). The detection of cloud-free snow-covered areas using AATSR measurements. *Atmospheric Measurement Techniques*, 3, 1005–1017. doi: 10.5194/amt-3-1005-2010.

Istomina, L., Nicolaus, M., and Perovich, D. (2013). Spectral albedo of sea ice and melt ponds measured during POLARSTERN cruise ARK XXII/3 (IceArc) in 2012. *PANGAEA Dataset*, Bremen: PANGAEA, Institut für Umweltphysik, Universität Bremen. doi: 10.1594/PANGAEA.815111.

Ivanova, N., Pedersen, L.T., Tonboe, R. T., Kern, S., Heygster, G., Lavergne, T., Sørensen, A., Saldo, R., Dybkjaer, G., Brucker, L. and Shokr, M. (2015). Inter-comparison and evaluation of sea ice algorithms: towards further identification of challenges and optimal approach using passive microwave observations. *The Cryosphere*, 9, 1797–1817. doi: 10.5194/tc-9-1797-2015.

Kaleschke, L., Lüpkes, C., Vihma, T., Haarpaintner, J., Bochert, A., Hartmann, J., and Heygster, G. (2001). SSM/I sea ice remote sensing for mesoscale ocean-atmosphere interaction analysis. *Canadian Journal of Remote Sensing*, 27, 5, 526–537, 2001.

Kaleschke, L., Tian-Kunze, X., Maaß, N., Beitsch, A., Wernecke, A., Miernecki, M., Müller, G., Fock, B. H., Gierisch, A. M. U., Schlünzen, K. H., Pohlmann, T., Dobrynin, M., Hendricks, S., Asseng, J., Gerdes, R., Jochmann, P., Reimer, N., Holfort, J., Melsheimer, C., Heygster, G., Spreen, G., Gerland, S., King, J., Skou, N., Søbjærg, S. S., Haas, C., Richter, F. and Casal, T. (2016). SMOS sea ice product: Operational application and validation in the Barents Sea marginal ice zone. *Remote Sensing of Environment*, 180, 264–273, http://dx.doi.org/10.1016/j.rse.2016.03.009

Karvonen, J. (2016) Virtual radar ice buoys – a method for measuring fine-scale sea ice drift. *The Cryosphere*, 10, 29–42. doi: 10.5194/tc-10-29-2016.

Kern, S., Ozsoy-Cicek, B. and Worby, A. P. (2016). Antarctic sea-ice thickness retrieval from ICESat: inter-comparison of different approaches. *Remote Sensing*, 8(7), 538, 2016. doi: 10.3390/rs8070538

Komarov, A. S. and Barber, D. G. (2014). Sea ice motion tracking from sequential dual-polarization RADARSAT-2 images. *IEEE Transactions on Geoscience and Remote Sensing*, 52(1), 121–136. doi: 10.1109/TGRS.2012.2236845

Kræmer, T., Johnsen, H. and Brekke, C. (2015). Emulating sentinel-1 Doppler radial ice drift measurements using Envisat ASAR Data. *IEEE Transactions on Geoscience and Remote Sensing*, 53(12), 6407–6418. doi: 10.1109/TGRS.2015.2439044

Kwok, R. (2002). Sea ice concentration estimates from satellite passive microwave radiometry and openings from SAR ice motion. *Geophysical Research Letters*, 29, 1311. doi: 10.1029/2002GL014787.

Kwok, R. (2014). Declassifed high-resolution visible imagery for Arctic sea ice investigations: An overview. *Remote Sensing of Environment*, 142, 44–56. doi: 10.1016/j.rse.2013.11.015

Kwok, R. and Cunningham, G. F. (2008). ICESat over Arctic sea ice: Estimation of snow depth and ice thickness. *Journal of Geophysical Research*, 113, C08010. doi: 10.1029/2008JC004753.

Kwok, R., Curlander, J. C., McConnel, R. and Pang, S. (1990). An ice motion tracking system at the Alaska SAR Facility. *IEEE Journal of Oceanic Engineering*, 15, 44–54. doi: 10.1109/48.46835

Lavergne, T., Eastwood, S., Teffah, Z., Schyberg, H. and Breivik, L. A. (2010). Sea ice motion from low-resolution satellite sensors: An alternative method and its validation in the Arctic. *Journal of Geophysical Research*, 115, C10032. doi: 10.1029/2009JC005958.

Liu, Y. H., Ackerman, S. A., Maddux, B. C., Key, J. R. and Frey, R. A. (2010). Errors in cloud detection over the Arctic using a satellite imager and implications for observing feedback mechanisms. *Journal of Climate*, 23(7), 1894–1907.

Liu, Y., Key, J. R., Frey, R. A., Ackerman, S. and Menzel, P. (2004). Nighttime polar cloud detection with MODIS. *Remote Sensing of Environment*, 92, 181–194. doi: 10.1016/j.rse.2004.06.004.

Maaß, N., Kaleschke, L., Tian-Kunze, X. and Drusch, M. (2013). Snow thickness retrieval over thick Arctic sea ice using SMOS satellite data. *The Cryosphere*, 7, 1971–1989. doi: 10.5194/tc-7-1971-2013.

Markus, T. and Cavalieri, D. J. (1998). Snow depth distribution over sea ice in the Southern Ocean from satellite passive microwave data, in: *Antarctic Sea Ice Physical Processes, Interactions and Variability*. Antarctic Research Series, 74, edited by M.O. Jeffries, pp. 19–40, Washington, DC: AGU, 1998.

Markus, T., and Cavalieri, D. J. (2000). An enhancement of the NASA team sea ice algorithm. *IEEE Transactions on Geoscience and Remote Sensing*, 38, 1387–1398.

Maslanik, J., Agnew, T., Drinkwater, M., Emery, W., Fowler, C., Kwok, R. and Liu, A. (1998). *Summary of Ice-Motion Mapping Using Passive Microwave Data*, Special Publication 8, Boulder, CO: National Snow and Ice Data Center.

Meiners, K. M., Vancoppenolle, M., Thanassekos, S., Dieckmann, G. S, Thomas, D. N., Tison, J. -L., Arrigo, K. R., Garrison, D. L., McMinn, A., Lannuzel, D., van der Merwe, P., Swadling, K. M., Smith Jr., W. O., Melnikov, I. and Raymond, B. (2012). Chlorophyll *a* in Antarctic sea ice from historical ice core data. *Geophysical Research Letters*, 39(21), L21602, 2012. doi: 10.1029/2012GL053478.

Meissner, T., and Wentz, F. (2012). The emissivity of the ocean surface between 6 – 90 GHz over a large range of wind speeds and Earth incidence angles. *IEEE Transactions on Geoscience and Remote Sensing*, 50(8), 3004 pp.

Melsheimer, C., Heygster, G. and Pedersen, L. T. (2009). Retrieval of Sea Ice Emissivity and Integrated Retrieval of Surface and Atmospheric Parameters over the Arctic from AMSR-E data. *Journal of the Remote Sensing Society of Japan*, 29(1), 236–241.

Muckenhuber, S., Korosov, A. A. and Sandven, S. (2016). Open-source feature-tracking algorithm for sea ice drift retrieval from Sentinel-1 SAR imagery. *The Cryosphere*, 10, 913–925. doi: 10.5194/tc-10-913-2016.

Mätzler, C. (1987). Applications of the interaction of microwaves with the natural snow cover. *Remote Sensing Reviews*, 2(2), 259–391.

Mätzler C. and Wiesmann, A. (1999). Extension of the microwave emission model of layered snowpacks to coarse-grained snow. *Remote Sensing of Environment*, 70, 317–325.

Ninnis, R. M., Emery, W. J. and Collins, M. J. (1986). Automated extraction of pack ice motion from advanced very high-resolution radiometry. *Journal of Geophysical Research*, 91, 10725–10734. doi: 10.1029/JC091iC09p10725.

Pedersen, L.T. (1994). Merging microwave radiometer data and meteorological data for improved sea ice concentrations. *EARSeL Advances in Remote Sensing*, 3, 81–89.

Perovich, D., Richter-Menge, J., Elder, B., Arbetter, T., Claffey, K. and Polashenski, C. (2013). Observing and understanding climate change: Monitoring the mass balance, motion, and thickness of Arctic sea ice. http://imb.erdc.dren.mil.

Ramseier, R. O. (1991). Sea Ice Validation, in: *DMSP special sensor microwave/imager calibration/validation*, edited by: Hollinger, J. P., Washington, DC: Naval Research Laboratory, 1991.

Ricker, R., Hendricks, S., Perovich, D. K., Helm, V. and Gerdes, R. (2015). Impact of snow accumulation on CryoSat-2 range retrievals over Arctic sea ice: An observational approach with buoy data. *Geophysical Research Letters*, 42, 4447–4455. doi: 10.1002/2015GL064081.

Rösel, A., Kaleschke, L. and Birnbaum, G. (2012). Melt ponds on Arctic sea ice determined from MODIS satellite data using an artificial neural network. *The Cryosphere*, 6, 431–446. doi: 10.5194/tc-6-431-2012.

Schanda, E. (1986, 2012). *Physical Fundamentals of Remote Sensing*, Springer, 1986 & 2012, 187 pp.

Shi, T., Clothiaux, E. E., Yu, B., Braverman, A. J. and Groff, D. N. (2007). Detection of daytime arctic clouds using MISR and MODIS data. *Remote Sensing of Environment*, 107 (1–2), pages 172–184, ISSN 0034-4257, http://dx.doi.org/10.1016/j.rse.2006.10.015.

Schmetz, J., Holmlund, K., Hoffman, J., Strauss, B., Mason, B., Gærtner, V., Koch, A. and Berg, L. V. D. (1993). Operational cloud-motion winds from Meteosat infrared images. *Journal of Applied Meteorology*, 32, 1206–1225.

Scott, K. A., Buehner, M., Caya, A. and Carrieres, T. (2012) Direct assimilation of AMSR-E brightness temperatures to estimate sea ice concentration. *Monthly Weather Review*, 140, 997–1013.

Scott, K. A., Ashouri, Z., Buehner, M., Pogson, L. and Carrieres, T. (2015). Assimilation of ice and water observations from SAR imagery to improve estimates of sea ice concentration. *Tellus A*, 67, 27218, http://dx.doi.org/10.3402/tellusa.v67.27218.

Shokr, M. E. (1998). Field Observations and model calculations of dielectric properties of Arctic sea ice in the microwave C-band. *IEEE Transactions on Geoscience and Remote Sensing*, 36(2), 463–478.

Shokr, M. and Sinha, N. K. (2015). *Sea Ice – Physics and Remote Sensing*, Hoboken, NJ: AGU/Wiley & Sons, 600 pp, ISBN: 978-1-119-02789-8.

Smith, D. M. (1996). Extraction of winter total sea ice concentration in the Greenland and Barents Seas from SSM/I data. *International Journal of Remote Sensing*, 17, 2625–2646.

Spreen, G., Kaleschke, L. and Heygster, G. (2008). Sea ice remote sensing using AMSR-E 89 GHz channels. *Journal of Geophysical Research*, 113, C02S03. doi: 10.1029/2005JC003384.

Sumata, H., Kwok, R., Gerdes, R., Kauker, F. and Karcher, M. (2015). Uncertainty of Arctic summer ice drift assessed by high-resolution SAR data. *Journal of Geophysical Research*, 120, 5285–5301. doi: 10.1002/2015JC010810.

Svendsen, E., Matzler, C. and Grenfell, T. (1987). A model for retrieving total sea ice concentration from a spaceborne dual-polarized passive microwave instrument operating near 90 GHz. *International Journal of Remote Sensing*, 8(10), 1479–1487.

Thomas, M., Geiger, C. A. and Kambhamettu, C. (2008). High resolution (400 m) motion characterization of sea ice using ERS-1 SAR imagery. *Cold Regions Science and Technology*, 52, 207–223.

Tonboe, R. T., Eastwood, S., Lavergne, T., Sørensen, A. M., Rathmann, N., Dybkjær, G., Pedersen, L. T., Høyer, J. L. and Kern, S. (2016). The EUMETSAT sea ice concentration climate data record. *The Cryosphere*, 10, 2275–2290. doi: 10.5194/tc-10-2275-2016.

Tonboe, R. T., Heygster, G., Pedersen, L. T. and Andersen, S. (2006). Sea ice emission modelling, in: *Thermal Microwave Radiation: Applications for Remote Sensing*, edited by: Mätzler, C., IET Electromagnetic Wave Series 52, London, UK, 382–400.

Tonboe, R. T., Pedersen, L. T. and Haas, C. (2010). Simulation of the Cryosat-2 satellite radar altimeter sea ice thickness retrieval uncertainty. *Canadian Journal of Remote Sensing*, 36(1), 55–67.

Toyota, T., Haas, C. and Tamura, T. (2010). Size distribution and shape properties of relatively small sea-ice floes in the Antarctic marginal ice zone in late winter. *Deep–Sea Research II*. doi: 10.1016/j.dsr2.2010.10.034.

Tsang, L. and Kong, J. A. (2001). *Scattering of Electromagnetic Waves: Advanced Topics*, 432 pages, Wiley, ISBN: 978-0-471-38801-2.

Tucker, W. B., Grenfell, T. C., Onstott, R. G., Perovich, D. K., Gow, A. J., Shuchman, R. A. and Sutherland, L. (1991). Microwave and physical properties of sea ice in the winter marginal ice zone. *Journal of Geophysical Research*, 96(C3), 4573–4587.

Ulaby, F. (2013). *Microwave Radar and Radiometric Remote Sensing*, University of Michigan Press, 1116 pages, ISBN-13: 978-0472119356.

Untersteiner, N. (Ed.) (1986). *The Geophysics of Sea Ice*. NATO ASI series, series B: Physics Vol. 146, New York & London: Plenum Press.

Wadhams, P. (2000). *Ice in the Ocean*. Gordon & Breach Science Publishers.

Wang, L., Scott, K. A., Xu, L. and Clausi, D. A. (2016a). Sea ice concentration estimation during melt from dual-pol SAR scenes using deep convolutional neural networks: A case study. *IEEE Transactions on Geoscience and Remote Sensing*, 54(8), 4524–4533. doi: 10.1109/TGRS.2016.2543660.

Wang, L., Scott, K. A. and Clausi, D. (2016b). Improved sea ice concentration estimation through fusing classified SAR imagery and AMSR-E data. *Canadian Journal of Remote Sensing*, 42(1). doi: 10.1080/07038992.2016.1152547.

Warren, S. G., Rigor, I. G., Untersteiner, N., Radionov, V. F., Bryazgin, N. T., Alexandrov, I. and Colony, R. (1999). Snow depth on Arctic sea ice. *Journal of Climate* 12, 1814–1829.

Wentz, F. and Meissner, T. (2000). *AMSR Ocean Algorithm Theoretical Basis* Document (Version 2), RSS Tech. Proposal 121599A-1, Santa Rosa, CA: Remote Sensing Systems, December 1999, revised 2000.

Wentz, F. and Meissner, T. (2007). *AMSR Ocean Algorithm Theoretical Basis Document Supplement-1*, RSS Tech. Rpt. 051707, Santa Rosa, CA: Remote Sensing Systems, May 2007.

Wiesmann, A. and Mätzler, C. (1999). Microwave emission model of layered snowpacks. *Remote Sensing of Environment*, 70, 307–316.

Worby, A. and Allison, I. (1999). A technique for making ship-based observations of Antarctic sea ice thickness and characteristics: PART I observational technique and results. *Research Report* 14, pp. 1–23, Hobart, Tasmania, Australia: Antarctic Cooperative Research Centre, University of Tasmania.

Worby, A. P., Geiger, C. A., Paget, M. J., Van Woert, M. L., Ackley, S. F. and DeLiberty, T. L. (2008). Thickness distribution of Antarctic sea ice. *Journal of Geophysical Research: Oceans*, 113(C5), C05S92. doi: 10.1029/2007JC004254.

Zygmuntowska, M., Rampal, P., Ivanova, N. and Smedsrud, L. H. (2014). Uncertainties in Arctic sea ice thickness and volume: new estimates and implications for trends. *The Cryosphere*, 8, 705–720, 2014. doi: 10.5194/tc-8-705-2014.

4

Sea Ice Data Assimilation

Mark Buehner, Laurent Bertino, Alain Caya, Patrick Heimbach, Greg Smith

4.1 Introduction

Data assimilation is a critical component of any automated prediction system (APS) used to forecast the weather, ocean or sea ice with lead times from hours to weeks. In the context of sea ice prediction, it enables the vast amount of information from all available in situ and remote sensing sea ice observations (described in Chapter 3) and forecasts from large-scale sea ice models (described in Chapter 2) to be optimally combined. Several existing sea ice APS (described in Chapter 5) use data assimilation techniques for this purpose. The merging of the information from observations and model forecasts results in more accurate and useful estimates of the sea ice conditions than could otherwise be obtained using either the observations or model forecasts in isolation. The goal of this chapter is to provide an overview of data assimilation with an emphasis on the particular challenges posed by its application to operational sea ice prediction.

The chapter begins with a general, non-mathematical explanation of data assimilation and the basic ideas that underpin it. It is hoped that this can serve as a useful introduction, even to those readers without a background in statistics or numerical modelling. Then, an overview is given of the most common data assimilation techniques that are used within existing operational systems for weather, ocean and sea ice prediction. This overview is necessarily brief. Readers interested in additional detail on these techniques should consult text books dedicated to the subject, such as Lahoz *et al.* (2010). Though these techniques span a wide range of mathematical and technical complexity, they are all based on the same basic theory. The differences between the techniques arise from different assumptions and simplifications that are necessary to be able to combine many thousands or even millions of observations with forecasts from a high-resolution forecast model within a short enough time to make the result useful for short-term prediction (i.e. usually in less than an hour of real time). Many of these techniques were originally applied to numerical weather prediction (NWP) or the assimilation of oceanographic observations. Compared with these applications, sea ice data assimilation poses several additional challenges that are also discussed, including: the highly non-linear nature of sea ice physics; the non-Gaussian (and range-bounded) distribution of errors for key sea ice model variables and observations; the use of categorical observations; and the simultaneous assimilation of observations with widely varying spatial resolution. Finally, the importance of coupling with the ocean and atmosphere for sea ice prediction is discussed along with the different data assimilation approaches for initializing coupled model forecasts.

4.2 What Is Data Assimilation?

In the context of an APS, data assimilation is generally used within an intermittent cycle in which a large number of observations are used to correct a short-term model forecast, referred to as the background state. The corrected model state, referred to as the analysis state, is then used to initialize the forecast model to produce the subsequent short-term forecast and the process is repeated. As part of an APS, this data assimilation cycle is repeated indefinitely in near-real time to produce updated analysis states at regular intervals (most commonly every six hours for global weather prediction). The analysis states are also used to initialize longer forecasts with lead times typically much longer than the period of the data assimilation cycle. A simple schematic representation of this data assimilation cycle is shown in Figure 4.1.

For sea ice prediction, the forecast model must be initialized with a complete gridded model state. Most importantly, this model state typically includes ice concentration, ice thickness (or a complete ice thickness distribution (ITD)), ice drift, variables describing the snow on top of the ice and thermodynamic information for the ice. These state variables must be provided at every grid point on the horizontal numerical grid used by the model. The assimilated observations can include both in situ and remote sensing

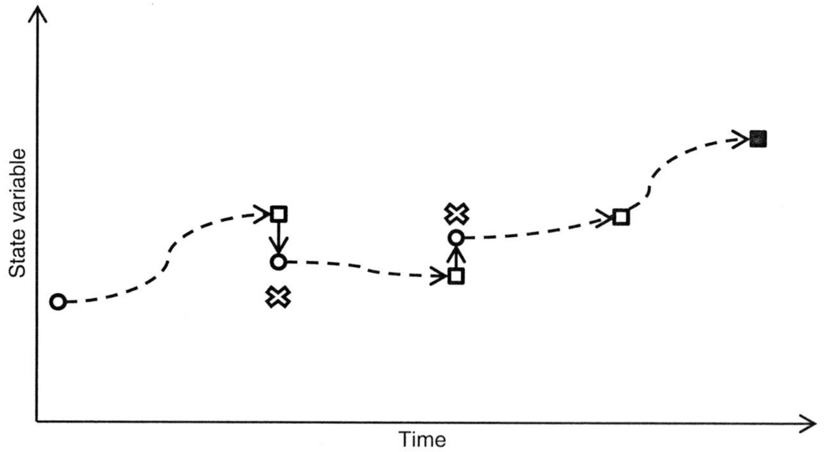

Figure 4.1: Schematic showing how a particular state variable is predicted to evolve by the forecast model (dashed arrows) and how the predicted value is corrected by the data assimilation procedure (solid vertical arrows) when assimilating observations (crosses). The forecast produces both the background state (hollow squares) and longer-term forecast (solid square) starting from the analysis state (circles).

observations, can be at various spatial resolutions of quantities that are either directly or indirectly related to the forecast model variables, and are distributed arbitrarily through space and time. The power of advanced data assimilation techniques is the ability to combine the information from such a diverse set of observations with that from the background state in a way that takes into account the relative uncertainty of each source of information. This is in contrast to simpler data assimilation approaches that either directly replace the values of model variables with values obtained from observations or gradually nudge the model variables towards observed values by including an ad hoc term in the model equations. Such simple approaches are not considered in this chapter since they are no longer used in most state-of-the-art sea ice prediction systems. With data assimilation, model variables that are not related (either directly or through statistical relationships with other variables) to any of the assimilated observations are not modified, and therefore their values in the analysis state equal those in the background state. For example, when only assimilating ice concentration observations, ice thickness will not be modified by the data assimilation procedure unless the data assimilation procedure makes use of a statistical or physical relationship between the background errors of ice concentration and thickness.

Data assimilation is also used for initializing coupled ice–ocean or ice–ocean–atmosphere models for either short-range forecasts or forecasts on the seasonal or longer timescale. Data assimilation with coupled systems can be used as a way of reconstructing the ocean and sea ice conditions over a period in the past. This application of data assimilation makes use of the forecast model to 'fill in' the gaps in the observations. These gaps are both in terms of the incomplete spatial coverage of observations and the lack of observations for some important variables, such as ice thickness and the ocean temperature and salinity below sea ice.

To be able to assimilate an observation, the so-called 'observation operator' (introduced in Section 3.9 in the context of retrieval algorithms for sea ice variables) must be known. This mathematical function relates the state vector (i.e. the gridded background or analysis state consisting of all model variables) to the observations. When used for assimilating observations that are directly related to the state variables, the observation operator may only involve a spatial interpolation from the surrounding model grid points to the observation location (and possibly also a temporal interpolation from the nearest model time levels to the valid time of the observation). Other observations are much less directly related to the state variables, for example a passive microwave satellite radiance observation at the top of the atmosphere. In this case, the observation operator must transform the state variables into an equivalent radiance value using an appropriate radiative transfer model that simulates the real observing process for this type of radiance (see Figure 3.21).

Many approaches have been developed for assimilating observations to correct a background state. Even though the errors associated with both the background state and each observation being assimilated are unknown, most data assimilation approaches require knowledge of the *statistics* of these errors. In addition, most approaches assume that the background errors are statistically independent from the errors of the observations being assimilated and that the errors in the observations are uncorrelated in time. While these assumptions are usually sufficiently satisfied, for certain types of observations or analysis variables they may not be. In the simplest case of a single observation assimilated to correct a single collocated background state variable, the resulting analysis state from most data assimilation approaches is a weighted average of the observed and background values. The weights given to the observed and background values are computed in a way that accounts for their relative uncertainty, such that the resulting analysis state will be closer to the quantity with lower uncertainty. Two- and three-dimensional data assimilation approaches additionally use information that characterizes the spatial statistical relationships of these errors (usually in the form of spatial background error covariances between all possible pairs of the gridded state variables and observation error covariances between all possible pairs of the assimilated observations) to spread the information from the observation location to correct the background state in the surrounding area. Inclusion of the statistical relationship between the background errors of different state variables (e.g. ice concentration and ocean surface temperature) also allows observations of one variable type to produce corrections to the background values of another variable type. Four-dimensional data assimilation approaches include the time dimension such that observations distributed both spatially and over a time window are used to correct the three-dimensional background state either at a single time or at multiple times within the same time window (Figure 4.2).

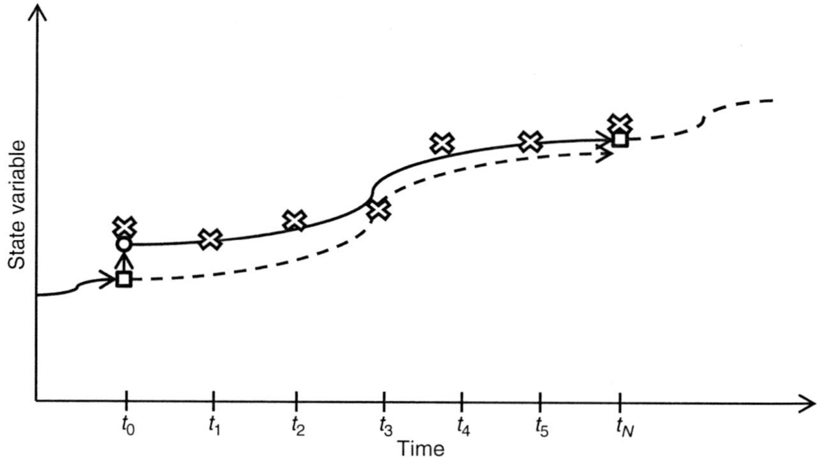

Figure 4.2: Schematic showing how a four-dimensional data assimilation procedure modifies the forecast (dashed line) initialized with the background state (square on the left of the figure) to better fit a series of assimilated observations (crosses) available at times $t_0, t_1, \ldots, t_N$ throughout a time window by iteratively computing a modified initial state (circle). Note that the final state of the modified forecast (solid line) becomes the initial background state for assimilating observations over the following time window (square on the right of the figure).

4.3 Data Assimilation Theory and Techniques

4.3.1 Bayes Theorem and the Kalman Filter

The presence of unknown errors associated with the background state and observations makes data assimilation a statistical estimation problem in which each quantity is formally represented not by a single value, but by a probability density function (PDF). These PDFs represent our imprecise knowledge regarding the true geophysical state of the system and the process of observing the system. In the hypothetical case of a perfectly known background state estimate or a perfectly accurate observation, the corresponding PDF would become a Dirac delta function centred on that value.

Bayes theorem can be considered as the fundamental recipe for updating the PDF of the background state by assimilating a set of observations for which its conditional PDF is known. This is presented for a general state vector, denoted as $\mathbf{x}$, that may contain some or all of the different types of model variables at all of the locations on the model grid. The resulting conditional probability that a particular value of the state vector equals the true state, denoted as $\mathbf{x}^{\text{true}}$, given a particular set of assimilated observations, denoted by $\mathbf{y}^{\text{o}}$, is

$$P(\mathbf{x} = \mathbf{x}^{\text{true}} | \mathbf{y} = \mathbf{y}^{\text{o}}) = \frac{P(\mathbf{y} = \mathbf{y}^{\text{o}} | \mathbf{x} = \mathbf{x}^{\text{true}}) P(\mathbf{x} = \mathbf{x}^{\text{true}})}{P(\mathbf{y} = \mathbf{y}^{\text{o}})}, \tag{4.1}$$

where $\mathbf{y}$ denotes the random variable of the 'true' observations (which is random because the unknown observation error is treated as a random variable). The second term in the product of the right-hand side numerator represents the background state PDF, that is, our knowledge of the state vector before assimilating the observations.

This update is equivalent to the procedure described earlier of correcting the background state by assimilating observations. When only a single forecast is produced by the prediction system as the background state, the background PDF can be thought of as the distribution of the forecasted value plus a random variable that represents the unknown background error. Similarly, the conditional PDF on the right-hand side of Equation 4.1 is the distribution of the observed value plus a random variable that represents the unknown observation error. Using the background state, a set of observations and an estimate of the PDF for the background and observation errors, Bayes theorem (as expressed in Equation 4.1) can then be applied. The result is the analysis state PDF from which, if required, the mean or mode (i.e. the most likely particular state given the background state and observations) of the PDF can be computed along with a measure of its associated uncertainty, which is related to the width of the PDF. In geographical regions with a large number of high-quality observations, the resulting analysis state uncertainty will be significantly less than the uncertainty in the background state. In other words, the width or variance of the analysis state PDF will be less than that of the background state PDF. The mean and mode of this PDF will also generally be different from the mean and mode of the background state PDF. In contrast, in regions with no nearby observations or only very few poor quality

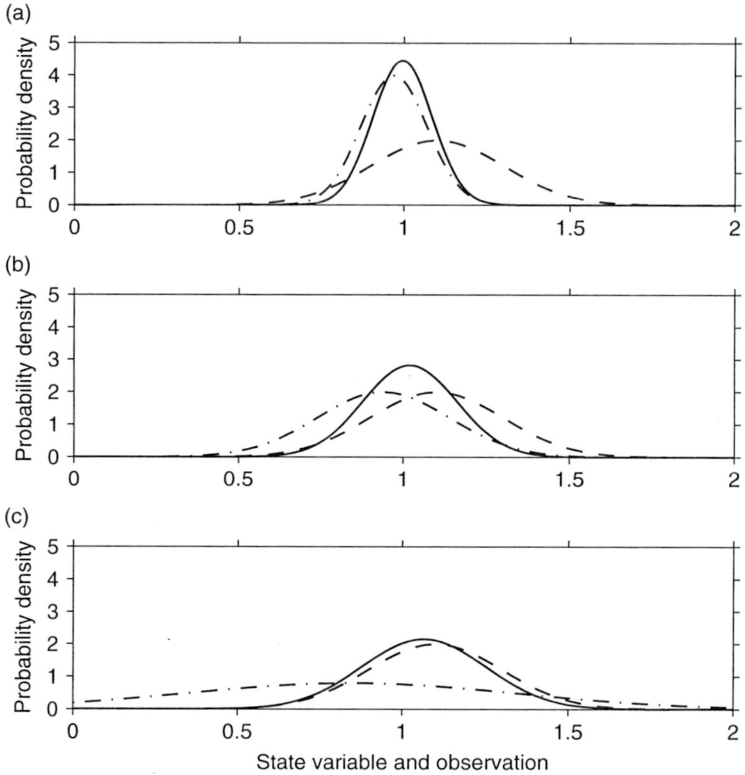

Figure 4.3: Examples of applying Bayes theorem to update the background state PDF (dashed) to produce the analysis state PDF (solid) of a single state variable when assimilating a single observation of the same quantity as the state variable (dashed-dotted). For all examples, the true value of the state variable is 1 and the background error standard deviation is 0.2. The observation error standard deviation is 0.1, 0.2 and 0.5, respectively, in panels (a), (b) and (c). Gaussian distributions are assumed for both the background and observation errors. The background value is 1.1 and the observed value is 0.97, 0.94 and 0.85 in panels (a), (b) and (c), respectively.

observations, the analysis and background state PDFs will be very similar to each other, in terms of the mean, mode and width. Figure 4.3 shows some simple examples of applying Bayes theorem to update the PDF of a single state variable by assimilating a single observation of varying accuracy.

For high-dimensional systems, the assumption that the error PDFs are both Gaussian and unbiased allows significant simplification of Bayes theorem. In this case, both the background and observation PDFs are characterized simply by their respective mean and covariance matrix. Note that the covariance matrix is a symmetric matrix with the variance of each variable of the state vector along the diagonal and the covariance between every pairwise combination of these variables in the corresponding off-diagonal positions. It should also be noted that the mean and mode are equal for a Gaussian distribution, which is generally not true for an arbitrary distribution. The resulting PDF of the analysis state (denoted as $\mathbf{x}^{\mathrm{a}}$) is expressed as

$$P(\mathbf{x}^{\mathrm{a}} = \mathbf{x}^{\mathrm{true}}|\mathbf{y} = \mathbf{y}^{\mathrm{o}}) \propto \exp\left[-\frac{1}{2}\left(\mathbf{y}^{\mathrm{o}} - H(\mathbf{x}^{\mathrm{a}})\right)^{\mathrm{T}}\mathbf{R}^{-1}\left(\mathbf{y}^{\mathrm{o}} - H(\mathbf{x}^{\mathrm{a}})\right)\right] \times \exp\left[-\frac{1}{2}\left(\overline{\mathbf{x}}^{\mathrm{b}} - \mathbf{x}^{\mathrm{a}}\right)^{\mathrm{T}}\mathbf{B}^{-1}\left(\overline{\mathbf{x}}^{\mathrm{b}} - \mathbf{x}^{\mathrm{a}}\right)\right], \tag{4.2}$$

where $\mathbf{H}$ denotes the observation operator, $\mathbf{R}$ is the observation error covariance matrix, $\overline{\mathbf{x}}^{\mathrm{b}}$ is the mean of the background state PDF, $\mathbf{B}$ is the background error covariance matrix and the superscript T indicates transposition of a vector or matrix. Since it is usually only the relative probability for different values of the state vector that is needed for data assimilation, the constant normalization factor of the Gaussian distributions and the probability of the observed values (i.e. the denominator in Equation 4.1) are ignored and the proportionality symbol is used in Equation 4.2. If only the most likely value is sought as the analysis state, then this can be conveniently obtained by finding the value of $\mathbf{x}$ that maximizes this probability or, equivalently, minimizes the cost function that is proportional to -1 times the natural logarithm of the analysis state PDF:

$$J(\mathbf{x}) = \frac{1}{2}\left(\mathbf{y}^{\mathrm{o}} - H(\mathbf{x})\right)^{\mathrm{T}}\mathbf{R}^{-1}\left(\mathbf{y}^{\mathrm{o}} - H(\mathbf{x})\right) + \frac{1}{2}\left(\overline{\mathbf{x}}^{\mathrm{b}} - \mathbf{x}\right)^{\mathrm{T}}\mathbf{B}^{-1}\left(\overline{\mathbf{x}}^{\mathrm{b}} - \mathbf{x}\right). \tag{4.3}$$

Furthermore, in the case of the observation operator being linear (and therefore simply denoted by the matrix $\mathbf{H}$), the resulting analysis state PDF is also Gaussian with mean denoted as $\overline{\mathbf{x}}^{\mathrm{a}}$ and the analysis error covariance matrix denoted as $\mathbf{P}^{\mathrm{a}}$. Explicit expressions for these are obtained by solving for the state vector that satisfies the gradient of $J(\mathbf{x})$ being zero, resulting in the Kalman filter analysis equations for the analysis state mean and error covariance:

$$\overline{\mathbf{x}}^{\mathrm{a}} = \overline{\mathbf{x}}^{\mathrm{b}} + \mathbf{K}[\mathrm{y}^{\mathrm{o}} - H(\overline{\mathbf{x}}^{\mathrm{b}})] \tag{4.4}$$

$$\mathbf{P}^{\mathrm{a}} = (\mathbf{I} - \mathbf{KH})\mathbf{B}, \tag{4.5}$$

where $\mathbf{I}$ is the identity matrix and $\mathbf{K}$ is the Kalman gain matrix, defined by

$$\mathbf{K} = \mathbf{BH}^{\mathrm{T}}\left(\mathbf{HBH}^{\mathrm{T}} + \mathbf{R}\right)^{-1}. \tag{4.6}$$

These equations can be easily understood for the case of a single state variable and a single collocated observation directly related to the state variable (and therefore the observation operator is simply the identity). In this case the Kalman gain is a scalar, denoted by K, given by the ratio of the background error variance to the sum of the background and observation error variances,

$$\begin{aligned}\overline{x}^{\mathrm{a}} &= \overline{x}^{\mathrm{b}} + K\left(y^{\mathrm{o}} - \overline{x}^{\mathrm{b}}\right) \\ &= \overline{x}^{\mathrm{b}} + \frac{\sigma_{\mathrm{b}}^2}{\sigma_{\mathrm{b}}^2 + \sigma_{\mathrm{obs}}^2}\left(y^{\mathrm{o}} - \overline{x}^{\mathrm{b}}\right).\end{aligned} \tag{4.7}$$

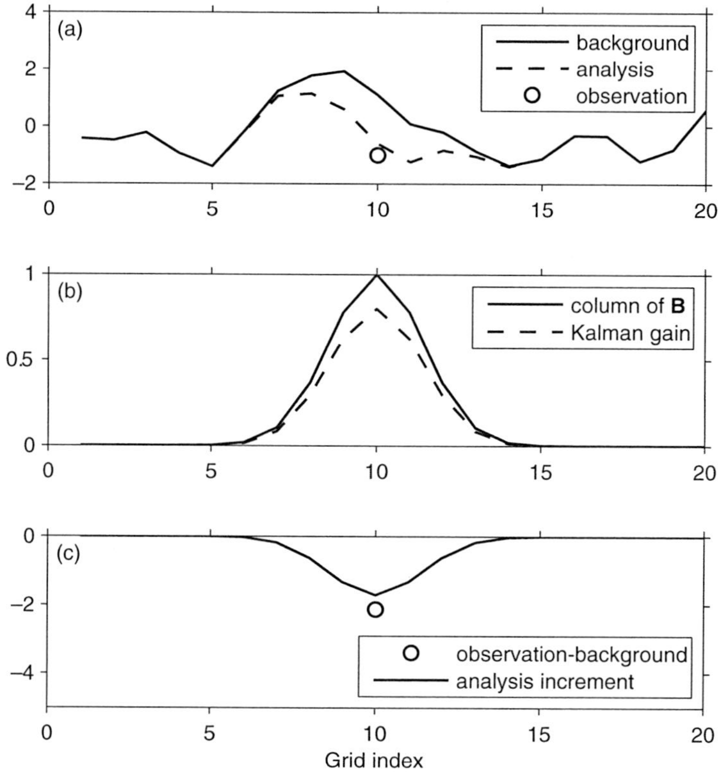

Figure 4.4: Simple data assimilation example with one spatial dimension and a single observation. The background state, analysis state and observed value are all shown in panel (a). The column of the background error covariance matrix (**B**), representing the covariance of the background error between the observed grid point and all of the others, together with the Kalman gain matrix are shown in panel (b). The observation-minus-background difference and the resulting analysis increment are shown in panel (c).

The value of the Kalman gain is close to zero when the background state has a much lower uncertainty than the observation and consequently the analysis state nearly equals the background state. In the opposite case, when the observation has a lower uncertainty than the background state, the Kalman gain is close to one and the analysis state will nearly equal the observed value. When the background state and observation have equal uncertainty, the Kalman gain equals 0.5, and the analysis state is the simple average of the background and observed values.

A simple example is shown in Figure 4.4 of correcting a state vector consisting of variables located at 20 evenly spaced grid points by assimilating a single observation located at grid index 10. The background error at each grid point is assumed to be positively correlated among nearby grid points, but this correlation decreases with increasing separation distance such that it becomes nearly zero beyond four grid points away. The observation (with error standard deviation equal to 0.5) is assumed to be more accurate than the background state (with error standard deviation equal to 1.0). Consequently, the analysis state at the observation location is closer to the observed value than the background value (Figure 4.4a) and the Kalman gain is greater than 0.5 at the observation location (Figure 4.4b). In such a simple example with only a single assimilated observation, the Kalman gain is simply a scaled version of a column of the background error covariance matrix (as shown in Figure 4.4b). The correction to the background state (also referred to as the analysis increment) is the product of the Kalman gain with the observation-minus-background difference (as shown in Figure 4.4c). Because the background error covariances decrease with separation distance, the impact of the observation on modifying the background state is greatest for the grid points closest to the observation location and becomes nearly zero starting at four grid points away. In cases with multiple assimilated observations, the analysis increment would not simply equal the sum of the increments obtained by assimilating each in isolation. Instead, the combined information from all of the observations is used to correct the background state by appropriately accounting for redundancy between the observations. This redundancy typically reduces the impact of each individual observation, while still producing an overall reduction in the uncertainty of the resulting analysis state.

Most data assimilation techniques are based on either an iterative approach for finding the state that minimizes the cost function (Equation 4.3) or by approximately solving the Kalman filter analysis equations (Equations 4.4–4.6). As described below, both types of data assimilation approaches require additional simplifications to make them useful for most practical applications.

4.3.2 Optimal Interpolation

Optimal interpolation (OI) is a data assimilation technique that explicitly solves the Kalman filter analysis equation for the mean of the analysis state PDF (Equation 4.4 and Equation 4.6) by making further simplifications. The background error covariance values are assumed to only depend on the distance between locations through a known spatially continuous function. Several such functions are possible (e.g. exponential, Gaussian, power functions) and can be easily evaluated for any pair of locations. While this leads to a covariance function that is homogeneous and isotropic, extensions to the standard OI approach have been developed that allow these constraints to be partially relaxed. The standard OI approach was originally developed to assimilate observations that are closely related to the state variables. In some cases, however, the observation may represent vastly different spatial scales than the state variable. For example, when the observation measures a quantity at a single location and time, but the state variable represents a quantity averaged over a large model grid cell. In such a case, modification of the standard approach is again required.

As the number of assimilated observations in typical applications of data assimilation is normally greater than several thousand, the computational cost of inverting the matrix $(\mathbf{HBH}^{\mathrm{T}} + \mathbf{R})$, required to solve the Kalman filter analysis equation, can become prohibitive. To overcome this, the model domain is first divided into geographical regions such that the number of observations near each is sufficiently small. Then, assuming the background error covariance approaches zero at some known distance, only the subset of observations located within this distance of a particular region is used to compute the analysis for the grid points within the region. Consequently, a smaller matrix is inverted when solving for the analysis state separately for each region instead of requiring the inversion of a much larger matrix when directly solving for the entire analysis state.

Optimal interpolation was one of the first data assimilation techniques to be successfully used in both atmospheric and oceanographic applications. However, due to the already mentioned limitations related to the background error covariances and the types of observations that can readily be assimilated, OI is no longer widely used compared with the techniques described in the remainder of this section.

4.3.3 Ensemble Kalman Filter

Monte Carlo simulation provides a means to approximate the PDF resulting from the application of a complex non-linear transformation (e.g. the integration of a large-scale sea ice model) to a high-dimensional vector of random inputs (e.g. the uncertain wind and ocean forcing fields). This is accomplished by numerically generating a pseudo-random sample of the input variables from their known PDF and applying the non-linear transformation separately to each realization of inputs. The resulting ensemble of outputs can be used to approximate its PDF (e.g. the uncertain sea ice concentration). Consequently, any statistic of the PDF, including its mean and covariance, can be estimated from the ensemble. The randomness of the ensemble ensures that any such estimator is unbiased and converges asymptotically as the ensemble size N_{ens} increases. The Monte Carlo simulation approach is used in the ensemble Kalman filter (EnKF) to generate ensembles of both background and analysis states that approximate their PDFs by simulating the uncertainties within all components of the complete data assimilation cycle. The main sources of uncertainty are due to the imperfections and unresolved processes in the forecast model when producing the background state and the errors associated with the assimilated observations. For sea ice prediction, an important source of model-related uncertainty is the error in the specified ocean and atmospheric forcing fields.

In practice, an ensemble of background states is obtained by propagating each member of an ensemble of analysis states using the forecast model $\mathcal{M}$. This is denoted as

$$\mathbf{x}_i^{\mathrm{b}}(t_n) = \mathcal{M}[\mathbf{x}_i^{\mathrm{a}}(t_{n-1}), \boldsymbol{\varepsilon}_i(t_{n-1})], \tag{4.8}$$

where $i = 1, \ldots, N_{\mathrm{ens}}$, is the ensemble member index and $\boldsymbol{\varepsilon}_i(t_{n-1})$ represents the i-th realization of the random model errors between the assimilation times t_{n-1} and t_n. The mean and covariance of the background state PDF can then be approximated by (after dropping the time level variable for clarity):

$$\overline{\mathbf{x}}^{\mathrm{b}} = \frac{1}{N_{\mathrm{ens}}} \sum_{i=1}^{N_{\mathrm{ens}}} \mathbf{x}_i^{\mathrm{b}} \tag{4.9}$$

$$\mathbf{B} = \frac{1}{N_{\mathrm{ens}} - 1} \sum_{i=1}^{N_{\mathrm{ens}}} \left(\mathbf{x}_i^{\mathrm{b}} - \overline{\mathbf{x}}^{\mathrm{b}}\right)\left(\mathbf{x}_i^{\mathrm{b}} - \overline{\mathbf{x}}^{\mathrm{b}}\right)^{\mathrm{T}}. \tag{4.10}$$

The EnKF approach can be extended to include 4D covariances by simply using ensembles of background states output by the forecast model at several times within the time window over which the assimilated observations are distributed. This captures both the time evolution of the covariances within the assimilation window and also the relationship between errors at different times within the window. The result is a true 4D data assimilation procedure, similar to what is shown in Figure 4.2.

The original stochastic EnKF (also referred to as the perturbed-observation EnKF) applies the Kalman filter analysis equation to each member of the background ensemble individually to assimilate observations that have been randomly perturbed,

$$\mathbf{x}_i^{\mathrm{a}} = \mathbf{x}_i^{\mathrm{b}} + \mathbf{K}(\mathbf{y}^{\mathrm{o}} + \boldsymbol{\eta}_i - \mathbf{H}(\mathbf{x}_i^{\mathrm{b}})), \tag{4.11}$$

with **K** obtained as in Equation 4.6, but using the ensemble estimate of the background error covariance matrix, **B**, given by Equation 4.10. The vector $\boldsymbol{\eta}_i$ is the *i*-th computed realization of the random observation errors with covariance equal to **R**. The analysis ensemble mean and covariance matrix converge to the Kalman filter expressions given in Section 4.3.1 as the number of ensemble members tends to infinity. Alternatively, several approaches known as square-root filters have been proposed that are similar to the approach described above, but avoid the need to perturb the observations (e.g. Tippett *et al.*, 2003).

Similar to the OI approach already described, the Kalman filter analysis equation is solved directly in all of the EnKF approaches, but using an estimate of the background error covariances derived from the ensemble of background states. Again, similar to the OI approach, the various EnKF approaches can only be made computationally feasible by reducing the size of the matrix that needs to be inverted when computing the Kalman gain matrix, Equation 4.6. This is typically accomplished by either of two general strategies. The first strategy is the same as used in OI: that is, dividing the spatial domain into a large number of small regions (possibly as small as a single grid point) and assimilating only the nearby observations expected to have an influence on the analysis state within each region. An advantage of this approach is that the calculation of the analysis states for each region can be done in parallel with those in all of the other regions. The second strategy assimilates small batches of observations (with batches being as small as a single observation) sequentially, one after the other. The entire ensemble of background states is modified by assimilating the first batch of observations, and this modified ensemble is used to estimate the background error covariances for computing the Kalman gain matrix for assimilating the second batch. The process is repeated, such that the influence of assimilating each batch of observations affects how the subsequent observations are assimilated, until all observation batches have been assimilated.

Most other data assimilation techniques require that the background error covariances must somehow already be specified, usually using an ad hoc approach. The specification of these covariances is very challenging since they depend on the accumulated effect, within the data assimilation cycle, of the accuracy and spatial distribution of the observations, the dynamics of error growth during the background forecasts, and the additional uncertainties introduced by errors in the forecast model. The EnKF approach accounts for all of these effects on the background error covariances by applying Monte Carlo simulation to the complete data assimilation cycle. Consequently, a major advantage of the EnKF approach is that a realistic estimate of the background error covariances can be produced using the resulting ensemble of background states. However, this assumes that the computational cost of generating a sufficient number of ensemble members is not prohibitive. The use of this background error covariance matrix when assimilating observations can lead to improved accuracy for the resulting analysis states as compared with using a less realistic estimate of the background error covariances obtained with an ad hoc approach. The background error covariances can include important relationships between different model variables that allow observations of one variable to produce corrections for another variable. For example, Lisæter *et al.* (2007) used an EnKF applied to a coupled ice–ocean model to demonstrate how the assimilation of simulated ice thickness observations results in direct improvements to the ocean salinity, surface temperature and ice concentration fields.

4.3.4 Variational Approaches

The data assimilation techniques presented so far are based on the direct solution provided by the Kalman filter analysis equation. A theoretically equivalent alternative is provided by variational data assimilation techniques. Instead of starting from the direct solution, the variational technique uses a standard iterative optimization algorithm to obtain the state vector that minimizes the cost function, Equation 4.3. This function is the sum of two terms. The first is a scalar measure of the normalized squared difference between the analysis state and the observations. The second term is a scalar measure of the normalized squared difference between the analysis state and the background state. By finding the analysis state that minimizes the sum of these two terms, the information on the sea ice conditions provided by both the observations and the short-term forecast initialized from the previous analysis state are optimally combined according to their relative accuracy. An important difference with the EnKF technique, which provides the mean and covariance of the analysis state distribution, is that variational approaches are typically used only to obtain a single analysis state, considered to be the most likely state, without any estimate of its associated uncertainty (though, in theory, uncertainty information can be obtained from the shape of the cost function, specifically the inverse of its Hessian matrix). The three-dimensional variational data assimilation (3DVar) approach is similar to the other techniques in that all observations assumed to be valid at the same time as the background state are assimilated to compute a correction to the background state. However, as mentioned previously, the background error covariances must be specified before this technique can be applied. As with OI, very simple representations of these covariances are often used, such as assuming a spatially constant variance and a spatially homogeneous and isotropic correlation function of a specific functional form. However, it is also possible to use covariances estimated from an ensemble of states that approximates a random sample of the background error PDF. This approach, referred to as ensemble-variational data assimilation (EnVar), can make use of ensembles of background states produced by an EnKF. A 4D version of EnVar makes use of ensembles of background states at several times within the time window of the assimilated observations.

Another difference compared to the EnKF technique, which uses an ensemble of pre-computed short-term model forecasts, is that the forecast model itself can be directly incorporated within the data assimilation procedure in the approach known as four-dimensional variational data assimilation (4DVar). In 4DVar, the cost function involves the state vector at many different times over a specified time window. The assimilated observations can then be arbitrarily distributed throughout this time window (as shown in Figure 4.2). The goal of 4DVar is to determine the series of state vectors through time that simultaneously satisfy the forecast model equations and are consistent with the observations throughout the time window while also being constrained to be 'close' to the background estimate of the state at the beginning of the time window. The 4DVar cost function that explicitly represents the time dependence of the state vector and the assimilated observations in the time window between times t_0 and t_N can be written as

$$\begin{aligned} J[\mathbf{x}(t_0), \mathbf{x}(t_1), \ldots, \mathbf{x}(t_N)] = & \frac{1}{2} \sum_{0 \le n \le N} [\mathbf{y}^{\mathrm{o}}(t_n) - \boldsymbol{H}(\mathbf{x}(t_n))]^{\mathrm{T}} \mathbf{R}(t_n)^{-1} [\mathbf{y}^{\mathrm{o}}(t_n) - \boldsymbol{H}(\mathbf{x}(t_n))] \\ & + \frac{1}{2} \left[\mathbf{x}^{\mathrm{b}}(t_0) - \mathbf{x}(t_0)\right]^{\mathrm{T}} \mathbf{B}^{-1} \left[\mathbf{x}^{\mathrm{b}}(t_0) - \mathbf{x}(t_0)\right], \end{aligned} \tag{4.12}$$

where $\mathbf{x}(t_n)$ is the state vector and $\mathbf{y}^{\mathrm{o}}(t_n)$ is the vector of observations, both at time t_n. Note that the background state is now denoted as $\mathbf{x}^{\mathrm{b}}$, instead of $\overline{\mathbf{x}}^{\mathrm{b}}$ used when describing the EnKF technique. The state vector at all times after t_0 is obtained from the initial state $\mathbf{x}(t_0)$ by integrating the non-linear forecast model,

$$\mathbf{x}(t_{n+1}) = \mathcal{M}[\mathbf{x}(t_n)], \tag{4.13}$$

forward in time for $n = 0, 1, \ldots, N-1$. Consequently, the cost function can be considered a function of the initial state only, that is, $J[\mathbf{x}(t_0)]$. The value of the initial state that minimizes this cost function can be efficiently obtained by using a gradient-based optimization algorithm, thus requiring calculation of the cost function gradient with respect to the initial state, $\partial J/\partial \mathbf{x}(t_0)$. By taking advantage of the chain rule of differentiation, the gradient is computed through a backward propagation of information from each observation back to the initial time. This can be seen by considering the gradient with respect to the state at time level n written in terms of the gradient with respect to the state at time level $n+1$ for $n = N-1, N-2, \ldots, 1$:

$$\frac{\partial J}{\partial \mathbf{x}(t_n)} = \frac{\partial \mathbf{x}(t_{n+1})}{\partial \mathbf{x}(t_n)} \frac{\partial J}{\partial \mathbf{x}(t_{n+1})} - \mathbf{H}^{\mathrm{T}} \mathbf{R}_n^{-1} \left[\mathbf{y}^{\mathrm{o}}(t_n) - H\left(\mathbf{x}(t_n)\right) \right], \tag{4.14}$$

where $\partial \mathbf{x}(t_{n+1})/\partial \mathbf{x}(t_n)$ denotes the Jacobian of the forecast model (i.e. a matrix containing the first derivatives of all elements of the vector $\mathbf{x}(t_{n+1})$ with respect to all elements of the vector $\mathbf{x}(t_n)$), usually referred to as the adjoint model. Then for the initial time, the gradient of the cost function term involving the background state also contributes to the gradient with the term: $-\mathbf{B}^{-1}[\mathbf{x}^{\mathrm{b}}(t_0) - \mathbf{x}(t_0)]$.

A major technical challenge for implementing 4DVar is the requirement to develop the adjoint model for the sea ice forecast model. As just shown mathematically, the adjoint model is used in 4DVar to propagate information backwards through time from the time of each observation back to the initial time. For comprehensive Earth system models whose complexity are continually increasing as the forecast models are being improved, the required work of developing and maintaining an adjoint model is similar to that of developing the forecast model itself. Alternatively, the path taken by two sea ice modelling groups (Kauker *et al.*, 2009 and Heimbach *et al.*, 2010) is to use algorithmic differentiation (AD). With AD, the source code of the original forecast model is provided to a software tool that analyses the code and automatically produces source code of the adjoint model. In addition, applying 4DVar to forecast models with highly non-linear processes (such as sea ice models) can cause practical challenges for efficiently obtaining the cost function minimum. This is because such non-linearities can result in a cost function that is not smooth and can even exhibit a discontinuous gradient.

Since the forecast model solution that minimizes the cost function provides an optimal fit to the observations distributed in time, the 4DVar approach is ideally suited for the reconstruction of the state over a past period. This is especially the case when applying 4DVar to a very long time window. For example, the sea ice–ocean state was estimated over a one-year period for the Labrador Sea and Baffin Bay by Fenty and Heimbach (2013a, b) and the global ocean state was estimated over a six-year period by Stammer *et al.* (2002).

4.3.5 Estimation and Modelling of Background and Observation Error Statistics

The background and observation error statistics used for assimilating observations have a large impact on the resulting analysis state. The use of particularly poor estimates of these statistics can produce an analysis state with a larger real error than that of the background state, even when the assimilated observations are highly accurate. As already mentioned, the relative magnitude of the background and observation uncertainty (as often measured by their error covariance) strongly influences the relative weight given to the observation when computing the correction to be applied to the background state. The spatial correlations of both the background and observation error affect how these corrections are distributed over different spatial scales. For example, large positive background error correlation with the surrounding grid points and no observation error correlation results in corrections being made mostly to the largest spatial scales. In contrast, small background error correlation and large positive observation error correlation results in corrections being made mostly to much smaller spatial scales. Of course, zero correlations for both the background and observation errors results in only correcting those state variables that directly affect the observation through the observation operator (usually those state variables located closest to the observation).

To estimate the background error covariances, ensembles of model states (such as those produced by an EnKF) are increasingly used. Due to the necessarily small number of ensemble members as compared with the dimension of the state vector being estimated, additional assumptions must be applied to obtain a useful covariance estimate. Spatial covariance localization is used in many data assimilation approaches that make direct use of ensembles. This relies on the assumption that the true background error covariance between spatially separated state variables becomes increasingly small and reaches zero beyond some separation distance. Alternatively, the parameters of an assumed covariance model can be estimated from the ensembles. These approaches allow relatively small ensembles (~100 members) to be used to assimilate a large volume of observations (10^3–10^6) to initialize a high-dimensional forecast model (with up to ~10^8 variables on the model grid). Without such approaches, the use of background error covariances estimated directly from an ensemble of, say, 100 members would only allow at most 99 independent linear combinations of the state variables to be corrected. Consequently, the assimilation of a single isolated observation would likely generate a non-zero correction for all other state variables, regardless of the distance or physical connection between the observation and the state variable (e.g. for global NWP, a temperature observation over Canada would modify the temperature over Australia and everywhere else on the Globe). The selection of only the subset of nearby observations when computing the analysis state for a relatively small geographical region of the state vector, as in OI and some EnKF approaches, has a similar effect as spatial covariance localization.

For several reasons, observation errors are usually assumed to be uncorrelated between measurements at different locations or at different times at the same location. This is due to both the computational challenge of including non-zero observation error correlations in many data assimilation algorithms and a lack of knowledge concerning these correlations. For some types of high-resolution observations used in NWP, improved results are obtained by first spatially thinning them before assimilation. By increasing the spacing between observations, thinning reduces the spatial correlation of the errors that is not taken into account when they are assimilated.

Even though most data assimilation approaches are derived under the assumption that the errors are unbiased, systematic errors associated with both observations and the background state are often present and can even be of equal importance as the random error. It is usually difficult to separate the systematic errors in the observations from those caused by the model. To make progress, it is usually necessary to assume one is unbiased to be able to estimate and correct the bias in the other. In the case of observation error bias, once estimated it can simply be subtracted from the observations prior to assimilation. In some cases, such as satellite radiance observations assimilated in NWP systems, the estimation and correction of systematic observation error is essential to be able to extract useful information from the observations. The application of such approaches may have potential benefits for sea ice data assimilation, but has not yet been extensively evaluated.

4.4 Adaptation of Data Assimilation Techniques for Sea Ice

When applying data assimilation techniques to sea ice observations and models, several important considerations arise. This section describes some of these, including possible strategies for addressing the associated challenges.

4.4.1 Spatially Discontinuous Nature of Sea Ice

Sea ice can be highly discontinuous in space, made up of discrete floes of various sizes and thicknesses with sharp transitions between ice-covered and open water areas. This is quite different from the fluid properties represented by oceanographic and atmospheric fields, such as temperature, velocity, salinity and humidity, that are the focus of conventional applications of data assimilation for ocean and weather prediction. These oceanographic and atmospheric fields vary continuously through space, often in a smooth way over relatively large distances. The spatially discontinuous nature of sea ice is also in contrast to how sea ice is represented in many of the currently available sea ice models we wish to initialize through data assimilation. These models typically represent the average ice conditions over relatively large model grid cells in a highly simplified way. Some of the observation types now available for assimilation are from satellite sensors that also measure the average sea ice conditions over fairly large footprints with a size comparable to the grid cell of a typical sea ice model (greater than ~5km). This is large enough to often include a high number of discrete ice floes separated by open water and therefore representing the ice cover as an average concentration is reasonable. Conventional techniques designed for assimilating data to correct relatively smooth fields of continuous variables can therefore be applied for assimilating such observations.

It is a serious challenge to effectively assimilate in situ observations that measure the sea ice conditions at a single location and time that are likely not representative of the average conditions over the grid cell. Increasingly, very high resolution satellite observations have become available for assimilation that can, in some cases, resolve individual ice floes. Similarly, particle-based sea ice models that resolve individual ice floes are being developed (see Section 2.8.5). The assimilation of such high-resolution observations to correct forecast models that can resolve individual ice floes will represent a fundamental change for data assimilation and require adaptation of the approaches currently used. At very high spatial resolution, the ice concentration will be expected to only have values of 0 per cent and 100 per cent. Consequently, the observation and background errors for ice concentration will have discrete probability distributions for which only 0 per cent and 100 per cent concentrations would have non-zero probabilities. In contrast to this, most data assimilation approaches are designed to estimate the values of continuous state variables from observations of continuous variables, both having continuous probability distributions. These challenges at such high resolutions suggest that it may be more effective to use a completely different set of continuous variables to represent the sea ice conditions, such as the location, horizontal size and thickness of each individual ice floe.

Even with current, relatively low-resolution sea ice models and satellite observations, the sea ice concentration can be highly spatially discontinuous. For example, near the ice edge, the concentration can change rapidly over the distance of a single model grid cell or observation footprint from values near 100 per cent to values near 0 per cent. An incorrect ice edge location in the background state can therefore result in very large positive or negative ice concentration errors close to the ice edge. In adjacent areas over the ice pack or open water, however, the errors are typically much smaller. The errors may also be highly correlated between grid cells along the ice edge, but have little correlation with the errors nearby within the ice pack or in the region of open water. This suggests that the variances and spatial correlations of the background error for ice concentration and other sea ice variables strongly depend on the location of the ice edge. Similar large gradients in the background error covariances can occur near land where the winds and ocean currents can cause openings in the ice cover as the ice drifts away from shore or when ice motion becomes impeded upstream of narrow passages. Approaches that allow such strongly state-dependent background error statistics to be accurately estimated and used when assimilating observations would likely produce improved analyses as compared with approaches that use spatially homogeneous and stationary statistics. Figure 4.5 shows an example of the mean and standard deviation for ice concentration obtained from a 60-member ensemble generated from an ensemble of 3DVar data assimilation systems in which multiple sources of uncertainty are simulated (more details given by Shlyaeva *et al.*, 2016). Note how the error standard deviation (Figure 4.5b) is close to zero where the background ice concentration (Figure 4.5a) is zero and it is highest near the ice edge.

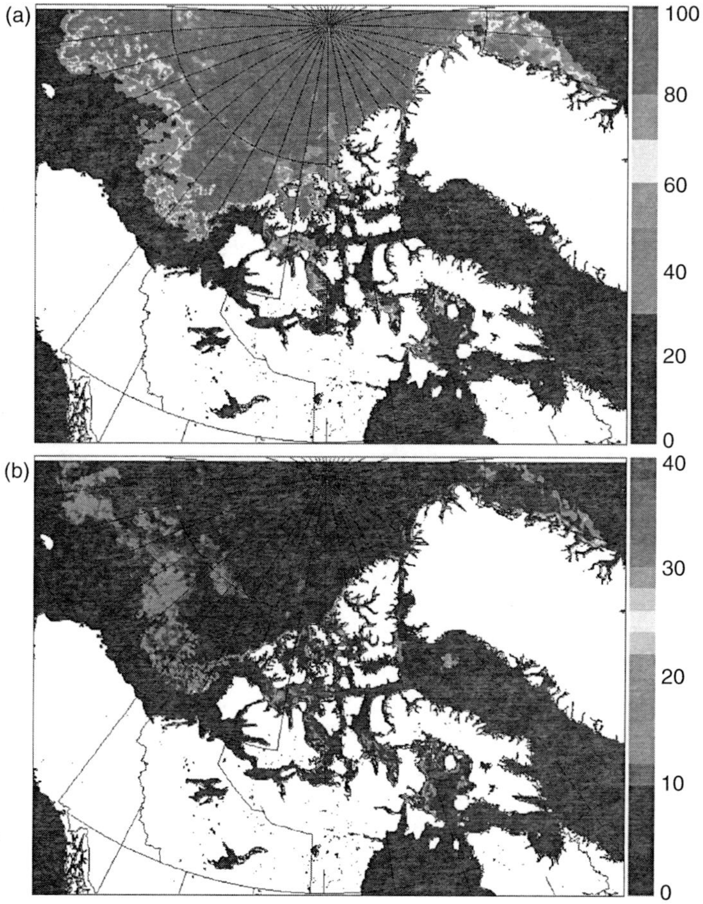

Figure 4.5: The background (a) mean sea ice concentration and (b) standard deviation of sea ice concentration computed from a 60-member ensemble. The ensemble was constructed using six-member ensembles of six-hour forecasts from 10 consecutive assimilation times, separated by six hours (from 15 August 2011 at 1200 UTC to 17 August 2011 at 1800 UTC). Note that the units are in per cent (similar to Figure 13 from Buehner and Shlyaeva, 2015). (A black and white version of this figure will appear in some formats. For the colour version, please refer to the plate section.)

4.4.2 Initialization of Important Unobserved State Variables

Sea ice thickness is an important sea ice property for which there are currently few sources of accurate observations suitable for operational sea ice prediction (though some sources, such as from L-band passive microwave and altimeter instruments can provide limited information). Moreover, several sea ice models require more than a single thickness value at each grid cell since they represent thickness using an ITD (see Chapter 2). Providing this level of detailed information through data assimilation is a significant challenge. This is especially difficult at locations where the background state has no ice, but the observations indicate the presence of ice. One approach to correct the ITD is to use the multivariate covariances within the data assimilation system, such as the ensemble covariances produced by an EnKF, to compute these corrections based on ice concentration observations. In this case, the EnKF relies on the ensemble of short-term model forecasts to represent the linear relationship between errors in total ice concentration and the errors in partial concentration for all thickness categories. Alternatively, more ad hoc approaches have been shown to be effective. For example, Smith *et al.* (2015) compared two methods for adjusting the background state ITD based on the ice concentration analysis. In the first, the entire ITD is simply rescaled by a constant chosen to give a total ice concentration that is consistent with the analysis. The second, which resulted in slightly better forecasts, applies a correction to the ITD that is proportional to the temporal tendencies over the most recent model forecast. The use of model tendencies is a relatively crude method, but it has the advantage of providing a means to enhance or reverse non-linear processes captured by the sea ice model.

While sea ice motion can strongly affect the ice concentration, thickness and the location of the ice edge in the background state, the assimilation of ice motion observations for the purpose of improving the sea ice analysis state and subsequent forecast is not straightforward. This is because the ice velocity within a sea ice model, as discussed in Chapter 2, is mostly determined by the balance between the rheology term, the wind stress and the water stress. If changes are made by the data assimilation system only to ice velocity, these will be forgotten in the subsequent forecast as the balance is quickly re-established. Ice motion errors can result from errors in numerous variables and parameters: e.g. surface winds, surface ocean currents, drag coefficients, ice strength and ice thickness. By modifying any of these, the relative balance between the rheology and stresses in the momentum equation would be changed and therefore result in a more sustained change to the ice motion in the forecast. Consequently, the most beneficial approach for assimilating ice drift observations is to use them to correct the causes of the ice motion error, most of which are not directly observed. This requires a multivariate assimilation approach that effectively maps the corrections for ice motion into corrections for these other model variables and parameters. For example, Massonnet *et al.* (2014) used the assimilation of ice drift to only correct three ice model parameters (the ice strength parameter and two drag coefficients) using an EnKF, leading to improved ice drift forecasts. Alternatively, when using a data assimilation technique that does not provide the needed multivariate relationships, a simpler approach can be adopted. For example, an additional ice stress term can be included in the sea ice model that is diagnosed from ice motion errors, as proposed in the study by Stark *et al.* (2008).

4.4.3 Direct Assimilation of Satellite Observations Versus Assimilation of Retrieved Quantities

Many types of satellite observations are only indirectly related to the sea ice model variables. For example, passive microwave radiances and synthetic aperture radar (SAR) backscatters are indirectly related to sea ice concentration. No simple transformation exists between the radiance or backscatter and ice concentration, though the ice concentration is one important factor affecting both types of observation. Two basic strategies are possible for extracting information concerning the sea ice conditions from such observations through data assimilation. For a given type of observation and depending on which sea ice variables are to be estimated, each strategy has its own relative advantages and disadvantages.

In the first strategy, a stand-alone retrieval algorithm is used to produce an estimate of sea ice concentration or another sea ice variable from the observation (examples of retrieval algorithms were discussed in Chapter 3). This estimate is then assimilated directly in a straightforward way using any of the data assimilation techniques described earlier. In this case, the observation operator only includes a spatial interpolation or averaging procedure to map the gridded sea ice variables to the location of the observation. The advantage of this strategy is that the retrieval algorithm can be highly non-linear, including threshold processes that are not differentiable. Also, it is straightforward to ensure physical values of the retrieved quantity before assimilation, such as ensuring ice concentration is between 0 per cent and 100 per cent or ice thickness is positive. For some types of satellite observation it may be possible to obtain a relatively accurate retrieval for categorical variables, whereas the estimation of continuous sea ice variables, such as ice concentration or thickness, may be more difficult. Examples of categorical variables include the ice type (i.e. new ice, first-year ice, multi-year ice, etc.) and the presence of ice versus open water. Approaches for assimilating categorical variables are discussed in Section 4.4.4.

The second strategy involves modelling the complete observing process as part of the observation operator within the data assimilation system. For passive microwave radiances, for example, this means that the gridded ice concentration along with information on all other geophysical factors affecting the microwave radiance measured at the top of the atmosphere (ice and ocean emissivity; ice and ocean surface temperature; atmospheric temperature, humidity, clouds, etc.) are input into a complete radiative transfer model to compute a corresponding predicted value of brightness temperature for the particular frequency band measured by the instrument. Not only ice concentration, but some or all of the other variables affecting the brightness temperature can be estimated as part of the data assimilation procedure, and therefore included as part of the analysis state. Compared to the first strategy, the complexity of the data assimilation procedure is increased, since additional inputs for the observation operator are required that are either fixed to assumed values or estimated together with ice concentration. For each additional variable to be estimated as part of the data assimilation procedure, a background estimate and its error covariances are also required. However, any ambiguities between multiple inputs to the observation operator do not need to be resolved by relying on just one type of observation, as they often would when using a stand-alone retrieval algorithm. Such ambiguities exist for many types of satellite observations. For example, at certain incidence angles high wind speed over open water can produce a SAR backscatter similar to 100 per cent sea ice cover, whereas at low wind speed the ice and open water may have very distinct backscatter.

Ambiguities can be resolved during the simultaneous assimilation of all available observations and also by taking into account information in the background state and the statistics of the errors of both the background state and the observations. In principle, all of this additional information, both at the location of the ambiguous observation and within the surrounding area, should work together in the data assimilation system to produce the most likely value of ice concentration and other related variables. However, in addition to the added complexity and input data that need to be specified, the direct assimilation of such satellite observations may also be problematic due to the observation operator being highly non-linear or even discontinuous, a problem for variational techniques that rely on a linearization of the observation operator. It may also be more difficult to prevent the assimilation of an observation from resulting in physically unrealistic values for the analysis state variables (e.g. ice concentration outside the range of 0 per cent to 100 per cent, negative ice thickness).

An idealized numerical example of C-band SAR backscatter with HH polarization (i.e. horizontal transmit-horizontal receive) is now presented to demonstrate how ambiguities in the relationship between the observation and the related geophysical variables can be treated by directly assimilating the backscatter within a data assimilation system. For simplicity, it is assumed that the surface consists of a combination of first-year (FY) ice, multi-year (MY) ice and open water, and that the observed backscatter is a weighted sum of typical backscatter values from each of these surface types. Furthermore, it is assumed that the backscatter from the open water varies linearly with the surface wind speed. (Note: these simple assumptions may not be entirely realistic, since backscatter from areas with intermediate ice concentrations in the marginal ice zone (MIZ) may actually be higher than areas with ice concentration close to either 0 per cent or 100 per cent due to the higher surface roughness of deformed ice and from the edges of individual floes in the MIZ.) Assuming a linear dependence with respect to the analysis state variables FY ice concentration (A_{fyi}), MY ice concentration (A_{myi}) and wind speed ($|\mathbf{u}_{\mathrm{a}}|$) results in the following idealized observation operator for backscatter (denoted as σ_0):

$$\sigma_0\left(A_{\mathrm{fyi}}, A_{\mathrm{myi}}, |\mathbf{u}_{\mathrm{a}}|\right) = A_{\mathrm{fyi}} \times \sigma_{0\mathrm{fyi}} + A_{\mathrm{myi}} \times \sigma_{0\mathrm{myi}} + \left(1 - A_{\mathrm{fyi}} - A_{\mathrm{myi}}\right) \times \left(\sigma_{0\mathrm{ow}} + \frac{d\sigma_0}{d|\mathbf{u}_{\mathrm{a}}|} \times |\mathbf{u}_{\mathrm{a}}|\right). \tag{4.15}$$

For an incidence angle of 35°, the following typical backscatter values for MY ice, FY ice and open water are respectively assumed: $\sigma_{0\mathrm{myi}} = -10\mathrm{dB}$, $\sigma_{0\mathrm{fyi}} = -18\mathrm{dB}$, $\sigma_{0\mathrm{ow}} = -25\mathrm{dB}$ and the slope of the relationship between backscatter and wind speed over open water is $d\sigma_0/d|\mathbf{u}_{\mathrm{a}}| = 0.75\mathrm{dB}$ per m s^{-1}. According to this relation, it is clear that the same backscatter value can be produced by many different combinations of FY and MY ice concentration and wind speed (Figure 4.6a, b). Therefore, it would be impossible to obtain a reliable estimate of the sea ice conditions by using a retrieval algorithm based on the backscatter alone. For an observed backscatter value of −20 dB, the resulting observation cost function equals the same minimum value for many combinations of the three analysis

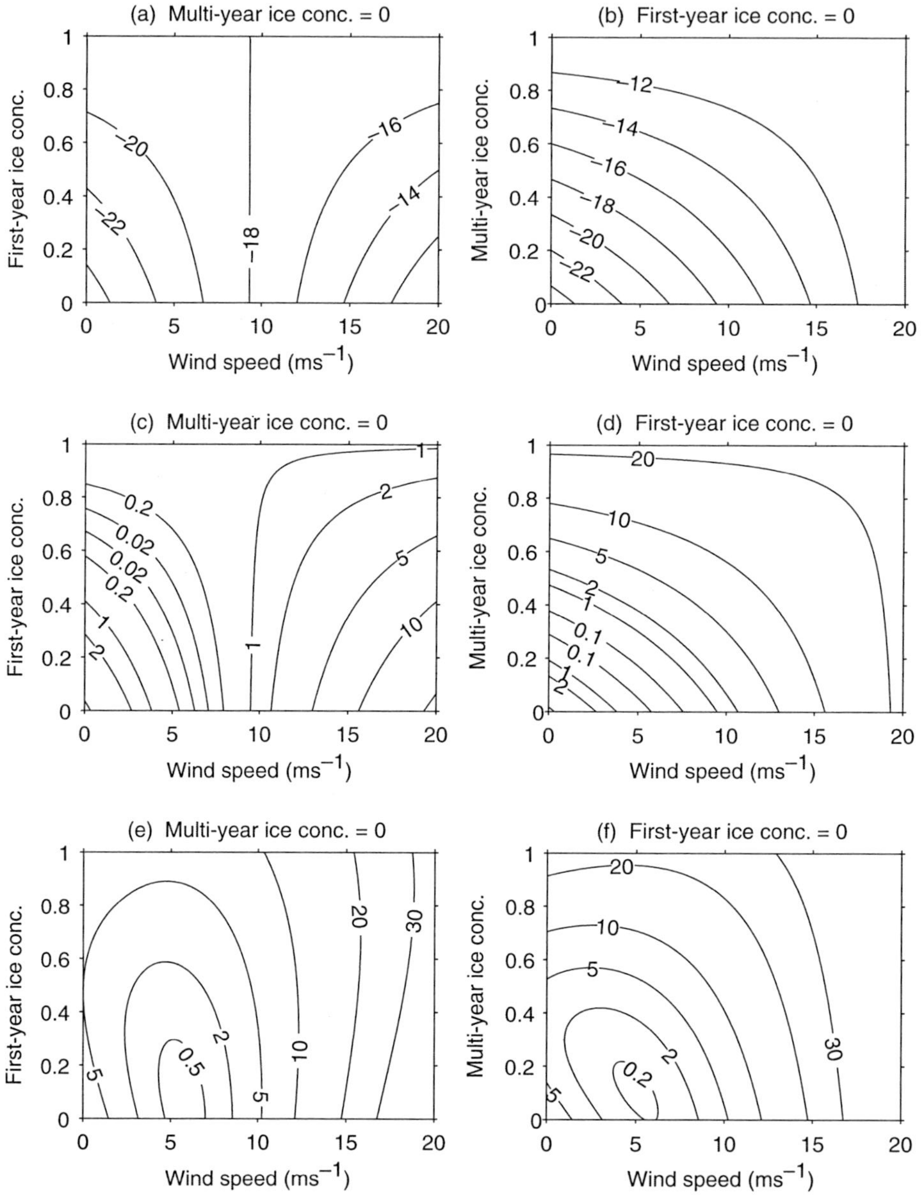

Figure 4.6: Idealized C-band SAR HH backscatter (a) as a function of wind speed and FY ice concentration when MY ice concentration is zero and (b) as a function of wind speed and MY ice concentration when FY ice concentration is zero. The resulting observation cost function (first term on the right-hand side of Equation 4.3) for an observed HH backscatter of −20 dB and an observation error standard deviation of 1.5 dB (c) as a function of wind speed and FY ice concentration when MY ice concentration is zero and (d) as a function of wind speed and MY ice concentration when FY ice concentration is zero. The result of adding to the observation cost function a background cost function (second on the right-hand side of Equation 4.3) with background values for FY ice concentration and wind speed equal to zero and 5 m s^{-1}, respectively, is shown in (e) and (f).

state variables (Figure 4.6c, d). For example, if the total ice concentration is close to zero, then the observation indicates that the wind speed is approximately 7 m s^{-1} (Figure 4.6c). Similarly, if just the FY ice concentration is zero, then the observed backscatter of −20 dB is also consistent with a wind speed close to zero and a MY ice concentration of 35 per cent (Figure 4.6d). On its own, the backscatter observation reduces the possible combinations of these three analysis state variables that are likely, but is not sufficient to provide a single estimate. However, by assimilating the backscatter in combination with a reasonably accurate background state and other types of observation that do not have the same ambiguities, a unique estimate can be obtained. For example, if the background state contains information that the FY ice concentrations is close to zero (with error standard deviation of 30 per cent) and the wind speed is 5 m s^{-1} (with error standard deviation of 2 m s^{-1}), then the ambiguity is significantly reduced, indicating a MY ice concentration near 10 per cent, even though no information on MY ice concentration was provided by the background state (Figure 4.6f).

4.4.4 Assimilation of Categorical Observations

In some cases, it may be impossible to extract accurate information related to a continuous analysis state variable from a type of observation using either of the strategies described in the previous section. However, it may be possible to estimate the discrete value of a categorical variable, such as ice type or simply the presence of ice or open water. For example, the Interactive Multisensor Snow and Ice Mapping System (IMS) currently operated by the US National Ice Center produces a daily gridded field over water in the Northern Hemisphere that simply indicates the presence of ice or open water (Helfrich *et al.*, 2007). The IMS product is manually generated by visually analysing multiple sources of high-resolution satellite data. Automated approaches based on image processing and pattern recognition approaches have also been applied for obtaining categorical information from satellite observations. For example, the MAp-Guided Ice Classification (MAGIC) system produces estimates of the presence of ice or open water from SAR imagery (Leigh *et al.*, 2014). Related to this are observation types that accurately provide a continuous value only up to a known maximum value. For example, ice thickness retrievals from passive microwave or infrared data can reliably provide thickness estimates only up to relatively low values (typically less than ~0.5 m, depending on the sensor and the ice conditions) beyond which the measurement saturates, causing the accuracy of the estimated thickness to significantly decrease. Therefore, an observed value at this saturation limit can be considered a categorical value since it only indicates that the thickness can be any value greater than the thickness associated with the saturation limit. In this section, we explore several possible methods for using categorical information, generated either with manual or automated approaches, within a sea ice data assimilation system to estimate fields of continuous analysis state variables, such as ice concentration or thickness.

To illustrate, we use an idealized numerical context to assimilate a binary quantity that indicates at each location if sea ice or open water is present. We take as the ‘true’ state the sea ice analysis state from an experimental version of the Canadian Regional Ice Prediction System (RIPS) valid on December 31 at 1800 UTC, 2014 (Figure 4.7a). This version of RIPS is similar to that described by Buehner *et al.* (2014), except that binary retrievals of ice and open water from high-resolution AVHRR observations have been assimilated in addition to ice concentration from AMSR2 and CIS regional ice charts. Observations are simulated from this true state at a spatial resolution of approximately 5 km by first adding Gaussian noise with a standard deviation of 10 per cent to the ice concentration (to represent the uncertainty due to observation error) and then applying a threshold of 50 per cent to generate the binary observations. The background ice concentration to be corrected by assimilating the binary observations is obtained by taking the background state from a version of RIPS that did not assimilate the high-resolution AVHRR observations (Figure 4.7b). Though the spatial domain of RIPS covers most of the ice-affected regions in the Northern Hemisphere, all figures and results related to this numerical example only consider the region consisting of Hudson Bay, Foxe Basin, Hudson Strait and parts of the Labrador Sea and Davis Strait.

The simplest approach to assimilate a binary ('ice' or 'open water') observation would be to assign the ice observations an ice concentration of 100 per cent and the open water observations a value of 0 per cent and assimilate them as any other type of ice concentration observation. In locations where the background state is very different from the true ice concentration, this approach will likely result in an analysis state that is closer to the truth than the background state. In many other locations, however, when the background state is already relatively close to the true ice concentration and is not close to 0 per cent or 100 per cent, this procedure could often lead to an increase in error for the analysis state relative to the background state. For example, a true ice concentration of 60 per cent would be classified by this observation type as ice and therefore its assimilation as an ice concentration observation of 100 per cent will result in an increase to the background ice concentration, even when the background ice concentration is already higher than the true concentration.

To greatly reduce the number of cases for which the error is increased by assimilating binary observations, a simple check can be made to eliminate, and therefore not assimilate, observations that are already consistent with the background state. This approach is similar to the treatment of 'out-of-range' observations in the so-called partial-EnKF proposed by Borup *et al.* (2015). Since the binary observation was originally generated using a concentration threshold of 50 per cent, an ice observation can be considered consistent with the background state wherever the background concentration is above 50 per cent. Similarly, open water observations would be consistent with the background state, and therefore not assimilated, wherever the background concentration is below 50 per cent. For the observations that remain, they are assimilated as ice concentration observations of 50 per cent. Borup *et al.* (2015) showed that this is equivalent to making the reasonable assumption of equal probability for all values beyond the threshold. In other words, an ice observation tells us only that the ice concentration could equally likely be any value above 50 per cent. Similarly, an open water observation indicates an equal likelihood for any concentration below 50 per cent.

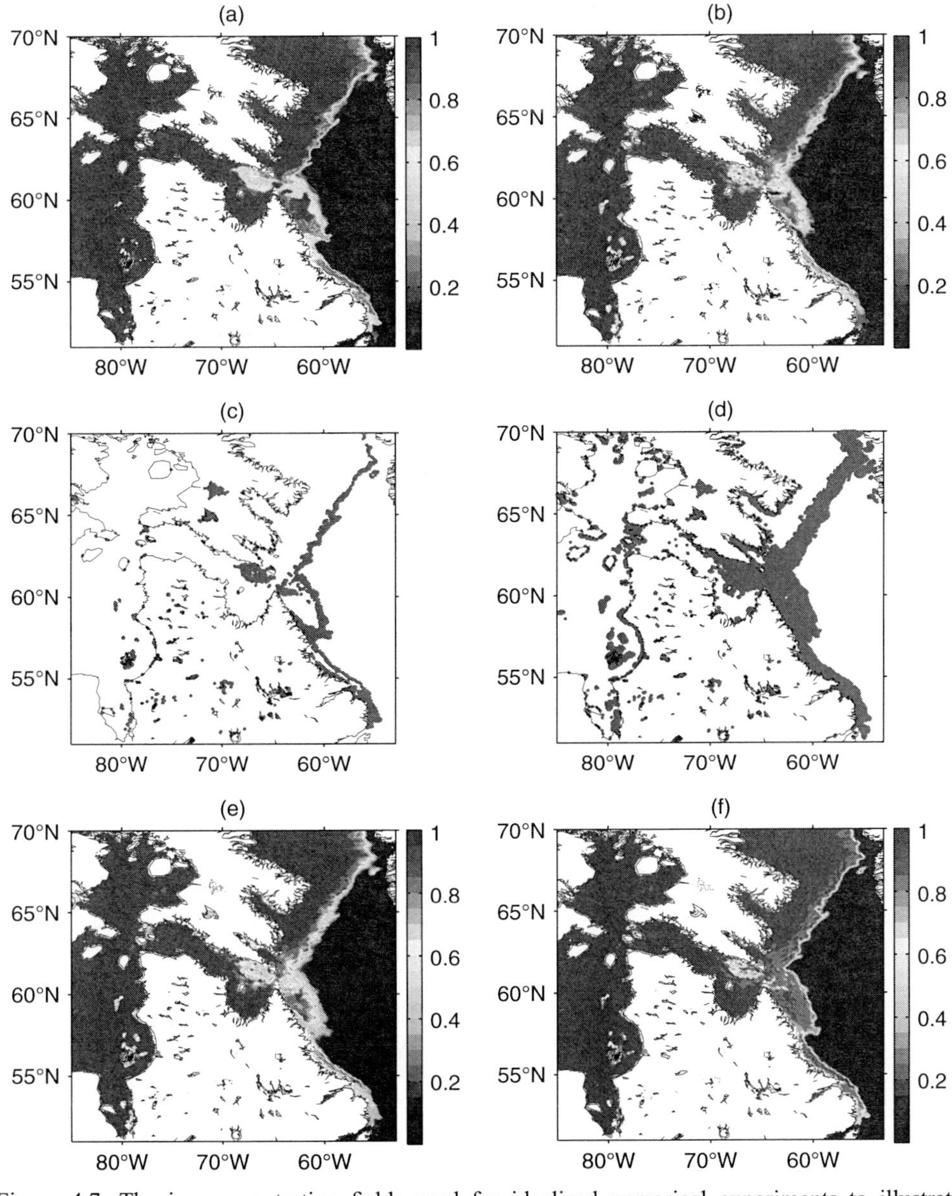

Figure 4.7: The ice concentration fields used for idealized numerical experiments to illustrate different approaches for assimilating categorical observations: (a) the 'true' ice concentration, (b) the background ice concentration to be corrected by assimilating observations. Locations of the ice and open water observations assimilated using two different sets of concentration threshold values: (c) 50 per cent for both ice and water, (d) 90 per cent for ice and 10 per cent for water. The case of threshold values 100 per cent for ice and 0 per cent for water are not shown, since the observations cover the entire domain. The resulting analysed ice concentration after assimilating high-resolution ice and open water observations using different concentration threshold values: (e) 50 per cent for both ice and water, (f) 90 per cent for ice and 10 per cent for water. (A black and white version of this figure will appear in some formats. For the colour version, please refer to the plate section.)

Figure 4.7c shows the locations of all binary observations that remain after removing the observations already consistent with the background concentration after applying a threshold of 50 per cent to the background concentration for both ice and open water observations, as just described. Note how the remaining observations are mostly only close to the ice edge or near land. To increase the number of assimilated observations, while also increasing the number of cases where assimilation of binary observations increases the error, a less stringent criterion can be used. Figure 4.7d shows the binary observation locations after eliminating the ice observations only where the background concentration is above 90 per cent and eliminating the open water observations where the background concentration is below 10 per cent. The use of this less stringent criteria results in nearly five times more binary observations being assimilated.

The ice concentration obtained from assimilating binary observations using the two different sets of thresholds are shown in Figure 4.7e, f. By comparing these analysis ice concentration fields, it is clear that the presence of intermediate ice concentration values is decreased by assimilating more binary observations with more extreme values. The relative accuracy of the ice concentration analysis state will therefore depend on the prevalence of intermediate concentration values in the true state. If intermediate values occur only rarely, then the advantage of assimilating more binary observations will outweigh the errors caused by incorrectly forcing the analysis towards more extreme concentration values. As the number of locations increases where the true ice concentration differs significantly from 0 per cent or 100 per cent, then the elimination of observations where the background concentration is to some extent already consistent with the observation will be beneficial.

Continuing with this example, Figure 4.8 shows the difference between the background and true ice concentrations (Figure 4.8a) in addition to the differences between the analysis state and true concentrations for three threshold values (Figure 4.8b, c, d). The use of the threshold values of 90 per cent for ice and 10 per cent for open water observations results in the lowest error, though only slightly lower than when using all observations (root-mean-square error, RMSE, values are given in the caption of Figure 4.8). The use of threshold values of 50 per cent for both ice and open water observations still reduces the error in the analysis state relative to the background state, though it is not as low as when using the less stringent criteria. It is interesting to note that when using the threshold values of 50 per cent (Figure 4.8b), which are consistent with how the binary observations were originally produced, the analysis error is reduced relative to the background error, but generally has the same pattern of positive and negative errors. That is, the error at every location is either reduced towards zero or unchanged, but almost always has the same sign as the background error. Using the less stringent criteria of 90 per cent and 10 per cent for ice and open water observations, respectively, or assimilating all of the observations, results in a greater overall reduction in the analysis error, though the spatial pattern of the error is significantly changed relative to the error in the background concentrations. Specifically, in locations where the background error is predominantly negative (near the coast of Labrador and along the ice edge in Davis Strait), the analysis error becomes mostly positive.

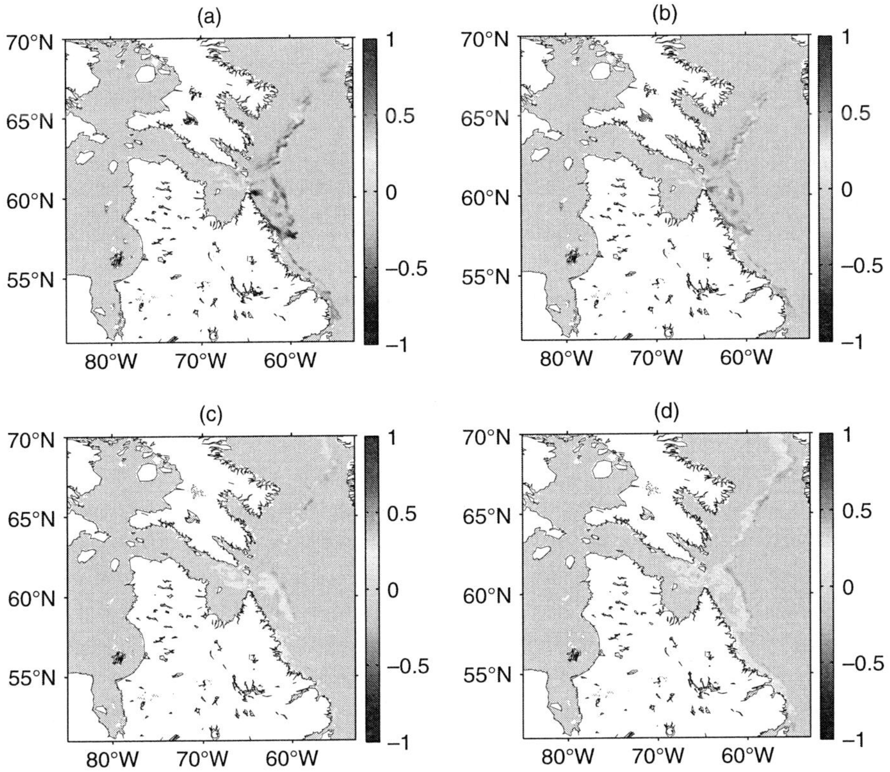

Figure 4.8: The ice concentration error for (a) the background state (RMSE of 15.5 per cent) and the analysis state after assimilating high resolution ice and open water observations using various concentration threshold values: (b) 50 per cent for both ice and open water (RMSE of 14.1 per cent), (c) 90 per cent for ice and 10 per cent for open water (RMSE of 12.8 per cent), (d) 100 per cent for ice and 0 per cent for open water (RMSE of 12.9 per cent). (A black and white version of this figure will appear in some formats. For the colour version, please refer to the plate section.)

A similar approach, proposed by Scott *et al.* (2015), can be used within a variational data assimilation system that avoids the need to explicitly eliminate categorical observations that are already consistent with the background state. Instead, a non-linear observation operator is employed that has nearly the same effect.

4.4.5 Simultaneous Assimilation of both High- and Low-Resolution Observations

As described in detail in Chapter 3, sea ice observations from different types of satellite-based sensors can have very different spatial resolutions. For example, some types of passive microwave observations (e.g. SSMI/S, SMOS) measure the brightness temperature averaged over a footprint with a diameter of approximately 50 km at the surface. Since this is much larger than the grid cell size of most sea ice models used for short-term forecasting, these observations must be treated carefully to avoid removing small-scale information from the model-generated background state. Other observation types are available with significantly higher resolution of approximately 1 km or better (e.g. Visible/Infrared, SAR). Ideally, both high- and low-resolution observations would be assimilated simultaneously with each only used to correct the background state at the spatial scales that they resolve. This can be accomplished by including, as part of the observation operator, a spatial averaging over the model grid cells that fall within the footprint of each low-resolution observation. When computing the difference between the observation and background state, this averaging removes from the background state the spatial scales not resolved by the observation. In contrast, high-resolution observations are assimilated using an observation operator that simply interpolates the background state from the nearest grid points. In this way, the inclusion of satellite observations with large footprints does not affect the ability of the assimilation procedure to correct the small scales (sharp features) by simultaneously assimilating high-resolution observations.

The same idealized numerical context used previously to demonstrate the assimilation of categorical observations is used here for the assimilation of an ice concentration retrieval from a low-resolution satellite sensor (such as SSMI/S), both individually and in combination with a high-resolution binary observation (indicating 'ice' and 'open water'). The low-resolution observations are simulated by averaging the true state over footprints of 50 km diameter followed by the addition of random observation error with a standard deviation of 18 per cent. The ice concentration background error (Figure 4.9a) is reduced by assimilating either the high-resolution binary observation using threshold values of 50 per cent for both ice and open water (Figure 4.9b) or an ice concentration observation with a footprint 50 km in diameter (Figure 4.9c). Note how the assimilation of the low-resolution observations can create fine scale analysis errors. This is seen mostly by the introduction of positive analysis errors along the ice edge parallel with the Labrador coast. This is because, without additional assimilated observations in the same area, the mean concentration over the footprint will be changed by modifying the concentration by a similar amount for all grid points within the footprint. However, by simultaneously assimilating higher-resolution observations, even a very limited number of binary observations (whose spatial distribution is shown in Figure 4.7c), the two types of observations can work together to reduce the analysis error more than when either is assimilated independently (Figure 4.9d).

4.4.6 Non-Gaussian Error Distributions

The use of standard data assimilation techniques for sea ice applications can often result in unrealistic or unphysical values that need to be modified before they are used to initialize a sea ice forecast model. For example, non-zero ice concentration and thickness, including negative values, over the large areas where no ice exists in reality often occurs due to the unavoidable random and systematic errors in the assimilated observations or the observation operator. Such unrealistic values of sea ice variables can also be caused by the inappropriate spatial propagation, by the background error covariances, of the corrections to the background state from nearby ice-covered areas. In a similar way, the analysis state can contain ice concentrations above 100 per cent in regions with high ice concentration.

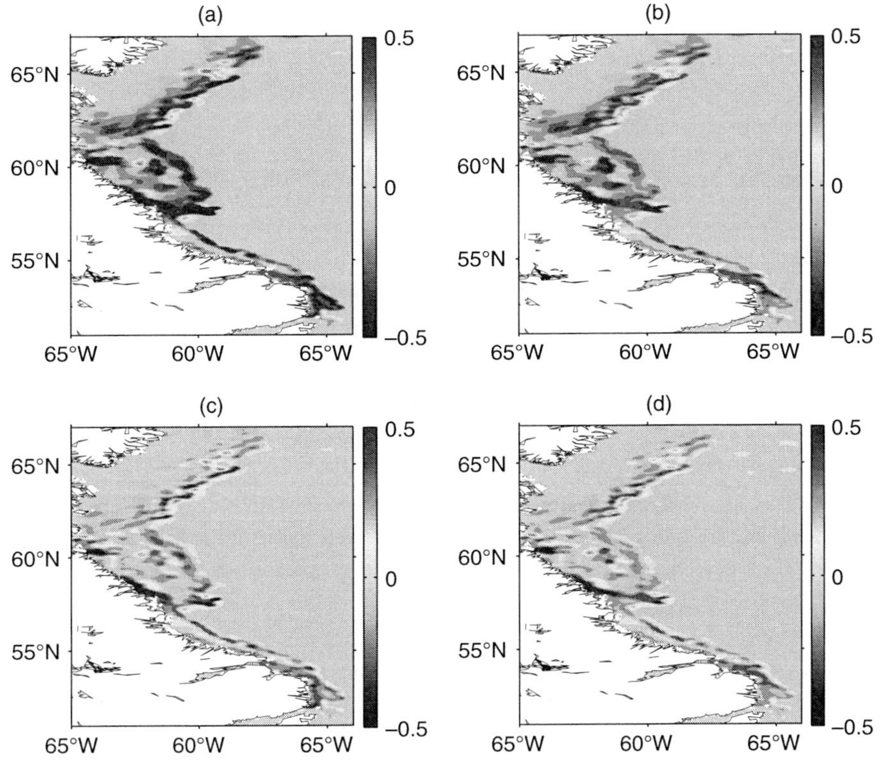

Figure 4.9: Idealized example showing the impact of assimilating low-resolution retrievals of ice concentration from passive microwave data in combination with high-resolution binary retrieval of ice and open water. The ice concentration error in (a) the background state (RMSE of 14 per cent) is shown in addition to the error in the analysis states after assimilating (b) only high-resolution binary observations with concentration threshold values of 50 per cent for both ice and open water (RMSE of 11 per cent), or (c) only low-resolution ice concentration observations (RMSE of 9.8 per cent), or (d) both low-resolution ice concentration observations and high-resolution binary observations (RMSE of 8.7 per cent). (A black and white version of this figure will appear in some formats. For the colour version, please refer to the plate section.)

Several possible strategies can be applied to avoid such unrealistic results. The simplest approach is to reset ice concentration values outside the range of 0 per cent to 100 per cent back to either 0 per cent or 100 per cent after the data assimilation procedure and before the sea ice model is initialized. Similarly, negative ice thickness values would be reset to either zero or some small specified minimum thickness. Avoiding the spurious introduction of small amounts of sea ice over areas that should have none is a challenging problem. This requires additional information that can be used to determine when and where an observation that indicates a very low concentration corresponds with zero ice concentration in reality. One approach is to set the assimilated ice concentration observations or the resulting ice concentration analysis to zero at all locations where the ocean surface temperature (obtained from an ocean model forecast or an analysis state based on satellite observations) is above some specified threshold.

Obtaining such unrealistic values for sea ice variables can often be caused by the observation and background error distributions being very different from the Gaussian distribution assumed by most data assimilation techniques. Therefore, an alternative approach is to apply a non-linear function to transform the original variables (such as ice concentration) into variables with distributions that are closer to being Gaussian. Such transformations (called Gaussian anamorphosis or, alternatively, normal score transform) can either be defined explicitly for variables following a known distribution or based on an empirical fit to a collection of observed or modelled data. The concept was applied to data assimilation by Bertino *et al.* (2003) in the context of the EnKF, since a random sample of the PDF provided by the EnKF ensemble makes the application of this technique more straightforward. The data assimilation procedure is applied to the transformed Gaussian variables and the resulting analysis state variables are then transformed back to the original physical variables using the inverse transform. This idea has been applied to the assimilation of sea ice concentrations by Barth *et al.* (2015) but with only moderate success. The effectiveness of the Gaussian anamorphosis technique to sea ice concentration is limited since its PDF has relatively high probability values at or close to the physical limits of 0 per cent and 100 per cent that must be represented continuously in the transformed Gaussian variable by arbitrary distribution tails.

4.5 The Challenge of Sea Ice Data Assimilation for Coupled Ice–Ocean and Ice–Ocean–Atmosphere Models

As discussed in Chapter 2, sea ice models predict the evolution of the ice conditions in response to the fluxes of heat, salt, moisture and momentum from both the atmosphere and ocean. Traditionally, sea ice models have been driven by atmospheric fluxes, either provided directly by atmospheric models or by using bulk formulae to calculate fluxes based on surface atmospheric fields. In contrast, due to the tightly coupled nature of the sea ice and underlying ocean, sea ice and ocean modelling components are often used in a fully coupled framework. In this context, the ocean component may range in complexity from a simple bulk mixed layer model to a full ocean general circulation model. Fully coupled ice–ocean–atmosphere models have been used for some time for climate simulations and seasonal prediction. The potential benefits for forecasts with a lead time of hours to weeks from including high-frequency coupled interactions between the atmosphere, ice and ocean models are becoming more widely recognized.

Initialization of a coupled model requires analysis states for all the various components (ice, ocean and atmosphere). The analysis states for each component are usually generated separately using systems tailored to the specifics of each system (e.g. timescales, observation types and density). As a consequence, significant inconsistencies can result between them. In an uncoupled forecasting framework, this can be overcome through offline manipulation, for example, by applying a static bias correction or specific tailoring of the analysis states and forecasts from one component when used as the forcing for another component. However, in a coupled model, inconsistencies between analyses can cause shocks to model fields and imbalances, resulting in the generation of an unphysical numerical response and degradation to the forecast accuracy.

4.5.1 Weakly Coupled Data Assimilation

Weakly coupled data assimilation involves using the coupled model to produce the background states for each component's separate data assimilation procedure. That is, within the data assimilation cycle, the forecast step is coupled, but the assimilation step is not. Therefore, the same coupled model configuration is used within the data assimilation cycle as for the forecasts. This approach has the advantage that an improved consistency between analyses can be obtained, since the background states are more likely to be consistent between components. This should result in a more balanced initial state and improved forecast skill.

In practice, implementing a weakly coupled data assimilation approach can be a challenging task. First, a number of technical issues need to be overcome. Data assimilation systems are complex pieces of software, and creating a framework that allows the different physical systems to be executed together in a coordinated way can be daunting. Moreover, in the context of operational prediction, the time delays involved in receiving various types of satellite and in situ observations that are assimilated in each system can be very different. As a result, it may be necessary to include both near-real time and delayed runs of the data assimilation systems to maintain the highest accuracy possible within the data assimilation cycle while also producing timely forecasts.

Additionally, each component may use a different time frequency for its data assimilation cycle. For example, it is common practice in NWP to assimilate observations every six hours, while sea ice and ocean data assimilation cycles often use a frequency of one day or even one week. Consequently, it may be necessary to either modify the data assimilation systems to use the same frequency, or develop another strategy to combine various systems.

While a weakly coupled data assimilation system is more complex than an uncoupled system, there is still no guarantee that fields will be consistent between the components. The observations assimilated in each component may provide inconsistent information resulting in physical inconsistencies between them. For example, the sea ice assimilation system may have assimilated observations that increase the ice concentration in an area directly adjacent to an ice-free area where the ocean assimilation system assimilated observations that increase the sea surface temperature. Due to the horizontal spreading of the information from the observations by the background error covariances, the likely result would be an area with both increased sea ice concentration and warmer ocean surface temperatures in the analysis states of the two components. When used to initialize the coupled model forecast for producing the subsequent background state, this could cause rapid ice melting and increased surface stratification due to changes in salinity, possibly affecting mixed layer properties. While such a situation could occur in reality, it would more likely lead to forecast degradation. The effect on ocean salinity could be particularly damaging since currently available surface salinity observations are inadequate to constrain this type of error.

4.5.2 Strongly Coupled Data Assimilation

The sea ice, ocean and atmosphere exhibit a strong inherent physical coupling. Opening and closing of leads and polynyas have large non-linear effects on fluxes of heat, moisture and momentum. The presence of sea ice and its characteristics (e.g. melt ponds, ridges, ice strength, snow cover) have direct impacts on other components. As a result, observations from one component provide valuable information about conditions in other components.

In strongly coupled data assimilation, the assimilation step is performed as a single combined procedure for all components, significantly increasing the complexity of the system. This coupling allows exploitation of the strong interconnection between the ice, ocean and atmosphere in a data assimilation scheme through the cross-covariances of the background error between the variables in different component models. An obvious example is the covariance between ice concentration and sea surface temperature, which is typically negative. A coupled data assimilation strategy would operate on the joint state vector formed by concatenating the state vectors of each individual component system. Then, the combined set of observations from all component systems would be assimilated to produce the analysis state for the entire coupled system. For example, ice concentration observations would influence the resulting sea surface temperatures, and sea surface temperature observations would influence nearby sea ice concentration. However, any benefit from this approach relies on the accurate specification of the cross-covariances of the background error between the different components. These cross-covariances can be obtained from ensembles of coupled model forecasts within an ensemble data assimilation approach, such as the EnKF. Alternatively, cross-covariances could be obtained implicitly through the forecast and adjoint models of the coupled system within a 4DVar data assimilation system. In practice, the development of such coupled adjoint models represents a number of technical challenges. In a coupled data assimilation approach, the observations of sea ice could benefit the atmospheric and ocean state and vice versa. In principle, this should allow for a more optimal use of available observations and result in a more coherent and balanced coupled analysis state providing greater forecast skill.

Strongly coupled ice–ocean data assimilation has been successfully implemented in the past. This includes theoretical studies and operational implementations using both ensemble and variational data assimilation approaches (e.g. Sakov *et al.*, 2012; Massonnet *et al.*, 2013; Fenty and Heimbach, 2013a, b). However, strongly coupled ice–ocean–atmosphere data assimilation remains a topic of active research at the present time. Idealized studies have demonstrated its potential benefits and full-scale applications are expected over the coming decade. Some initial examples of coupled data assimilation using simplified approaches have been shown to be effective. For example, Toyoda *et al.* (2016) used a simple assumption about the relationship between ice concentration and both the near-surface atmospheric and oceanic temperatures to correct these temperatures as a result of ice concentration assimilation. Similarly, Barth *et al.* (2015) used a simple regression between ice drift and near-surface winds to obtain a correction to the winds derived from the assimilation of ice drift observations.

4.6 Chapter Summary

This chapter introduced the concept of using data assimilation to correct a background state, usually obtained from a short-term forecast, by extracting the useful information from a large number of diverse observations of the physical system of interest. Numerous techniques have been developed for this purpose, often originally developed for and applied to atmospheric and oceanographic prediction. The application of data assimilation to sea ice prediction is still relatively new and involves unique challenges, several of which were discussed.

Examples were mentioned in which the 4DVar data assimilation technique has been successfully applied to sea ice prediction. However, the challenge of developing and maintaining an adjoint version of complex and highly non-linear sea ice models limits their application. Instead, ensemble-based methods are becoming increasingly used for sea ice prediction, as they are for operational weather and ocean prediction. These methods require significantly more computational resources than simpler techniques, such as 3DVar, but are relatively simple to implement and can provide estimates of the complex spatial and multivariate statistical relationships between physically related analysis state variables. Using these statistical relationships when assimilating observations should lead to improved accuracy. Ensemble-based approaches are also more straightforward to implement in the context of strongly coupled data assimilation for coupled prediction, which is increasingly becoming a focus for the development of automated prediction systems.

References

Barth, A., Canter, M. and Van Schaeybroeck, B. (2015). Assimilation of sea surface temperature, sea ice concentration and sea ice drift in a model of the Southern Ocean. *Ocean Modelling*, **93**, 22–39. doi: 10.1016/j.ocemod.2015.07.011

Bertino, L., Evensen, G. and Wackernagel, H. (2003). Sequential data assimilation techniques in oceanography. *International Statistical Review*, **71**, 223–241. doi: 10.1111/j.1751-5823.2003.tb00194.x

Borup, M., Grum, M., Madsen, H. and Mikkelsen, P. S. (2015). A partial ensemble Kalman filtering approach to enable use of range limited observations. *Stochastic Environmental Research and Risk Assessment*, **29**, 119–129. doi: 10.1007/s00477-014-0908-1

Buehner, M., Caya, A., Carrieres, T. and Pogson, L. (2014). Assimilation of SSMIS and ASCAT data and the replacement of highly uncertain estimates in the Environment Canada Regional Ice Prediction System. *Quarterly Journal of the Royal Meteorological Society*, **142**, 562–573. doi: 10.1002/qj.2408

Buehner, M. and Shlyaeva, A. (2015). Scale-dependent background-error covariance localisation. *Tellus A*, **67**, 28027. doi: 10.3402/tellusa.v67.28027

Fenty, I. and Heimbach, P. (2013a). Hydrographic preconditioning for seasonal sea ice anomalies in the Labrador Sea. *Journal of Physical Oceanography*, **43**, 863–883. doi: 10.1175/JPO-D-12-064.1

Fenty, I. and Heimbach, P. (2013b). Coupled sea ice–ocean-state estimation in the Labrador Sea and Baffin Bay. *Journal of Physical Oceanography*, **43**, 884–904. doi: 10.1175/JPO-D-12-065.1

Heimbach, P., Menemenlis, D., Losch, M., Campin, J.-M. and Hill, C. (2010). On the formulation of sea-ice models. Part 2: Lessons from multi-year adjoint sea-ice export sensitivities through the Canadian Arctic Archipelago. *Ocean Modelling*, **33**, 145–158. doi: 10.1016/j.ocemod.2010.02.002

Helfrich, S. R., McNamara, D., Ramsay, B. H., Baldwin, T. and Kasheta, T. (2007). Enhancements to, and forthcoming developments in the Interactive Multisensor Snow and Ice Mapping System (IMS). *Hydrological Processes*, **21**, 1576–1586. doi: 10.1002/hyp.6720

Kauker, F., Kaminski, T., Karcher, M., Giering, R., Gerdes, R. and Voßbeck, M. (2009). Adjoint analysis of the 2007 all time Arctic sea-ice minimum. *Geophysical Research Letters*, **36**(3), L03707. doi:10.1029/2008GL036323

Lahoz, W. A., Khattatov, B. and Menard, R. (2010). *Data Assimilation: Making Sense of Observations*. Germany: Springer-Verlag Berlin Heidelberg. doi: 10.1007/978-3-540-74703-1

Leigh, S., Wang, Z. and Clausi, D. A. (2014). Automated ice-water classification using dual polarization SAR satellite imagery, *IEEE Transactions on Geoscience and Remote Sensing*, **52**, 5529–5539. doi: 10.1109/TGRS.2013.2290231

Lisæter, K. A., Evensen, G. and Laxon, S. (2007). Assimilating synthetic CryoSat sea ice thickness in a coupled ice-ocean model, *Journal of Geophysical Research Oceans*, **112**, C07023. doi: 10.1029/2006JC003786

Massonnet, F., Goosse, H., Fichefet, T. and Counillon, F. (2014). Calibration of sea ice dynamic parameters in an ocean-sea ice model using an ensemble Kalman filter. *Journal of Geophysical Research Oceans*, **119**, 4168–4184. doi: 10.1002/2013JC009705

Massonnet, F., Mathiot, P., Fichefet, T., Goosse, H., König Beatty, C., Vancoppenolle, M., and Lavergne, T. (2013). A model reconstruction of the Antarctic sea ice thickness and volume changes over 1980–2008 using data assimilation. *Ocean Modelling*, **64**, 67–75. doi: 10.1016/j.ocemod.2013.01.003

Sakov, P., Counillon, F., Bertino, L., Lisæter, K. A., Oke, P. R., and Korablev, A. (2012). TOPAZ4: an ocean-sea ice data assimilation system for the North Atlantic and Arctic. *Ocean Science*, **8**, 633–656. doi: 10.5194/os-8-633-2012

Scott, K. A., Ashouri, Z., Buehner, M., Pogson, L. and Carrieres, T. (2015). Assimilation of ice and water observations from SAR imagery to improve estimates of sea ice concentration. *Tellus A*, **67**, 27218. doi: 10.3402/tellusa.v67.27218

Shlyaeva, A., Buehner, M., Caya, A., Lemieux, J.-F., Smith, G. C., Roy, F., Dupont, F. and Carrieres, T. (2016). Towards ensemble data assimilation for the Environment Canada Regional Ice Prediction System. *Quarterly Journal of the Royal Meteorological Society*, **142**, 1090–1099. doi: 10.1002/qj.2712

Smith, G. C., Roy, F., Reszka, M., Surcel Colan, D., He, Z., Deacu, D., Belanger, J.-M., Skachko, S., Liu, Y., Dupont, F., Lemieux, J.-F., Beaudoin, C., Tranchant, B., Drévillon, M., Garric, G., Testut, C.-E., Lellouche, J.-M., Pellerin, P., Ritchie, H., Lu, Y., Davidson, F., Buehner, M., Caya, A. and Lajoie, M. (2015). Sea ice forecast verification in the Canadian Global Ice Ocean Prediction System. *Quarterly Journal of the Royal Meteorological Society*, **142**, 659–671. doi: 10.1002/qj.2555

Stammer, D., Wunsch, C., Giering, R., Eckert, C., Heimbach, P., Marotzke, J., Adcroft, A., Hill, C. N. and Marshall, J. (2002). Global ocean circulation during 1992–1997, estimated from ocean observations and a general circulation model. *Journal of Geophysical Research*, **107**(C9), 3118. doi: 10.1029/2001JC000888

Stark, J. D., Ridley, J., Martin, M. and Hines, A. (2008). Sea ice concentration and motion assimilation in a sea iceocean model. *Journal of Geophysical Research*, **113**, C05S91. doi: 10.1029/2007JC004224

Tippett, M. K., Anderson, J. L., Bishop, C. H., Hamill, T. M. and Whitaker, J. S. (2003). Ensemble square root filters. *Monthly Weather Review*, **131**, 1485–1490. doi: 10.1175/1520-0493(2003)131<1485:ESRF>2.0.CO;2

Toyoda, T., Fujii, Y., Yasuda, T., Usui, N., Ogawa, K., Kuragano, T., Tsujino, H. and Kamachi, M. (2016). Data assimilation of sea ice concentration into a global ocean–sea ice model with corrections for atmospheric forcing and ocean temperature fields. *Journal of Oceanography*, **72**, 235–262. doi: 10.1007/s10872-015-0326-0

5

Automated Sea Ice Prediction Systems

Tom Carrieres, Alain Caya, Pam Posey, E. Joseph Metzger, Laurent Bertino, Arne Melsom, Greg Smith, Michael Sigmond, Viatcheslav Kharin, Adrienne Tivy

Various centres around the world are developing their own sea ice automated prediction system (APS). Some of these are truly operational and are directly tied to a National Ice Information Service (NIIS) while others are used more as research tools. Similarly, the sophistication of numerical guidance provided and use of that guidance varies from centre to centre. This chapter provides a description of several systems being actively developed. The intention is to provide a few representative examples, not an exhaustive list.

A number of points should be noted when reviewing these systems. Coupling between models of each Earth system component varies from systems with two-way fully coupled atmosphere, ice and ocean components to those for which the atmospheric and ocean forcing are supplied as external forcing to the ice model from uncoupled forecasts. All of these operational systems include sea ice model components that are available as separate packages. The most commonly used sea ice models are those developed at Los Alamos (CICE) and Louvain-la-Neuve (LLN). These models were not specifically designed for producing short-term, operational ice forecasts, and therefore research is currently underway to identify and remedy related shortcomings.

The Nucleus for European Modelling of the Ocean-Océan Parallélisé (NEMO-OPA) and HYbrid Coordinate Ocean Model (HYCOM, Metzger *et al.*, 2014a) are two of the most common ocean models used in these systems. Atmospheric forcing or two-way coupling is mainly provided through national environmental prediction systems that are usually best suited for the corresponding NIIS primary area of interest. In practice, the duration (or lead time) of the sea ice forecast is usually limited by the numerical weather prediction (NWP) forecast.

The use of two-way coupling between component models can significantly influence the accuracy of the forecasts produced by these systems. Documentation on what and how often information is exchanged between components is important to understand system behaviour. It is also important to recognize that apparently minor differences between the systems, especially surface properties in ice covered regions, can have a large impact on the resulting sea ice analyses and forecasts.

Model initialization is also quite varied between these systems. In some systems gridded retrievals of sea ice concentration prepared elsewhere are directly used to initialize the forecast model, while others use a data assimilation method to combine information from numerous types of observations with varying accuracy, spatial resolution and coverage.

It should be noted that sea ice thickness observations are not normally assimilated, and therefore the forecast model is generally allowed to evolve the ice thickness without constraint. In any case, choices are required to ensure consistencies in ice fields such as concentration and thickness. For systems with two-way coupling, consistencies are also required between components, for example ice cover and sea surface temperature (SST).

5.1 Regional Ice Prediction System – Environment and Climate Change Canada

The Regional Ice Prediction System (RIPS) has been designed to provide operational numerical guidance for sea ice analysis and forecasting services as well as to provide lower boundary surface conditions for the Canadian Centre for Marine and Environmental Prediction (CCMEP) NWP models. Early development of the system began in 2002, with the first operational release in March 2011. Operational implementation updates take place once or twice per year to incorporate new or revised observation sources, to upgrade the data assimilation system and to upgrade the forecast model. RIPS provides an analysis and 0- to 48-hour forecasts every six hours. The RIPS domain is illustrated in Figure 5.1.

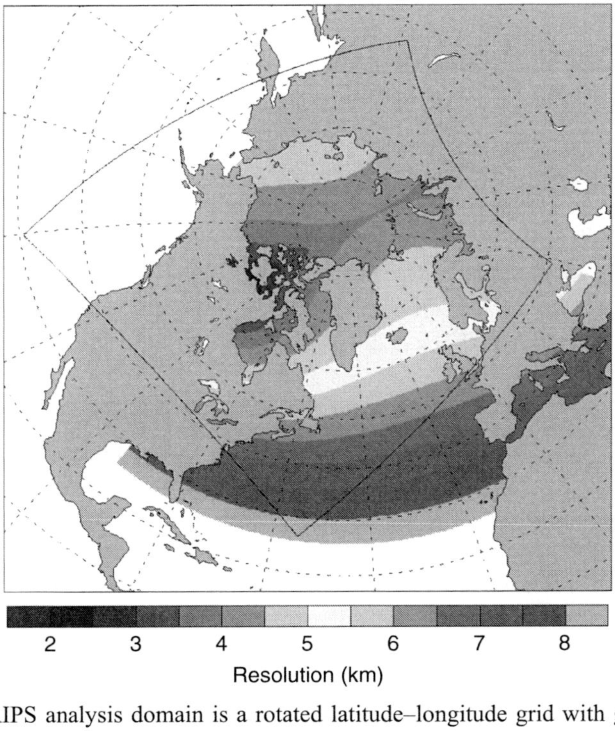

Figure 5.1: The RIPS analysis domain is a rotated latitude–longitude grid with grid points evenly spaced every 0.045° (area enclosed by black line). The colour contours indicate the average grid spacing of the RIPS forecast model grid which is a tri-polar grid with 12 grid points evenly spaced every degree. Note that most of the larger inland lakes are included in the analysis but not in the forecast system. (A black and white version of this figure will appear in some formats. For the colour version, please refer to the plate section.)

The analysis grid is on a rotated, nearly uniform 5 km resolution latitude–longitude projection while the forecast grid is on a tri-polar grid with 1/12° resolution (ORCA12).

5.1.1 System Description: Data Assimilation

The 3DVar (three-dimensional variational) data assimilation technique is used to assimilate a variety of types of ice concentration observations to compute corrections to the ice concentration background state, which is the analysis from six hours earlier. The background-error covariance matrix is assumed to be stationary in time and horizontally homogeneous, using a horizontal diffusion operator to model the spatial error correlations. The background-error standard deviation is set to 0.093 and the background-error horizontal correlation length scale is set to 10 km. Ice concentrations are modified after the data assimilation procedure to ensure they remain within the range of 0 to 100 per cent. A further quality control measure sets all ice concentrations to zero when the SST is greater than 4°C.

5.1.1.1 Observations

All observations with a valid time within 3 hours of the analysis valid time are assimilated. However, the data assimilation procedure is started 4.5 hours after the analysis time and therefore observations received after this time are not used. In general, application of the observation operator involves a spatial interpolation of the gridded state to each observation location. Some observation types are related to the average ice concentration over an area larger than the analysis grid cell. For these observations, a footprint operator is applied whereby the average concentration from all grid cells falling within a specified radius is computed instead of using spatial interpolation (discussed in Section 4.4.5). Observations are removed if any of these grid cells involved in the observation operator are identified as land. Details of the empirically derived observation errors and the assimilation method are described in Buehner *et al.* (2014). An impression of the relative impact of each observation type is given by the ratio of the background error variance in observation space ($\mathbf{HBH}^{\mathrm{T}}$) to the sum of the background and observation error variance ($\mathbf{HBH}^{\mathrm{T}} + \mathbf{R}$), since this equals the Kalman gain matrix in observation space for one observation (see Equation 4.6). Each observation set undergoes a gross error check to ensure the data has not been corrupted. Whenever the RMS (root mean square) value of the difference between the observations and the background state of an observation set is above the values specified in Table 5.1, the entire observation set is rejected. The different types of assimilated observations are described in the following list.

a) Canadian Ice Service (CIS) Ice Charts
   Total ice concentration is extracted on a 5 km grid from three different types of manually prepared charts.
   1) RADARSAT SAR (synthetic aperture radar) image analysis charts: these represent a snapshot of ice conditions at the valid time of the satellite overpass. These are considered the highest quality observations with relatively uniform accuracy across

Table 5.1: *Parameters used for the assimilation of each type of observation.* $\boldsymbol{HBH}^T$, *the background-error standard deviation projected into observation space, provides an impression of the relative impact of each observation type on the analysis. The observation error standard deviation is the estimated uncertainty in each observation. The RMS threshold is used in a quality control procedure to identify large errors in an observation set.*

Observation type	Horizontal interpolation	Footprint size/radius	$\mathbf{HBH}^\mathbf{T}$	Observation error standard deviation	RMS threshold
AMSR2	Footprint	11.0 km	0.071	0.2	0.6
SSM/I	Footprint	27.5 km	0.040	0.18	0.6
SSMIS	Footprint	29.0 km	0.038	0.05	0.6
ASCAT	Footprint	25.0 km	0.044	0.03[1]	Not applied
CIS Image Analysis Charts	Bilinear	–	0.087	0.1	0.7
CIS Daily Ice Charts	Bilinear	–	0.087	0.1	0.38
CIS Regional Ice Charts	Bilinear	–	0.086	0.15	0.6
CIS Lake Ice Data	Spatial average over lake	Lake or portion thereof	Size dependent	0.01	0.95

[1] This is the value for open water retrievals. For the anisotropy, the observation-error standard deviation is incidence angle- and season-dependent.

the image swath but they are only available in areas with marine activity and only when imagery is acquired.

2) Daily ice charts: these represent a nowcast of conditions at 1800 UTC incorporating any available and relevant information, including model forecasts, SAR images and image analysis charts. Accuracy of these products vary according to the information available in any given area and they are only available in regions that are operationally active.
3) Regional ice charts: these are prepared once per week for climate monitoring purposes. As such, they are prepared several days after the valid time, incorporating information available before and after the valid time and covering a much larger area than the other chart types. Because of this time delay, they cannot be assimilated at the analysis time closest to the valid time of the chart. Instead, the regional ice charts are assimilated at the analysis time following when the chart becomes available. The main advantage of these charts is that they provide information in areas that are not resolved by other observations. To minimize the errors that could be introduced due to the time delay, they are only assimilated where the estimated analysis uncertainty in the background state is high.

b) CIS Lake Ice Analyses

Mainly in support of numerical weather prediction, CIS prepares average ice concentration analyses for small lakes and lake areas in North America on a weekly basis. Information used to prepare these analyses is similar to that used for the daily ice charts. These are the only lake ice observations used for all but the Great Lakes and they are only produced once per week. To compensate for this low frequency and to ensure there is a reasonable impact on the analysis, the observation error standard deviation is set quite low.

c) Passive Microwave Satellite Data

Swaths of brightness temperatures observed by the DMSP (Defense Meteorological Satellite Program), F-15 SSM/I (Special Sensor Microwave/Imager), DMSP F-16, 17 and 18 SSMIS (Special Sensor Microwave Imager/Sounder) and AMSR2 (Advanced Microwave Scanning Radiometer2) instruments are used to retrieve ice concentration using the NASA (National Atmospheric and Space Administration) Team 2 (NT2) algorithm. The footprint operator size corresponds to the footprint of the coarsest channel used in NT2. Observations are removed over freshwater. To eliminate spurious retrievals, a number of quality control conditions are employed using data from the CCMEP NWP system: 1) to avoid the effects of wet snow and melt water, observations are removed when the near surface air temperature is 0°C or above; 2) potentially weather contaminated non-zero concentration observations are removed when the SST is 4°C or above; 3) observations are removed for climatological ice free areas; and 4) to avoid bias in ice concentration arising from wind roughened ocean, observations are removed when wind speeds are greater than 46 km h^{-1}.

d) Scatterometer Satellite Data

The anisotropy of the backscatter returns from the three look directions detected by the Advanced Scatterometer (ASCAT) instrument on the Meteorological Operational

(METOP) satellite is used to distinguish ice from water. The assumption is that the signal returned from ice is isotropic due to volume scattering and backscatter from randomly oriented features, while open water return is anisotropic due to the orientation of capillary waves in appropriate wind conditions. ASCAT data is assimilated in two ways: (1) 0 per cent concentration is specified in conditions of high anisotropy when the included European Centre for Medium Range Weather Forecasting (ECMWF) forecast wind speed is greater than 4 m s^{-1} and the observation quality has been flagged by the instrument as good; and, (2) using a linear observation operator that relates ice concentration to the backscatter anisotropy using tie-points values for open water and 100 per cent ice.

5.1.2 System Description: Forecast Model

The forecast model is a tailored implementation of CICE version 4.0 with the following options:

- ice strength is calculated using the Rothrock (1975) formulation;
- 10 ice thickness categories are used;
- one snow and four ice layers per ice category are used for thermodynamic calculations; and
- the Community Climate System Model 3 (CCSM3) scheme is used to calculate the albedo and the attenuation of the absorbed shortwave radiation.

Apart from the model parameters listed in Table 5.2, RIPS uses all the default parameters of CICEv4.0. Note that the surface ice roughness parameter and ice–ocean drag coefficient were adjusted in order to decrease the errors between the simulated sea ice drift and the observed drift from the International Arctic Buoy Programme dataset. Moreover, to reduce a positive bias in the simulated sea ice concentration (during the growth season), the specified thickness of the frazil ice is set to 8 cm (instead of default 5 cm in CICEv4.0).

Table 5.2: *Parameters used for the CICE implementation within RIPS 2.2*

Parameter	Value
timestep	1200 s
elastic damping timescale	432 s
number of subcycling steps	300
Ranges for the thickness categories	(0–10, 10–15, 15–30, 30–50, 50–70, 70–120, 120–200, 200–400, 400–600, ≥600 cm)
surface ice roughness parameter	1.6×10^{-4} m
ice–ocean drag coefficient	0.023
thickness of frazil ice	8 cm
ice salinity	0 psu

Atmospheric forcing is obtained from the 10 km resolution atmospheric forecasts from the Environment and Climate Change Canada (ECCC) Regional Deterministic Prediction System (RDPS) blended with the 25 km Global Deterministic Prediction System (GDPS) at the RDPS lateral boundaries. The atmospheric forcing fields are:

- winds at approximately 40 m above the surface,
- surface air temperature, humidity and precipitation and
- downward longwave and shortwave fluxes.

For initialization, the ice concentration estimated by the 3DVar is combined with the snow cover, ice thickness distribution and ice velocity from the previous short-term forecast. To ensure the system does not drift from realistic conditions, the Global Ice Ocean Prediction System (GIOPS, described later) provides these non-analysed fields every Wednesday for the 0000 UTC run. For grid cells with non-zero ice concentration and zero ice thickness in the previous forecasts, the thickness is set to 25 cm. Ice temperature is initially given a linear profile with the freezing temperature at the lower surface and the surface air temperature at the upper surface.

The ocean component of RIPS consists of a mixed-layer ocean model with GIOPS providing the initial ocean temperature, salinity and mixed-layer depth. The salinity and mixed layer depth are held constant during a forecast while the ocean temperature evolves. GIOPS also provides surface ocean currents as dynamic forcing for the ice model but these are not used for lateral ocean heat advection.

Results given in this book come from version 2.2 of RIPS. Note, however, that RIPS was recently replaced by the Regional Ice Ocean Prediction System (RIOPS). RIOPS includes an improved 3DVar sea ice concentration analysis by also assimilating high-resolution AVHRR data. Significant modifications to the forecast component were also done. The simple mixed-layer ocean model of RIPS was replaced by a 3-D ocean model (NEMO-OPA). Another major improvement relative to RIPS is the addition of explicit tides in RIOPS. RIOPS uses a spectral nudging approach to constraint ocean fields toward GIOPS in place of a true ocean data assimilation component. Finally, RIOPS includes a new grounding scheme and increased tensile strength for a better representation of land-fast ice.

5.1.3 Plans for Future Development

RIPS/RIOPS will continue to evolve on a variety of fronts including the addition of new types of observations and analysis variables, improved data assimilation techniques and improvements to the forecast model. The following developments are in progress and should be available within the next five years or so.

Satellite-borne SAR represents the backbone observation source for many NIIS. For data assimilation purposes, algorithms must be robust and truly automatic but it is likely better to reject data in areas that are problematic than to assimilate observations that can degrade the quality of the resulting estimates. One such algorithm that is near completion is a simple retrieval of the presence of ice and open water that only provides information in specific incidence angle and wind speed ranges.

While the RIPS ice concentration analysis feeds into the forecast model, the model does not yet provide the background state for the data assimilation system. This cycling process requires that the model short-term forecasts have nearly equivalent skill as compared with persistence. However, introducing cycling would allow the model to correct for some deficiencies in observations.

5.2 Global Ice Ocean Prediction System – Environment and Climate Change Canada

This section describes an ice–ocean analysis and forecasting system recently implemented operationally at the CCMEP called GIOPS. GIOPS was developed in response to a growing need in Canada for a multipurpose global marine core service. Such a system is needed to provide real-time information on the state of the ice–ocean system and its potential evolution over the medium-range timescale (up to 10-day forecasts). This information is needed to support fisheries and aquaculture management, marine operational needs (e.g. by the Canadian Coast Guard), marine emergency response and to provide guidance to the CIS. Additionally, this system was developed to provide an evolving marine boundary condition for coupling to numerical weather prediction systems. Here, we provide a description of the system and future directions.

GIOPS is built on four core components: the System d'Assimilation Mercator version 2 (SAM2) ocean data assimilation system, a 3DVar sea ice data assimilation system, NEMO-OPA ocean model and the CICE sea ice model. The 3DVar sea ice assimilation system is similar to that described for RIPS. SAM2 is a multivariate Singular Evolutive Extended Kalman (SEEK) filter that assimilates satellite altimetry and sea surface temperature as well as in situ temperature and salinity observations.

5.2.1 System Description: Forecast Model

The GIOPS model configuration was designed to provide a realistic representation of key physical processes responsible for the evolution of the state of the ice–ocean system. NEMO is actually a modelling framework, based on the OPA ocean model. OPA is a primitive equation z-level model making use of the hydrostatic and Boussinesq approximations. The current version has 50 levels in the vertical, with a spacing ranging from 1 m at the surface to 500 m in the deep ocean. NEMO exchanges momentum, heat and freshwater (salt) with the CICE sea ice model at every model time step (i.e. every 10 mins).

The NEMO-CICE coupled model is run on a 1/4° resolution global tri-polar grid called ORCA025, similar to that used for RIPS but at lower resolution. This provides an eddy-permitting resolution over most of the world's oceans with sufficient resolution to capture large-scale basin features (e.g. gyres, mode water formation) and to permit mesoscale ocean features (eddies and fronts). The resolution also allows realistic coastal currents and the exchange of water masses between subpolar and polar ocean basins, thereby affecting thermodynamic sea ice growth and the export of ice out of the Arctic.

The evaluation of GIOPS has focused on the quality of analyses and forecasts of surface fields, in particular sea surface temperature and sea ice concentration. Sea level anomalies and in situ temperature and salinity fields have also been verified as a proxy for ocean currents, which are not well observed in the ocean. These fields are also of relevance for ensuring accurate ocean heat content estimates for seasonal forecasting and for fisheries and national defence applications. Numerous multi-year 'forced' simulations have also been made to thoroughly evaluate and optimize model physics and reduce biases.

This model has been developed for a variety of applications ranging from climate simulations to high-resolution operational oceanography. Physical parameterizations have been developed to provide the most realistic representation of processes in the ocean. In particular, the conservation of energy, enstrophy, heat, mass and momentum have been the focus of past efforts. Additionally, a number of studies have focused on the representation of exchanges with the atmosphere and the evolution of the turbulent surface mixed layer.

Atmospheric forcing for NEMO-CICE is taken from the bottom level of CCMEP GDPS forecasts. The fields required are wind, air temperature, specific humidity, incident downward shortwave and longwave radiation fluxes, accumulated precipitation and surface atmospheric pressure. Fluxes are calculated using the Common Ocean Reference Experiment (CORE) bulk formula adapted for application at the height of the bottom GDPS model level. The surface roughness of ice is set to the same value used by the atmospheric forecast model for consistency (1.6×10^{-4} m).

5.2.2 System Description: Data Assimilation

GIOPS uses the operational global ice concentration analysis produced at CCMEP. This analysis is used as the surface boundary condition for most NWP applications at CCMEP. The analysis system is closely related to the 3D Var data assimilation system used in RIPS with two main differences: a lower resolution 10 km global grid is used, and fewer observations are assimilated. The global 3DVar analysis system only assimilates SSMI, SSMI/S and CIS products (except Regional Charts) using the same parameters as in Table 5.1. Innovations implemented in RIPS will soon be ported to the global system.

SAM2 has been developed with different NEMO configurations for the regional and global oceans by Mercator-Océan. It is intended to be a flexible tool for performing oceanic analyses and forecasts over a wide range of space and time scales. It is able to assimilate various data types such as satellite (2-D) and in situ data (3-D). The data assimilation technique is based on SEEK, a Reduced Order Kalman Filter.

SAM2 assimilates three types of observations: SST, sea level anomaly and vertical profiles (e.g. Argo). The assimilated SST observations consist of the CCMEP gridded SST analysis product at 0.2° resolution. Prior to assimilation, the CCMEP SST product is

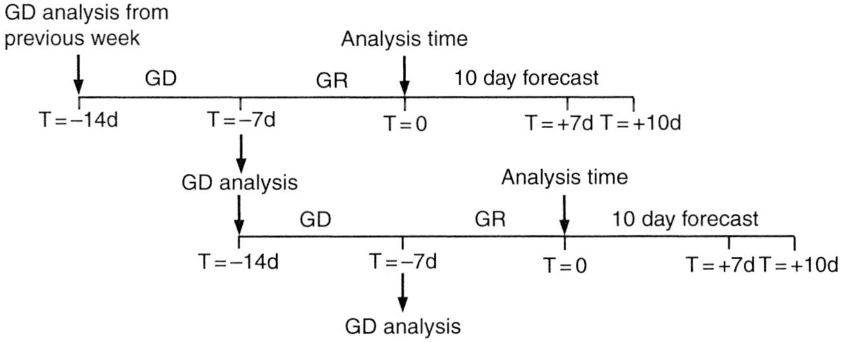

Figure 5.2: Schematic showing the functioning of the GIOPS delayed (GD) and real-time (GR) analyses. All analyses are valid at 0000 UTC. The GD is produced on Tuesdays and the GR on Wednesdays. GR analyses for other days are produced using a one-day assimilation window and assimilating SST only.

interpolated onto the ORCA025 grid. Observation measurement error is set to 0.2°C. The sea level anomaly observations are taken from the SSALTO/DUACS AVISO Near-Real-Time Along-Track product. As these observations have been filtered to remove effects such as tides and inverse barometer, a similar filtering is applied to the model sea surface height as part of the observation operator. Three satellite altimeters are being used: Jason2, Cryosat2 and Saral/Altika. Observation measurement errors for all three instruments are set to 2 cm. The vertical profile observations are obtained from Collecte Localisation Spatiale (CLS). At CLS, all available in situ temperature and salinity observations are taken from the Coriolis data centre and additional quality control, thinning and reformatting is applied.

In order to provide the best quality daily analysis given the inherent delays in receiving the observations, GIOPS has been implemented using a three-tier approach (Figure 5.2). The backbone of the system which provides the continuity in time is provided by the weekly GIOPS delayed (GD) model analysis produced each week on Tuesdays, valid for the previous Wednesday at 0000 UTC. The GD analysis is used to initialize the GIOPS real-time weekly analysis (GR), produced on Wednesdays and valid that day at 0000 UTC. A GR daily update is then run from Thursday to Tuesday to provide a daily analysis.

The GD and GR weekly analyses use all three data types with an assimilation window of seven days and produce an analysis valid at the end of the window. SST observations are only assimilated for day 7 of the window. The choice of seven days was made as it allows sufficient observations to provide a statistically optimal analysis while also taking into account the roughly seven-day delay to accumulate 90 per cent of the in situ vertical profile observations (see Figure 5.2). The GR-daily analyses assimilate only SST observations, as sea level anomaly and vertical profile data are not available. As such, the quality of the in-situ and sea surface height fields in GIOPS analyses are expected to have the lowest quality on Tuesdays prior to the GD and GR weekly update that occurs on Wednesdays.

As discussed in Section 4.5, it is extremely important to consider interactions between the ice and ocean data assimilation approaches to ensure as much consistency between analyses as possible. Accordingly, the 3DVar ice concentration fields are used to modify the gridded SST product that is assimilated into GIOPS. Moreover, this SST product itself uses the 3DVar ice analysis to create pseudo-observations of freezing point SST under the ice. While this improves the coherence between the three analysis systems, some degree of inconsistency remains.

5.2.3 Plans for Future Development

Overall, GIOPS sea ice evaluations demonstrate a consistent picture of skilful medium-range forecasts in both the Northern and Southern Hemispheres as compared to persistence. While these results support the use of GIOPS forecasts for coupled environmental prediction and marine applications, they should nonetheless be viewed as a baseline for further improvement. A particular weakness of GIOPS is the use of three separate analysis systems for the ocean, ice concentration and SST. Coupling these systems together should lead to better consistency and the potential to further exploit ice–ocean covariances to improve GIOPS initial conditions, and as a result, sea ice forecasts. The use of additional data sources to constrain GIOPS analyses should also improve forecasts, in particular, sea ice thickness, drift and water mass properties near and under the ice.

The GIOPS sea ice forecasts also show greater error growth as compared to trial fields. This demonstrates the important role of atmospheric forecasts in terms of influencing the evolution of the sea ice state. As such, improved sea ice forecasts should be possible by improving atmospheric fields used to force them. This could be possible by using coupled ice–ocean–atmospheric models to better represent the complex and non-linear interactions across the marine interface. Additionally, ensemble atmospheric forecasts (either coupled or uncoupled) could be used to provide an estimate of uncertainty related to the atmospheric forecasts. Current plans for GIOPS include both coupling to the atmosphere and an extension to probabilistic forecasting.

5.3 TOPAZ4 – Arctic Marine Forecasting Centre – European Union

The Arctic Monitoring and Forecasting Center (MFC) of the Copernicus Marine Environment Monitoring Service (CMEMS) is responsible for providing services above 62° N. The latest version of the prediction system is called 'Towards an Operational Prediction system for the North Atlantic European coastal Zones' version 4 (TOPAZ4). The global MFC in CMEMS employs the Mercator-Océan NEMO-LIM system. Copernicus is a European Programme for Earth Observation and Monitoring. Copernicus also has a substantial space component (the Sentinel series of satellites) which provides space based observations to the services. The TOPAZ4 domain is shown in Figure 5.3.

The developments of TOPAZ started in 2000 and TOPAZ has been running in near-real-time forecasting mode since 2003. From the first version of the system, ocean and sea ice remote sensing data were assimilated with a local version of the Ensemble Kalman Filter (EnKF) and a coupled ice–ocean model based on the HYCOM code. In 2008, the

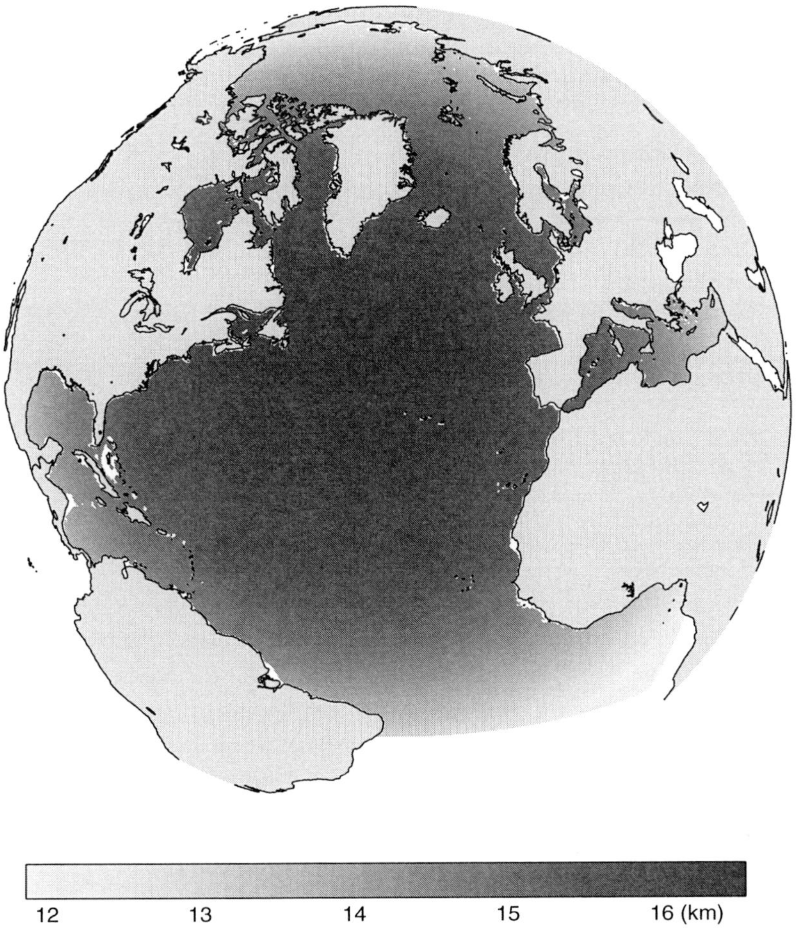

Figure 5.3: Map of the TOPAZ4 domain with grid resolution in km. (A black and white version of this figure will appear in some formats. For the colour version, please refer to the plate section.)

operational system was transferred to the Norwegian Meteorological Institute, where it became part of the operational suite. In 2012 TOPAZ was coupled on-line to the ecosystem model NORWECOM, developed by the Institute of Marine Research and the University of Bergen. A reanalysis was delivered both for the physics (Sakov *et al.*, 2012) and the biology.

5.3.1 System Description: Forecast Model

The TOPAZ4 system presently uses HYCOM 2.2.37 and is described in Sakov *et al.* (2012). It uses 28 hybrid z-isopycnal layers, and the top layer has a thickness of 3 m. The model grid has a horizontal resolution of 12–16 km, which is eddy permitting from the Equator to the Nordic Seas but is still far from being eddy-resolving in the Arctic. The lateral boundaries of

Table 5.3: *Parameters used for the sea ice model within TOPAZ4.*

Parameter	Value
Time step	7200 s
Number of subcycling steps	120
Ice thermodynamics	Zero-layer (Semtner II 1976)
Sea ice tracers advection scheme	WENO3
Sea ice strength	27.5 kN m^{-2}
Ratio of air–ice to ice–ocean drag coefficients	0.29
Maximum fractional ice cover	99.5%
Thickness of newly formed ice	10 cm
Sea ice salinity	6 psu

temperature and salinity are relaxed to a combination of the World Ocean Atlas of 2005 and version 3.0 of the Polar Science Center Hydrographic Climatology. The ocean bottom topography is based on the General Bathymetric Chart of the Oceans database at 1-min resolution. HYCOM is coupled to a sea ice model that employs the elastic–viscous–plastic (EVP) rheology and sea ice thermodynamics as described in Drange and Simonsen (1996). The surface momentum fluxes use a bulk formula parameterization (Kara *et al.*, 2000), and the related thermodynamic fluxes are computed as described in Drange and Simonsen (1996) but the cloud cover fields are updated every three hours in the computation of the shortwave radiation to better represent the diurnal cycle. The CICE model is implemented as a subroutine in HYCOM rather than through a coupler and it is running at the same horizontal resolution. A summary of the model parameters is given in Table 5.3. The following atmospheric fields are taken from the ECMWF operational forecasts and analyses at six-hour frequency:

- 10-metres winds,
- 2-metres temperature and dew point temperature,
- total cloud cover and
- precipitation.

A monthly climatological river discharge is estimated by applying the run-off estimates (Oki and Sud, 1998) from the ECMWF Reanalysis (ERA-interim) over the 20-year period 1989–2009.

The model has been initialized from the same climatology data as used at the boundaries. It was then spun up from 1973 onwards using the ERA-Interim data. The Pacific water inflow is imposed by a barotropic inflow of 0.7 Sv through the Bering Strait at the model boundary and balanced by an outflow at the southern boundary of the domain.

5.3.2 System Description: Data Assimilation

TOPAZ4 uses the Deterministic variant of the EnKF (DEnKF, Sakov and Oke, 2008) with 100 dynamical members integrated with random perturbations of their surface boundary

Table 5.4: *Ocean and ice observations assimilated into TOPAZ. 3-D/4-D corresponds to synchronous or asynchronous assimilation respectively.*

Observation	Assimilation Method	Format	Resolution	Provider
SLA	4-D	Track	7 km	Sea Level TAC
SST	3-D	Gridded	5 km	Ocean and Sea Ice TAC
In situ T/S	3-D	Point	–	In Situ TAC
Ice concentrations	3-D	Gridded	25 km	Ocean and Sea Ice TAC
Ice drift (winter only)	4-D	Gridded	62.5 km	Ocean and Sea Ice TAC

conditions in order to obtain an ensemble representation of the flow-dependent background error covariances. Compared to Sakov *et al.* (2012), the only modification is the removal of the 1 per cent multiplicative inflation, which becomes problematic when used with a spatially varying observational network (Anderson, 2001). Hence, multiplicative inflation in absence of observations (such as in the interior of the Arctic Ocean) leads to an exponential increase of the spread, which combined with a multivariate update projects the bias in the observed variable to other variables. Satellite sea ice concentration retrieval algorithms misinterpret melt ponds as open water, which are not considered in TOPAZ4. This bias in the observations, has led to a degradation of the stratification in the Arctic until the multiplicative inflation has been removed in January 2014.

The observations assimilated into the TOPAZ4 system are satellite derived SST, along-track sea level anomalies (SLA) from satellite altimeters, in situ temperature and salinity profiles, sea ice concentration and low-resolution sea ice drift data from satellites (see Table 5.4). The pre-processing, temporal averaging and observation errors are described in Sakov *et al.* (2012).

All assimilated observations come from the CMEMS Thematic Assembly Centers (TACs) and are available through the CMEMS portal.[2] The SST data are from the UK Met Office. The SLA data assimilated in this system are the along-track data with a resolution of 7 km from CLS. Assimilated sea ice concentration and two-day ice drift trajectories are prepared by the Ocean and Sea Ice Satellite Application Facility (OSI SAF). The temperature (T) and salinity (S) profiles, including Argo floats, Ice-Tethered Profilers (ITP) and all other platforms, are provided by the CMEMS Arctic In-situ Thematic Assembly Centre. The sea ice concentration error variance is a function of ice concentration value A set to $0.01 + (0.5-|0.5 - \mathrm{A}|)^2$. This is intended to moderate the effect of melt ponds and other wet signatures and yields the smallest error standard deviation of 10 per cent at the extreme ice concentration values of 0 per cent and 100 per cent, increasing to 50 per cent when the ice concentration is 50 per cent. The errors in the two components of the ice drift are both set to 4.67 km day^{-1} and are assumed to be independent of each other.

[2] http://marine.copernicus.eu/

5.3.3 Related Activities and Plans for Future Development

The best evaluation of forecasts is produced when results are compared to independent observations, i.e. observations that were not used during model assimilation. Since ice chart data are not used in TOPAZ, this product was chosen for the evaluations presented in Chapter 6. However, TOPAZ provides forecasts for the entire Arctic Ocean and adjacent seas, and large parts of this domain are not covered by ice charts. For this reason, evaluation of sea ice concentration will be expanded by including results using a CMEMS SSMIS product as the observational baseline. Even though these data are assimilated by TOPAZ, a dependent evaluation is preferred over no evaluation at all.

Subsequently, APS evaluation will be further expanded by comparing model results for sea ice thickness to observations, most likely using CryoSat2 observations. Finally, evaluation of sea ice stage of development forecasts (first-year ice vs. multi-year ice) will also be included using the daily analysis of sea ice types from OSI SAF. This product distinguishes between young ice, first-year ice and ice that has survived a summer melt season.

Evaluation results provide precious hints of model shortcomings: the seasonal cycle of the bias, the reduction of accuracy in early summer and the difficulties in forecasting the marginal ice zone make tempting targets for further upgrades of the sea ice thermodynamics. But those results alone do not draw a complete picture of the needs without additional evaluation results of other sea ice variables (ice drift, ice thickness, snow depths) and ocean surface variables.[3] They reveal local biases in surface ocean temperatures (for example, a cold bias in St Anna Trough, North of the Barents Sea) as well as both systematic and random errors in ice drift, similar in amplitude to the ice drift magnitude itself (about 6 km per day). Both could make large contributions to ice edge displacement errors and must be attributed to other modules of the TOPAZ system. Therefore, the following developments are planned for the TOPAZ5 prototype, due for operations in April 2018, given in their chronological order:

- moving to the coupled ESMF (Earth System Modeling Framework), including an upgrade of HYCOM and an upgrade of the sea ice thermodynamics to version 5.1 of the CICE model;
- a doubling of the ocean model resolution, both horizontally (from 12 km to 6 km) and vertically, which should improve the representation of currents, notably those that are topographically steered and improve the performance of the mixing scheme in HYCOM;
- inclusion of a variable floe size distribution following the wave-in-ice model of Williams *et al.* (2013a, b); and
- assimilation of ice thickness observations with the EnKF, in both thin and thick ice from SMOS and CryoSat-2 satellite missions, respectively.

Developments of alternative rheological models to the ubiquitous VP/EVP model seem also necessary at a time when those become available. The elastic–brittle

[3] Many of those can be found on http://cmems.met.no/ARC-MFC/V2Validation/index.html and are corroborated by the evaluation of the corresponding 23-years reanalysis (http://marine.copernicus.eu/documents/QUID/CMEMS-ARC-QUID-002–003.pdf).

rheology has proven adept at representing the multi-fractal properties of sea ice deformations, as well as for simulating the ice drift and the seasonal cycle of ice thickness (Rampal *et al.*, 2016); the next generation sea ice model (neXtSIM) model thus makes a desirable candidate for improving sea ice forecasts. This change will however necessitate the transition into a coupled framework (like the ESMF) that is able to manage the communication between a fixed ocean model grid and a Lagrangian sea ice model grid. This may come after the 2018 upgrade due to the technological challenges ahead.

Other upgrades are planned, but with only indirect effects expected on sea ice forecast skills: among others, improved mean dynamic topography and boundary conditions from the global Mercator-Océan model.

5.4 Arctic Cap Nowcast/Forecast System and Global Ocean Forecast System – US Navy

The Arctic Cap Nowcast/Forecast System (ACNFS)[4] is presently the US Navy's operational coupled sea ice and ocean model system used to provide ice conditions in the Northern Hemisphere. ACNFS has been developed by the Naval Research Laboratory (NRL), transitioned to the NAVal OCEANographic Office (NAVOCEANO) and was declared operational in 2013. The system runs once per day at NAVOCEANO and provides ice–ocean nowcast and forecast products (such as ice concentration, ice thickness, ice drift, lead opening rate, 3-D ocean temperature/salinity and ocean currents) with lead times up to 7 days. The horizontal resolution is approximately 3.5 km near the North Pole and 6.5 km near 40° N (Figure 5.4).

Along with ACNFS, the Navy also uses the Global Ocean Forecast System (GOFS)[5] 3.1 to predict ice/ocean conditions. GOFS 3.1 has been transitioned to NAVOCEANO and at the time of publication was in the final operational testing phase. GOFS 3.1 uses the same model and assimilation components as ACNFS. When GOFS 3.1 becomes operational, it will replace ACNFS and provide a global sea ice prediction capability that includes both the Arctic and Antarctic. The National Ice Center (NIC) uses ACNFS and GOFS 3.1 output to improve the accuracy and resolution of the analysed ice edge location.

5.4.1 System Description: Forecast Model

ACNFS is based on HYCOM coupled via ESMF to CICE and uses the Navy Coupled Ocean Data Assimilation (NCODA) system (Cummings and Smedstad, 2013). The ice and ocean models are coupled on the same horizontal grid and exchange information every hour.

[4] ACNFS graphical products are publically available from www7320.nrlssc.navy.mil/hycomARC.
[5] GOFS 3.1 graphical products are publically available from www7320.nrlssc.navy.mil/GLBhycomcice1-12.

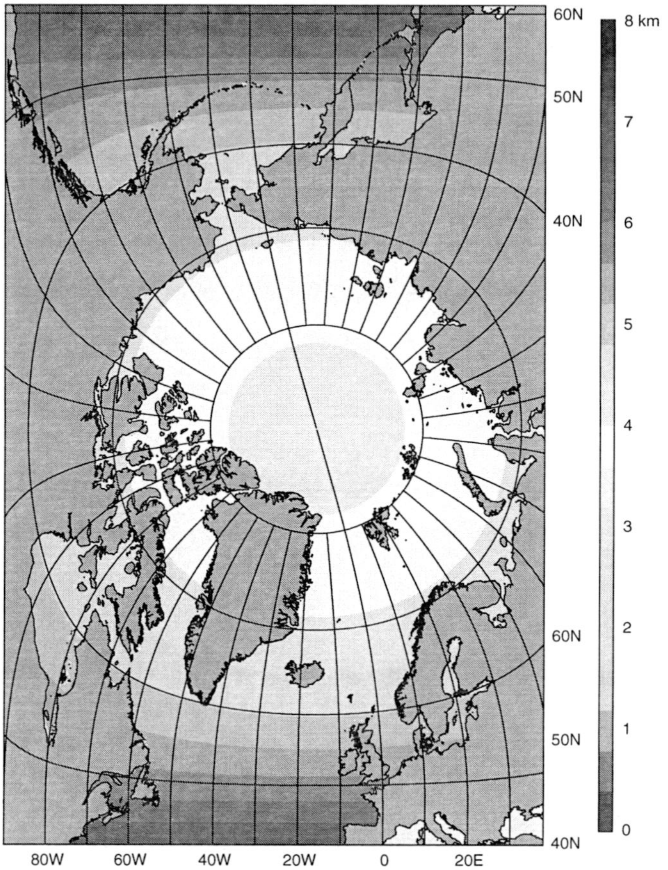

Figure 5.4: The Arctic Cap Nowcast/Forecast System (ACNFS) and Global Ocean Forecast System (GOFS 3.1) grid resolution (km). (A black and white version of this figure will appear in some formats. For the colour version, please refer to the plate section.)

Like RIPS, the ACNFS and GOFS 3.1 ice forecast models are tailored implementations of CICE version 4.0 with the following options used:

- ice strength is calculated using the Rothrock (1975) formulation;
- one snow and four ice layers are used for thermodynamic calculations; and
- the CCSM3 scheme is used to calculate the albedo and the attenuation of the absorbed shortwave radiation.

Other model parameters are listed in Table 5.5. Note that some of the parameters listed are the CICE default values, while others were tuned for the model thermodynamics and dynamics using optimizations of forecast errors compared to observations.

Atmospheric forcing used in ACNFS/GOFS 3.1 is from the Fleet Numerical Meteorology and Oceanography Center NAVy Global Environmental Model (NAVGEM). NAVGEM

Table 5.5: *Parameters used for the CICE implementation within ACNFS/ GOFS 3.1.*

Parameter		Value
time step		600 s
elastic damping timescale		216 s
number of sub-cycling steps		120
number of thickness categories	5	Cat 1: 0–0.64 m Cat 2: 0.64–1.39 m Cat 3: 1.39–2.47 m Cat 4: 2.47–4.57 m Cat 5: >4.57 m
surface ice roughness parameter		5×10^{-4} m
ice–ocean drag coefficient		0.00536 (ACNFS)/0.01072 (GOFS 3.1)
ice salinity		Profile with 0 psu as surface, 3.2 psu at ice/ocean interface

employs an eddy-diffusivity mass flux parameterization of turbulence and boundary layer mixing. The following atmospheric fields are used in CICE/HYCOM:

- air temperature at 2 m,
- specific humidity at 2 m,
- downward surface short-wave and long-wave radiation,
- precipitation,
- ground/sea temperature,
- zonal and meridional wind velocities at 10 m and
- mean sea level pressure.

5.4.2 System Description: Data Assimilation

ACNFS and GOFS 3.1 are both initialized daily via NCODA. Satellite observations provide sea surface height from altimeters, SST from AVHRR, Geostationary Operational Environmental Satellite (GOES), Meteosat Second Generation (MSG) and AMSR2 and ice concentration from SSMI, SSMIS and AMSR2 instruments (see Table 5.6 for ice observations currently assimilated into the forecast systems). Subsurface ocean data are assimilated via vertical profile observations from expendable bathy-thermographs (XBT), conductivity–temperature–depth (CTD) profilers, and profiling floats (e.g. Argo). Moored and drifting buoy observations of temperature and salinity, but not velocity, are also assimilated. Interactive Multisensor Snow and Ice Mapping System (IMS) data is used to mask these satellite-derived ice concentration products and makes a correction for the known problem associated with the interpretation of sea ice signatures in passive microwave data during summer months, with the presence of moist snow, wet ice surfaces and melt ponds.

Table 5.6: *Ice observations assimilated into ACNFS and GOFS 3.1.*

Observation type	Horizontal interpolation	Footprint size (radius)	Algorithm used to derive ice concentration
AMSR2	Footprint	11.0 km	Bootstrap Algorithm (Comiso and Nishio, 2008)
SSMI	Footprint	27.5 km	NT2
SSMIS	Footprint	29.0 km	NT2

NCODA generates both ocean and ice analyses using all available observations and the 24-hour forecast from the previous day as the background state. The ocean analysis variables include temperature, salinity, geopotential and vector velocity components that are all analysed simultaneously and provide corrections to the next model forecast in a sequential incremental update (Bloom *et al.*, 1996). The ice concentration analysis is inserted into ACNFS with a weighted average based on the model ice concentration and adjusts other fields (e.g. volume and energy of melting for both ice and snow) for consistency. The model also adjusts its ocean temperature to prevent ice from immediately melting or forming.

Recently, investigators at the National Snow and Ice Data Center (NSIDC), National Atmospheric and Space Administration (NASA), NIC and NRL developed a gridded ice concentration product that uses the daily observations from the IMS. This is blended with the AMSR2 passive microwave data to form a product with 4 km resolution. Posey *et al.* (2015) showed a significant improvement (56–62 per cent) in the predicted ice edge location at short forecast times when using this blended technique over assimilating SSMIS ice concentration alone. On 2 February 2015, these two new data sources (AMSR2 and IMS) were added to the operational ACNFS and pre-operational GOFS 3.1 job streams (i.e. swath SSMIS and ASMR2 data are assimilated into NCODA and the resulting ice concentration is masked using the IMS product). This masked product is then read into CICE.

5.4.3 Plans for Future Development

As noted above, GOFS 3.1 will replace ACNFS and provide an ice forecasting capability in both hemispheres. In the 2017–2018 timeframe, GOFS 3.5 is scheduled for transition to NAVOCEANO in which the model horizontal resolution will double to approximately 4.5 km at the equator and 1.75 km at the North Pole. This system will include tidal forcing in the ocean and the CICE code within GOFS will be updated to version 5 or above.

Developmental efforts are also underway to build a data-assimilative, fully coupled global atmosphere (NAVGEM), ocean (HYCOM), ice (CICE), wave (WAVEWATCH III™), land (NAVGEM-Land Surface Model), and aerosol (Navy Aerosol Analysis and Prediction System) system as part of the Earth System Prediction Capability (Eleuterio and

Sandgathe, 2012; Metzger *et al.*, 2014b). A demonstration of the first operational version of this system is presently scheduled for late 2018.

5.5 Long-Range Forecasting and the CANadian Seasonal to Inter-annual Prediction System – Environment and Climate Change Canada

With the increased marine accessibility of the Arctic waters in recent years, the interest in long-range forecasting of Arctic sea ice has increased as well. A large range of end users, from industries like transport, fishing and resource extraction to tourism, would greatly benefit from skilful predictions on seasonal timescales. Predictability studies suggest that Arctic sea ice is potentially predictable on seasonal to interannual timescales. The extent to which this potential predictability can be realized in practice is an active research area.

Until recently, long-range sea ice forecasts were solely based on statistical models, which rely on empirical relationships between cryospheric, oceanic and atmospheric predictors and sea ice concentration in subsequent months, derived from historical datasets. Such statistical models generally assume error stationarity which may be a poor assumption in the rapidly changing Arctic environment.

Dynamical models represent a new and promising tool for seasonal forecasting of Arctic sea ice. Although these models have been applied to weather and climate change predictions for a few decades, the widespread application of dynamical models to prediction on seasonal timescales is relatively new. The application to seasonal forecasting of sea ice is even more recent, with the first operational seasonal forecasting systems including prognostic sea ice appearing in the early 2010s. The CANadian Seasonal to Inter-annual Prediction System (CanSIPS), ECCC's operational seasonal forecast system since December 2011, was one of the first such systems.

5.5.1 System Description: Forecast Model

CanSIPS employs two global coupled climate models developed at the Canadian Centre for Climate Modelling and Analysis (CCCma), CanCM3 and CanCM4. A multi-model approach is taken as previous studies found generally greater skill of multi-model ensembles for a given number of ensemble members (e.g. Kharin *et al.*, 2009). The atmospheric components of the CanSIPS models are based on versions 3 and 4 of CCCma's atmospheric general circulation models, CanAM3 (Scinocca *et al.*, 2008) and CanAM4 (von Salzen *et al.*, 2013), with CanAM4 featuring a number of improvements related to aerosols, radiation, clouds and vertical resolution. These spectral models employ T63 triangular spectral truncation with physical tendencies calculated on a 128 × 64 (~2.8°) horizontal Gaussian grid. The models include the evolution of anthropogenic radiative forcings such as greenhouse gases and aerosols. CanCM3 and CanCM4 share the same ocean, land and sea ice model components. The ocean component (CanOM4) has approximately 100 km horizontal resolution and 40 vertical levels. The two-category (thickness–volume) sea ice component is formulated on the T63 atmospheric model grid. Sea ice dynamics is governed by cavitating fluid rheology (Flato and Hibler, 1992) and sea ice thermodynamics by a simple energy balance model.

5.5.2 System Description: Initial Conditions

CanSIPS forecasts consist of 12-month simulations with 20 ensemble members (10 for each model) initialized at the start of each calendar month. The initialization and ensemble generation procedure is illustrated in Figure 5.5. Initial conditions are obtained from a set of coupled assimilation simulations, one for each ensemble member, which are statistically consistent with observation-based products. The assimilation runs were initialized from slightly different initial conditions in 1948, allowing for more than three decades of assimilation input before the first seasonal forecasts initialized in 1979. For the atmosphere, larger spatial scales (T21 and larger) in moisture, temperature, vorticity and divergence fields were constrained to closely follow observation-based values in the form of gridded reanalyses, by means of a procedure that is effectively very similar to nudging. The strength of the atmospheric nudging is chosen so that the resulting spread between ensemble members is similar to the spread between different observation-based products, hence representing uncertainties in observations. Sea surface temperature and sea ice concentration are nudged towards observation-based products,

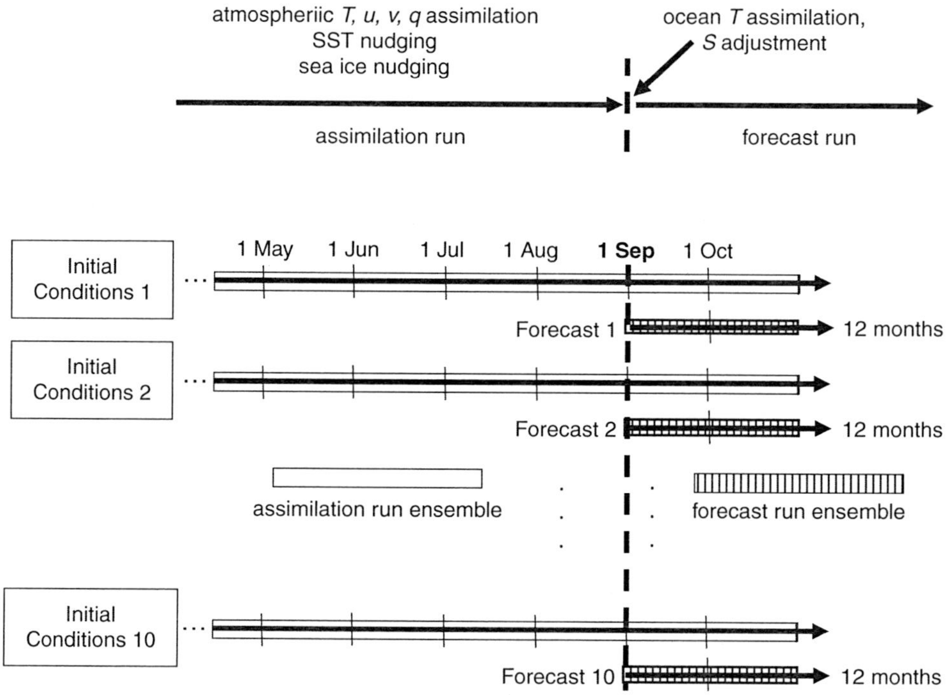

Figure 5.5: Schematic of the assimilation and ensemble generation procedure for initializing CanSIPS forecasts (adapted from Merryfield *et al.*, 2013).

whereas interior ocean temperatures are assimilated with an offline variational procedure, followed by adjustments to salinity to preserve stability in each grid column. Due to the lack of reliable long-term observations of sea ice thickness, the initialization of sea ice thickness is particularly challenging. As in other dynamical forecast systems, no observations are used to constrain sea ice thickness. Instead, in the current CanSIPS version sea ice thickness is initialized by relaxing to a seasonally varying model climatology.

5.5.3 Ongoing Research: Improved Sea Ice Thickness Initialization

The relatively simple initialization procedure of sea ice thickness fails to represent the long-term trend and interannual variations in sea ice thickness, which are both potentially important sources of forecast skill of sea ice area. An improved sea ice thickness initialization procedure has recently been developed and is available for application in future generations of CanSIPS. This procedure employs observed sea ice concentration to estimate sea ice thickness in real time by applying a statistical model (Dirkson *et al.*, 2016). First, the long-term trends in sea ice concentration and thickness are extrapolated using a linear fit, after which a simple proportionality relationship between de-trended anomalies in (known) sea ice concentration and (unknown) sea ice thickness is applied. An experimental set of hindcast simulations with CanCM3 using the improved thickness initialization shows significantly enhanced forecast skill, as illustrated in Figure 5.6. This enhancement is particularly large for total sea ice area anomalies (solid lines) for both 1 May and 1 June initial conditions, and de-trended sea ice area anomalies (dashed lines) for 1 May initial conditions.

5.5.4 Ongoing Research: Statistical Downscaling

Increased activity in the Arctic has increased demand for operational centres to provide temporally and spatially more detailed seasonal sea ice forecasts at longer lead times.

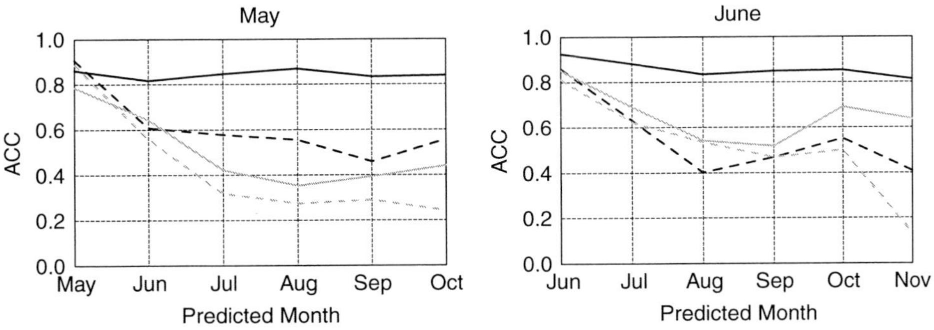

Figure 5.6: CanCM3 forecast skill (quantified by the anomaly correlation coefficient (ACC)) for sea ice area for forecasts initialized on (left) 1 May and (right) 1 June for the original hindcasts (grey lines) and in hindcasts with improved thickness initialization (black lines). Solid is for full anomalies and dashed for linearly de-trended anomalies.

The relatively coarse resolution of the current CanSIPS models prevents the direct usage of CanSIPS output. While this limitation will be partially addressed in future CanSIPS versions, statistical downscaling techniques may be applied to current CanSIPS output to improve the level of detail and accuracy. Two approaches for statistical downscaling are currently being studied: multiple linear regression and maximum covariance analysis.

A first technique presented here to downscale CanSIPS output to local, end user-relevant sea ice variables is multiple linear regression (MLR). MLR is a multivariate technique that quantifies the linear relationship between two or more explanatory variables (predictors) and a response variable (predictand). In numerical weather prediction, MLR techniques are commonly employed to provide subgrid-scale forecasts under a perfect prognosis (PP) or model output statistics (MOS) approach. Both MOS and PP exploit strong empirical relationships between predictors (model or observed atmospheric fields) and a predictand (subgrid-scale variable). In MOS the predictors are taken from the model to be downscaled; the main advantage of MOS is that the method accounts for systematic model errors.

The MOS approach could be applied to CanSIPS, taking full advantage of the 31-year hindcasts to train MLR equations and apply a bias correction. This may inform operational forecasts such as the seasonal outlook of sea ice events that the CIS issues every June. These sea ice events are relevant for shipping and community activities in northern Canadian Waters. Since 1959, CIS has issued a seasonal outlook for the summer shipping season. The outlook contains forecasts for the specific dates of over 40 regional sea ice events. Examples of these events are the fracture date of the sea ice around Pond Inlet and Lancaster Sound and the first day the shipping route from Davis Strait to Thule is ice-free. The current CIS outlook is based on a combination of analogue techniques, heuristic techniques and regression based predictions. Statistically downscaled CanSIPS output may provide an additional source of guidance for the production of such outlooks.

An alternative to the multivariate regression approach is based on the multivariate statistical technique known as the Maximum Covariance Analysis (MCA). While multivariate regression methods are optimal in minimizing mean square error when very long samples are available, MCA may be statically more robust when only short samples are available, such as in the case of seasonal forecasting (with sample sizes typically ranging from 20 to 40). MCA seeks pairs of orthonormal patterns in two related fields, maximizing the squared covariance of the corresponding time components. The solution to this problem is given by singular value decomposition (SVD) of the cross-covariance matrix of the two fields. In the application of MCA to statistical downscaling, these two fields would be the output from a coarse-resolution forecast model and the (high-resolution) observations.

The MCA-based downscaling works as follows. First, MCA is performed on coarse-resolution model hindcasts and corresponding high-resolution observations in the hindcast period, resulting in two sets of SVD patterns: one for the model hindcasts and the other for the corresponding observations. The top two panels in Figure 5.7 illustrates the first pair of SVD patterns obtained for monthly mean sea ice concentrations in a Canadian Arctic domain in March 1979–2012 based on the CanSIPS ensemble mean hindcasts initialized on March 1 (corresponding to a 0-lead forecast) and the NSIDC sea ice concentration

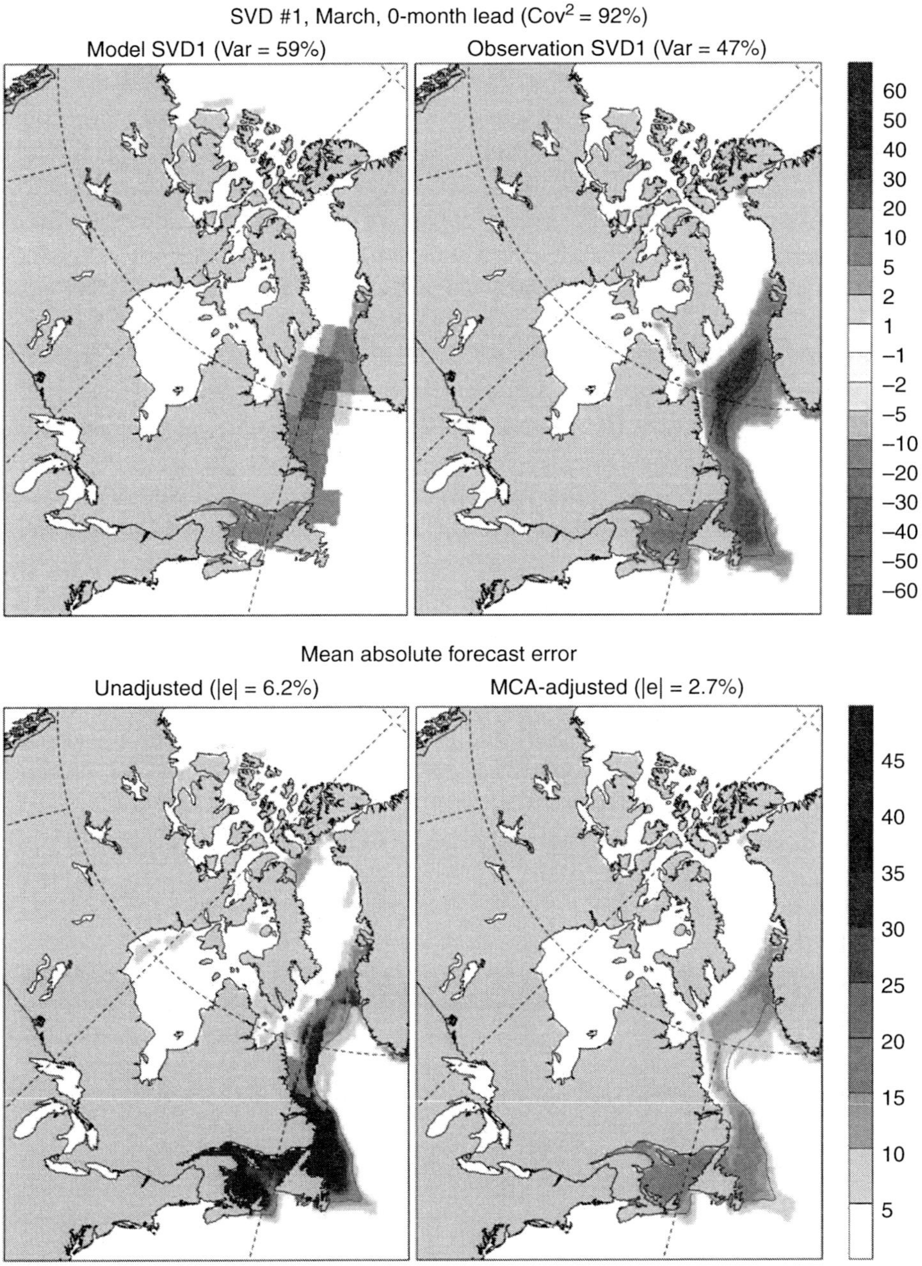

Figure 5.7: The first pair of SVD patterns (top) obtained for monthly mean CanSIPS predictions of March sea ice concentrations initialized on 1 March (lead = 0) in 1979–2012 (left) and the corresponding observed (NSIDC) field (right). This pair of patterns explains more than 90 per cent of the squared covariability in the two fields. The mean absolute error (bottom) of 0-lead CANSIPS predictions of March sea ice concentration for the original unadjusted (left) and MCA-adjusted (right) forecasts, with domain-averaged error indicated in the panel titles. (A black and white version of this figure will appear in some formats. For the colour version, please refer to the plate section.)

observational products. This pair explains more than 90 per cent of the total squared covariability in the two fields. A new model forecast (for future conditions) is then expanded in terms of the obtained model SVD patterns. The MCA-downscaled forecast is then obtained by replacing in this expansion the coarse-resolution model SVD patterns with the corresponding high-resolution observed counterparts. The resulting forecast is therefore available at the spatial resolution of the high-resolution observations. This MCA-based downscaling approach can be seen as a 'pattern correction' as it does not attempt to rescale the magnitude of the corresponding time components.

The bottom panels in Figure 5.7 display the mean absolute error of the 0-lead CanSIPS predictions of March sea ice concentration of the MCA downscaling approach applied to monthly mean sea ice concentrations in the Canadian Arctic domain. Relative to the mean error of the original unadjusted CanSIPS forecasts (left panel), the error is greatly reduced in the MCA-adjusted forecasts (right panel) with the domain-averaged error value reduced by more than a factor of two.

The same MCA-based adjustments have been applied to CanSIPS sea ice concentration predictions for each initialization and target month. The results are summarized in Figure 5.8 which presents the lead time dependence of the domain-mean absolute errors averaged over all 12 target months. The original unadjusted forecasts and the MCA-adjusted forecasts are compared with: persistence, in which the monthly mean anomaly observed in the month prior to the forecast initial month is persisted throughout the whole forecast period; the climatology forecast, in which the 1979–2012 monthly climatology is used as a prediction; and the bias-corrected forecast, in which the model climatology is replaced with the observed climatology. It is evident that the errors of the raw unadjusted forecasts substantially exceed those of all other forecasts, while the MCA-adjusted forecasts perform best. The improvements in the MCA-adjusted forecasts are larger for shorter

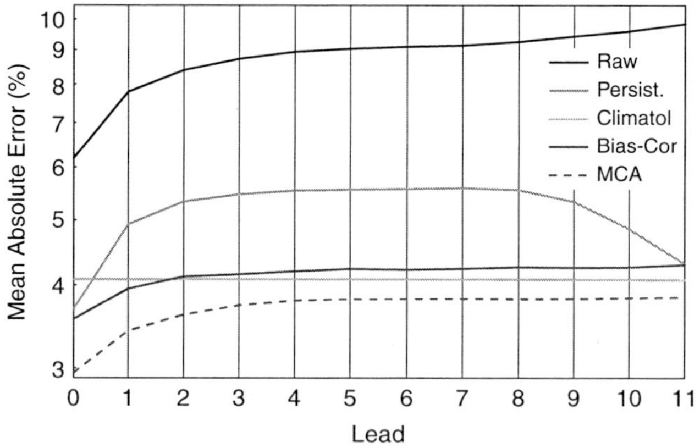

Figure 5.8: The domain-averaged mean absolute error of CanSIPS hindcasts averaged over all target months as a function of lead time (in months), for different forecasting methodologies (details in text). (A black and white version of this figure will appear in some formats. For the colour version, please refer to the plate section.)

lead times when predictive skill is higher and statistical relationships between model forecasts and corresponding observations are more robust.

5.5.5 Plans for Future Development

As described above, the sea ice model currently employed in CanSIPS has relatively coarse resolution such that Arctic coastlines and passages are not well resolved. In addition, it uses relatively simple physical parameterizations relative to the standards of present-generation climate models. Both aspects will be improved considerably in two new models currently under development: CanESM5, which couples an updated version of CCCma's atmospheric model to a NEMO-based ocean component, and GEM-NEMO, which couples ECCC's Global Environmental Multiscale (GEM) model also to a NEMO-based ocean component. Both models represent ocean and sea ice on the NEMO ORCA1 grid, which has horizontal resolution of a few tens of kilometres throughout the Arctic, and both employ multilayer, multi-category sea ice models, LIM for CanESM5 and CICE for GEM-NEMO.

5.6 Discussion

As will be seen in Chapter 6, it is not feasible to directly compare the accuracy of APS output for each of the systems described above and, even if it were possible, it would not necessarily be useful since each system serves different types of clients or needs. Nevertheless, it is worthwhile to note some of the advantages and possible disadvantages associated with the different approaches employed in these systems.

In general, it is beneficial to use a collaborative approach for the development of an APS and its components. A good example is the fairly widespread use of CICE, with many groups and researchers contributing to both the development and evaluation of the model. Additionally, the team at the Los Alamos National Laboratory provides services in terms of source code maintenance and testing as well as documentation. Disadvantages could include keeping track of new releases and APS adaptation and testing when implementing new versions.

In the real world, the evolution of the atmosphere, ice and ocean is affected by the exchange of heat, momentum, freshwater and salt. So, in an ideal world, an environmental APS would explicitly include all of these parts of the natural environment that would have an impact on predictions. This is not presently feasible for all APSs due to limitations of system development and operational resources. In order to capture the most crucial ice–ocean fluxes, all of the above systems include an ice model component and some form of an interactive ocean. Some already include an atmospheric model and others may have development plans to be coupled to an existing atmospheric APS. The only APSs that are currently operational and resolve all interactions between the atmosphere, ocean and sea ice are those used for long-range predictions, such as CanSIPS. The advantage of such coupled APSs is that the inclusion of explicit exchange of fluxes should improve predictions, especially for longer lead times as the effect of interactions accumulate. One method to overcome the significant resource issues related to the development and maintenance of atmospheric and ocean APS is to couple the sea ice component to an existing ocean APS (such as GOFS) or an existing atmospheric APS (as in the plans for GIOPS). There are other challenges that have to be considered in developing a fully coupled APS. One of these is that it requires understanding of the methodology to link the component models and possibly use a coupler that is not ideally suited to the model used. Another challenge is that full APS simulations require significantly greater computer resources. Lastly, evaluation of new APS versions must consider all components to ensure changes to one component do not negatively impact the others.

The initialization of an APS is another area where the above systems differ fairly significantly both in terms of data assimilated as well as the assimilation technique employed. For sea ice initialization, some APSs assimilate satellite observations directly (e.g. RIPS and GIOPS, to some extent) while others use or combine pre-existing analyses and possibly adjust them as necessary. The advantage of assimilating multiple sources of observations directly is that it is possible to account for temporal and spatial variability in observation accuracy and the resulting analysis should be more robust and hopefully more accurate. However, this approach requires more resources to effectively use each observation type. Assimilating pre-existing analyses requires fewer resources as the expectation is that the information provider has optimized the analysis. It is difficult though to account for spatial and temporal varying observation errors.

Since sea ice, ocean and atmospheric conditions are related, it is beneficial to adjust the respective initializations to be physically consistent. Most of the APSs make some adjustments to a greater or lesser extent but it is fair to say that more research is required in this area. Similarly, assimilation techniques vary quite significantly. Part of this is related to the availability of resources that can be devoted to data assimilation. Another reason is that there may already be existing data assimilation systems at a given APS facility and it is most efficient to adapt these for the sea ice APS (e.g. 3DVar for RIPS/GIOPS and NCODA for GOFS).

The resolution of an APS may be determined by the physical processes that affect clients. It is also affected by both the operational resources available to run the system as well as limitations in observation resolution. For example, ocean eddies may have a significant impact on ice conditions in a given area, but the APS may not resolve these eddies or the observations may be insufficient to correctly initialize them in the APS. Beyond this, grid projections may also affect the ability of an APS to resolve key ice covered areas. For example, with the tri-polar ORCA grid projection that has a pole near the Canadian Arctic Archipelago, some channels near that numerical pole must be masked out because they would have such high resolution that a very small model time-step would be required. Other projections, such as a rotated latitude–longitude grid can provide fairly uniform resolution along the grid equator but singularities would occur at the numerical poles.

5.7 Chapter Summary

A number of sea ice APSs are currently operational in the sense that they are providing output to national ice service organizations. In some areas there is great consistency between these APSs, such as the fairly widespread use of the CICE sea ice model. For other areas, such as ocean forcing and model initialization, there is considerably less consistency. The recent implementation of these systems shows the growing maturity of the field, but there remains an opportunity to compare methods and identify best practises.

RIPS is a system mainly designed to support the CIS. It includes a sophisticated data assimilation system that uses a variety of observation types. A number of other innovations have been implemented to adapt the system to support the regional ice conditions and operational support for which the system is targeted.

GIOPS is a global ice–ocean prediction system developed in collaboration with Mercator-Océan. This system satisfies a growing need in Canada for a multipurpose global marine core service. It will also become part of a global environmental prediction system.

TOPAZ is part of the larger CMEMS development (referred as Arctic MFC) and is its official system north of 62° N. It takes advantage of the existing CMEMS ice information both to initialize the APS and for evaluation, as will be shown in Chapter 6. Although it does use the HYCOM ocean model, it follows a different route than ACNFS/GOFS for data assimilation, sea ice modelling, ecosystem modelling and it will remain a regional system receiving boundary conditions from the global MFC at Mercator-Océan.

The ACNFS/GOFS system supports the NIC and its global mandate. These systems build on a history of operational ocean forecasting through the inclusion of CICE and sea ice initialization. The convergence of ice and ocean modelling systems provides an opportunity for consistent and efficient ongoing development.

CanSIPS targets extended range forecasts through the use of an ensemble approach. Unlike climate modelling systems, initial ice conditions have a significant impact on forecast accuracy. Although the current system resolution is coarse, statistical downscaling of CanSIPS output may provide important information for NIIS clients.

References

Anderson, J.L. (2001). An ensemble adjustment Kalman filter for data assimilation. *Monthly Weather Review*, **129**, 2884–2903.

Bloom, S.C., Takacs, L.L., da Silva, A.M. and Ledvina, D. (1996). Data assimilation using incremental analysis updates. *Monthly Weather Review*, **124**, 1256–1271.

Buehner, M., Caya, A., Carrieres, T. and Pogson, L. (2014). Assimilation of SSMIS and ASCAT data and the replacement of highly uncertain estimates in the Environment Canada Regional Ice Prediction System. *Quarterly Journal of the Royal Meteorological Society*, **142**, 562–573. doi: 10.1002/qj.2408.

Comiso, J.C. and Nishio, F. (2008). Trends in the sea ice cover using enhanced and compatible AMSR-E, SSM/I, and. SMMR data. *Journal of Geophysical Research*, **113** (C02S07). doi: 10.1029/2007JC004257.

Cummings, J.A. and Smedstad, O.M. (2013). Variational data assimilation for the global ocean. In S.K. Park and L. Xu, eds., *Data Assimilation for Atmospheric, Oceanic and Hydrologic Applications* (Vol. II), Springer-Verlag, Berlin Heidelberg, 303–343. doi: 10.1007/978-3-642-35088-7_13.

Dirkson, A., Merryfield, W.J. and Monahan, A. (2016). Impacts of sea ice thickness initialization on seasonal Arctic sea ice predictions. *Journal of Climate*, in press.

Drange, H. and Simonsen, K. (1996). Formulation of air-sea fluxes in the ESOP2 version of MICOM. *Technical Report No. 125*, Nansen Environmental and Remote Sensing Center.

Eleuterio, D.P. and Sandgathe, S. (2012). The Earth System Prediction Capability program. *Proceedings of OCEANS 12 MTS/IEEE*, October 14–19, 2012, Hampton Roads, Virginia, 1–3. doi: 10.1109/OCEANS.2012.6404895.

Flato, G.M. and Hibler, W.D. (1992). Modeling pack ice as a cavitating fluid. *Journal of Physical Oceanography*, **22**(6), 626–651.

Kara, A.B., Rochford, P.A. and Hurlburt, H.E. (2000). An optimal definition for ocean mixed layer depth. *Journal of Geophysical Research*, **105**, 16803–16821. doi: 10.1029/2000JC900072. ISSN: 0148–0227.

Kharin, V.V., Teng, Q., Zwiers, F.W., Boer, G.J., Derome, J. and Fontecilla, J.S. (2009). Skill assessment of seasonal hindcasts from the Canadian Historical Forecast Project. *Atmosphere-Ocean*, 47(3), 204–223.

Merryfield, W.J., Lee, W.S., Boer, G.J., Kharin, V.V., Scinocca, J.F., Flato, G.M., Ajayamohan, R.S., Fyfe, J.C., Tang, Y. and Polavarapu, S. (2013). The Canadian Seasonal to Interannual Prediction System. Part I: models and initialization. *Monthly Weather Review*, **141**, 2910–2945. doi: 10.1175/MWR-D-12-00216.1.

Metzger, E.J., Smedstad, O.M., Thoppil, P.G., Hurlburt, H.E., Cummings, J.A., Wallcraft, A.J., Zamudio, L., Franklin, D.S., Posey, P.G., Phelps, M.W., Hogan, P.J., Bub, F.L. and DeHaan, C.J. (2014a). US Navy operational global ocean and Arctic ice prediction systems. *Oceanography*, **27**(3), 32–43, dx.doi.org/10.5670/oceanog.2014.66.

Metzger, E.J., Ruston, B.C., Dykes, J.D., Whitcomb, T.R., Wallcraft, A.J., Smedstad, L.F., Chen, S. and Chen, J. (2014b). Operational implementation design for the Earth System Prediction Capability (ESPC): A first-look. *NRL Report* NRL/MR/7320-14-9498. (Available at www7320.nrlssc.navy.mil/pubs.php).

Oki, T. and Sud, Y.C. (1998). Design of the global river channel network for Total Runoff Integrating Pathways (TRIP), *Earth Interactions*, **2**, 1–37.

Posey, P.G., Metzger, E.J., Wallcraft, A.J., Hebert, D.A., Allard, R.A., Smedstad, O.M., Phelps, M.W., Fetterer, F., Stewart, J.S. and Helfrich, S.R. (2015). Improving Arctic sea ice edge forecasts by assimilating high horizontal resolution sea ice concentration data into the U.S. Navy's ice forecast systems. *The Cryosphere*, **9**, 1–11. doi: 10.5194/tc-9-1-2015.

Rampal, P., Bouillon, S., Ólason, E. and Morlighem, M. (2016). neXtSIM: a new Lagrangian sea ice model. *The Cryosphere*, **10**, 1055–1073. doi: 10.5194/tc-10-1055-2016.

Rothrock, D.A. (1975). The energetics of the plastic deformation of pack ice by ridging. *Journal of Geophysical Research*, **80**, 4514–4519.

Sakov, P. and Oke, P.R. (2008). A deterministic formulation of the ensemble Kalman Filter: an alternative to ensemble square root filters. *Tellus A*, **60**(2), 361–371. doi: 10.1111/j.1600-0870.2007.00299.x.

Sakov, P., Counillon, F., Bertino, L., Lisæter, K.A., Oke, P.R. and Korablev, A. (2012). TOPAZ4: an ocean-sea ice data assimilation system for the North Atlantic and Arctic. *Ocean Science*, **8**, 633–656. doi: 10.5194/os-8-633-2012.

von Salzen, K., Scinocca, J.F., McFarlane, N.A., Li, J., Cole, J.N.S., Plummer, D., Verseghy, D., Reader, M.C., Ma, X., Lazare, M. and Solheim, L. (2013). The Canadian Fourth Generation Atmospheric Global Climate Model (CanAM4). Part I: representation of physical processes. *Atmosphere-Ocean*, **51**, 104–125. doi: 10.1080/07055900.2012.755610.

Scinocca, J.F., McFarlane, N.A., Lazare, M., Li, J. and Plummer, D. (2008). The CCCma third generation AGCM and its extension into the middle atmosphere. *Atmospheric Chemistry and Physics*, **8**, 7055–7074. doi: 10.5194/acp-8-7055-2008.

Williams, T.D., Bennetts, L.G., Squire, V.A., Dumont, D. and Bertino, L. (2013a). Wave–ice interactions in the marginal ice zone. Part 1: Theoretical foundations. *Ocean Modelling*, **71**, 81–91.

Williams, T.D., Bennetts, L.G., Squire, V.A., Dumont, D. and Bertino, L. (2013b). Wave–ice interactions in the marginal ice zone. Part 2: Numerical implementation and sensitivity studies along 1D transects of the ocean surface. *Ocean Modelling*, **71**, 92–101.

6

System Evaluation

Tom Carrieres, Barbara Casati, Alain Caya, Pam Posey, E. Joseph Metzger, Arne Melsom, Michael Sigmond, Viatcheslav Kharin, Frédéric Dupont

System evaluation is a fundamental component of the development and application of an automated prediction system (APS). Process-based and physically meaningful diagnostics can provide guidance for system improvements and robust and informative evaluation metrics are in demand to monitor progress and enable comparisons of different systems. Moreover, specific information on the strengths and weaknesses of the models can be valuable for forecast end users to make optimal decisions in their planning. In the pioneering work of Murphy (1993), the aims of forecast evaluation were classified as: scientific, administrative and economic. More recently the World Meteorological Organization (WMO) Joint Working Group in Forecast Verification (JWGFVR) has identified the three most important reasons for evaluation as monitoring, improving and comparing. All these are encompassed in the definition of Jolliffe and Stephenson (2012): evaluation aims to be informative on the forecast consistency, quality and value.

Prior to designing an evaluation strategy it is essential to outline the purposes and to identify to whom the evaluation is addressed. This enables the variables and physical processes of interest to be identified, and the attributes of the forecast that ought to be evaluated. Moreover, prior to designing an evaluation strategy, it is essential to understand the statistical characteristics of the variable to be evaluated, and the constraints that might be imposed by the available observations.

In the context of this book, evaluation is the process of measuring the quality of a prediction. The primary reason for APS evaluation is to inform a variety of interested communities about the system performance. Some specific goals are:

- Understanding system behaviour in a variety of conditions;
- Comparing new systems with previous versions and other systems;
- Documenting system performance;
- Detecting degradation of operational system quality; and
- Demonstrating progress.

While it is not necessary for a particular evaluation process to address each of the above, common themes arise. In this chapter, performance measures are introduced for different types of sea ice variables. Continuous, categorical and spatial evaluation methods are presented and their statistical characteristics are outlined in a rigorous but simplified fashion.

6.1 Evaluation Data

Any evaluation study requires a reference dataset that should ideally be independent of the APS. As outlined in Chapter 3, there are many sources of observations each with its own uncertainty. These uncertainties include observation errors resulting from instrument measurement errors and the inherent uncertainty in the assumptions of retrieval algorithms (for satellite-based observations). Representativeness errors may arise if there is a scale mismatch between observations and APS resolution. Similarly, sampling errors occur due to limitations in temporal and spatial coverage of the observations. If the magnitude of the observation uncertainty is much smaller than the APS error, then perhaps the former can be neglected. In theory, if the statistics of the error in the verifying observations are known, and these errors are independent of the errors in what is to be verified, then their uncertainty can be accounted for in the evaluation: i.e. the variance(forecast error) = variance(forecast – observation) – variance(observation error). Accounting for observation uncertainty in verification practices is still not resolved and has been identified as a key research challenge for the forthcoming decade. For many uses (e.g. measuring improvements to the system, or comparing two systems), the absolute measure of the error is not needed. Instead, a relative measure that indicates if the accuracy is improved or not is sufficient. So while it's not ideal, fairly uncertain observations can still be used for this purpose, although the independence of the error is even more important.

Another point to consider is how APS output is mapped to observations and how the grid-point value of the variable of interest is defined in the APS. For example, some variables may represent a grid-cell average, while others may represent point-values, and therefore interpolation is the more suitable approach. Different interpolation strategies affect the value of the interpolated variables in different ways. For example, bilinear interpolation reduces peak values but can also lead to discontinuous slopes whereas bi-cubic interpolation can introduce small yet unphysical values. The appropriate interpolation approach should be chosen to minimize its effects on the evaluation results. This depends on the characteristics of the variable to be interpolated and on the discrepancy between observation and model resolution.

The most straightforward means to evaluate forecasts is to compare them with analysis states produced by the data assimilation subsystem and valid at the same times. With this approach, there is no need to interpolate and the observation and representativeness errors are addressed in the data assimilation and gridding process. The major difficulty with this method is that areas with few or no observations will revert to the background field. The analysis is then too correlated with the forecast, leading to evaluations where errors appear smaller than they actually are. Continued use could result in APS development drifting away from reality. An evaluation strategy which could mitigate this effect, while still using a model-based analysis as reference, could be to verify solely against analysis grid-points that have assimilated an observation within a small time window from the forecast valid time. A refinement to this method would be to use estimates of the analysis uncertainty to filter out grid cells with uncertainty above a chosen threshold. An alternative approach, to be used in the context of multi-model ensembles, is to verify against an ensemble of analyses. A probabilistic evaluation approach is adopted in this case, and the observation uncertainty will be accounted for in the spread of the analysis ensemble.

It cannot be overstated that satellite-based observations are crucial for the evaluation of sea ice. In situ observations in Polar Regions historically have always been very sparse and they are spatially under-representative. They are also affected by seasonality, with fewer observations in winter, and suffer from large measurement errors, possibly due to the harsh environmental conditions. As a consequence, Polar Regions rely more on satellite observations than lower latitudes. Satellite observations, despite the limitations described in Chapter 3, enable large spatial and temporal coverage. Evaluation processes benefit enormously from this enhanced coverage, not only in terms of dealing with a more complete and large sample, but these observations also permit the use of spatial evaluation approaches, which have enhanced diagnostic power compared to point-by-point traditional approaches.

6.2 Evaluation Using Categorical Approaches

Methods based on categorical approaches treat variables defined as discrete values. As an example, sea ice pressure as reported by ship observations can be classified as 'slight', 'moderate/severe' and 'ship beset'.

6.2.1 Categorical Scores

For categorical evaluations a contingency table is useful for calculating a variety of error indices. Table 6.1 is a general contingency table for G categories of forecasts and H categories of observations, noting that G and H do not have to be equal. Each entry in the table represents the count of forecasts in category g when corresponding observations were in category h. The contingency table can be interpreted as the forecast–observation joint distribution (Murphy and Winkler, 1987). The contingency table entries divided by the total number of forecast/observed events (N) are the estimates of the joint probabilities, e.g.

$$n(F_1, O_2)/N = P(\text{forecast category 1, observed category 2}) \tag{6.1}$$

(i.e. the frequency of the joint occurrence of forecast category 1 and observed category 2). The marginal totals are associated with the marginal observation and forecast probabilities, e.g.

$$(n(F_1, O_1) + n(F_2, O_1) + \ldots + n(F_G, O_1))/N = n(O_1)/N = P(\text{observed category 1}) \tag{6.2}$$

and similarly

$$(n(F_1, O_1) + n(F_1, O_2) + \ldots n(F_1, O_H))/N = n(F_1)/N = P(\text{forecast category 1}) \tag{6.3}$$

Table 6.1: *General $G \times H$ contingency table used to evaluate models against observation categories. The cell values $n(F_g, O_h)$ indicate the number of occurrences when a forecast made in forecast category g corresponded to an observation made in observation category h. The column entries $n(F_g)$ are the total number of forecasts for forecast category g. The row entries $n(O_h)$ are the total number of observations in observation category h. N is the total number of forecast–observation pairs.*

		Observations				
		Category 1	Category 2	...	Category H	Totals
	Category 1	$n(F_1, O_1)$	$n(F_1, O_2)$	...	$n(F_1, O_H)$	$n(F_1)$
	Category 2	$n(F_2, O_1)$	$n(F_2, O_2)$	...	$n(F_2, O_H)$	$n(F_2)$
Forecasts	...	...	...	...	...	...
	Category G	$n(F_G, O_1)$	$n(F_G, O_2)$	...	$n(F_G, O_H)$	$n(F_G)$
	Totals	$n(O_1)$	$n(O_2)$	...	$n(O_H)$	N

and the frequency bias for category 1 is the ratio of these marginal probabilities, assuming forecast category 1 corresponds to observed category 1. Finally, the conditional per cent correct by category are the estimates of the conditional forecast probabilities, e.g.

$$PC_g = P(\text{forecast category } g|\text{observed category } h) \tag{6.4}$$

i.e. the probability that the forecast was category g, when the observed category h overlapped.

For many situations, there are two forecast categories and two corresponding observation categories and in this case, Table 6.1 collapses to a 2×2 distribution. Useful evaluation statistics can be defined following the format of Table 6.2. In addition to the error measures described above, often it is desirable to assess if the forecast is better (or worse) with respect to a baseline or a (null) reference forecast (e.g. persistence). This can be done by evaluating a skill score. Skill scores are defined as $(S - S_{REF})/(S_{BEST} - S_{REF})$, where S is a score measuring forecast accuracy, S_{REF} is the score obtained for a reference forecast and S_{BEST} is the score obtained for a perfect forecast. Skill scores range between $-\infty$ and 1; a positive skill score denotes that the APS forecast is better than the reference forecast, a negative skill score denotes that the APS forecast is worse than the reference forecast, zero skill score denotes that the APS forecast performs similarly to the reference forecast. The Heidke Skill Score (*HSS*), defined in Table 6.2, is the skill score obtained when using a randomized reference forecast. The *PCT* (proportion correct total) of this randomized forecast is denoted PCT_R.

Dichotomous (2×2) contingency tables, because of their relation with the forecast–observation joint distribution, can be fully described by three numbers. Categorical scores defined from the entries of a contingency table (e.g. as those listed in Table 6.2) are therefore often related, and sometimes different scores can exhibit very similar behaviours.

Table 6.2: *Contingency table statistics. The scores are ratios and should only be calculated when the denominator is greater than some threshold. Note that* PC_1 *and* PC_2 *are conditional proportion correct scores with respect to the corresponding observation category.*

Name	Definition	Range	Best score
Observation Count (N)		0 to ∞	
Proportion Correct Total (PCT)	$(n(F_1,O_1)+n(F_2,O_2))/N$	0 to 1	1
Proportion Error Total (PET)	$(n(F_2,O_1)+n(F_1,O_2))/N$	0 to 1	1
Proportion Correct – Category 1 (PC_1)	$n(F_1,O_1)/n(O_1)$	0 to 1	1
Proportion Correct – Category 2 (PC_2)	$n(F_2,O_2)/n(O_2)$	0 to 1	1
Frequency Bias – Category 1 (FB_1)	$n(F_1)/n(O_1)$	0 to ∞	1
Frequency Bias – Category 2 (FB_2)	$n(F_2)/n(O_2)$	0 to ∞	1
PCT for Random Forecast (PCT_R)	$(n(F_1)n(O_1)+n(F_2)n(O_2))/N^2$	0 to 1	1
Heidke Skill Score (HSS)	$(PCT-PCT_R)/(1-PCT_R)$	$-\infty$ to 1	1

6.2.2 Ice Extent

Ice extent may be defined as the portion of the earth's surface (or over a defined geographical area) covered by ice with a concentration above a certain threshold. So while ice concentration may be a continuous variable bounded between 0 and 1, the ice extent resulting from imposing a binary ice–no ice categorization may also be evaluated. The threshold value used to categorize the continuous ice concentration into ice and no-ice categories is usually chosen either as a function of the observation capability or physically relevant values. Interpretation of categorical scores must consider that small errors in forecasts or observations near the threshold concentration value have the same numerical effect as larger errors.

There also could be a mismatch between the threshold chosen for the observations and the prediction. For example, Smith *et al.* (2015) verify sea ice concentration from the Global Ice Ocean Prediction System (GIOPS) versus the Interactive Multisensor Snow and Ice Mapping System (IMS) ice extent product and show the sensitivity of categorical evaluation scores to the threshold choice. The IMS ice-extent is defined using a 40 per cent ice concentration threshold whereas, for GIOPS, a 20 per cent sea ice concentration threshold is more appropriate as this value has been used in the assimilation to provide bogus values of sea surface temperature at the freezing point.

The use of multi-categorical evaluation scores can partially help address the issues associated with the choice of different thresholds. In the example discussed above, a multi-category contingency table can be evaluated based on different thresholds. The entries of this table are then combined and weighted by the entries of a scoring matrix which are defined to highly reward correct ice and water for the extreme categories near 0 and 1, while penalizing mildly incorrect forecast for intermediate categories.

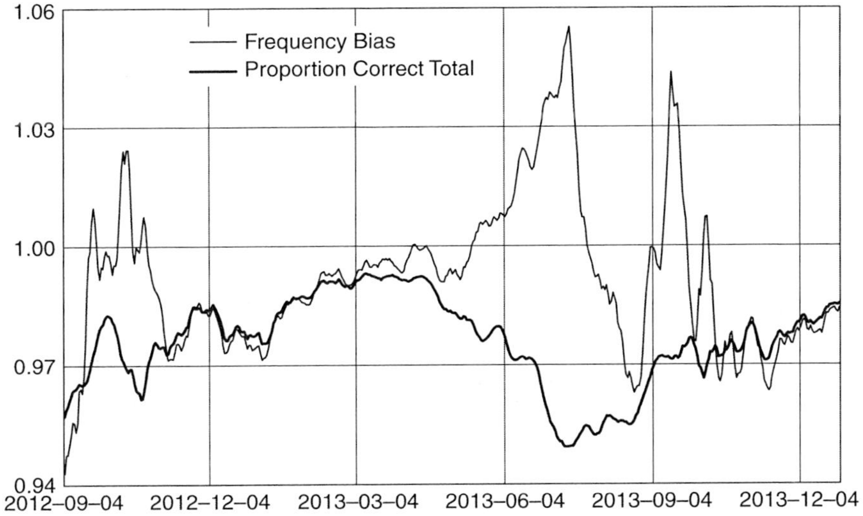

Figure 6.1: Time series of ice extent Frequency Bias and Proportion Correct Total from a comparison of the RIPS 18 UTC ice concentration analysis and IMS data. A threshold of 25 per cent has been used to determine the RIPS analysis extent.

Example: RIPS

A comparison of the Regional Ice Prediction System (RIPS) analysis to the manually prepared IMS analysis is shown in Figure 6.1. The published ice/no-ice threshold in IMS ice extent product documentation is 40 per cent, but an unpublished study (personal communication – Alain Caya) comparing IMS with Canadian Ice Service (CIS) Image Analysis Charts determined that the threshold may be closer to 25 per cent. For the evaluation process, RIPS output is first interpolated to the 4 km IMS observations and the latter threshold is used to determine the RIPS ice/no-ice category. Overall the RIPS analysis agrees with IMS to a very high degree but less so during the summer melt period (May–August) and during early stages of ice growth (September–October). In general, RIPS has a lower ice extent than IMS with exceptions in early summer and early fall.

6.2.3 Ice Pressure

While sea ice models predict ice pressure as a continuous variable, continuous observations do not exist. Most of the observations are reported by ships, and these tend to be subjective and related to the vessel capabilities. For example, a small fishing vessel can become beset in much lighter ice conditions than a heavy icebreaker. International standards for ships reporting ice pressure include the following categories: 'slight', 'moderate/severe' and 'ship beset'. Calibration and different thresholding of the modelled pressure values are needed, depending on the ship characteristics, in order to issue a forecast useful for navigation. Another issue is that models predict pressure over the dimensions of a grid cell, whereas ice pressure is localized (at subgrid-scales), so that evaluation is affected by representativeness issues. Evaluation against independent observations could then be for two categories (ice pressure or no ice pressure) or as many as five categories (beset, severe, moderate, light or no ice pressure) although there likely would be few observations for some or all of these categories.

Categorical scores from (multi-category) contingency tables are the natural approach for verifying sea ice pressure, when observations are expressed in categories. Like sea ice extent, evaluation scores are sensitive to the thresholds used for the predicted values, in order to define the forecast categories.

6.3 Evaluation of Continuous Variables

Continuous variables (as opposed to categorical variables) can take values in the continuum of real numbers. They can be bounded or unbounded. As an example, sea ice concentration can take continuous values in the range [0,1]; sea ice thickness can have values in the range $[0, +\infty)$. Note that these variables can be spatially discontinuous, in the sense that values of nearby points can be characterized by sharp changes (e.g. ice concentration in the vicinity of a lead).

6.3.1 Continuous Scores

Typical error indices of continuous variables include:

$$\text{Error}_i = E_i = F_i - O_i \tag{6.5}$$

$$\text{Bias} = \text{Mean Error} = \overline{E} = \frac{1}{N}\sum_{i=1}^{N} E_i \tag{6.6}$$

$$\text{Mean Absolute Error} = MAE = \frac{1}{N}\sum_{i=1}^{N} |E_i| \tag{6.7}$$

$$\text{Mean Square Error} = MSE = \frac{1}{N}\sum_{i=1}^{N} (E_i)^2 \tag{6.8}$$

$$\text{Mean Square Error Skill Score} = SS_{MSE} = 1 - \frac{MSE_F}{MSE_{REF}} \tag{6.9}$$

$$\text{Root Mean Square Error} = RMSE = \sqrt{\frac{1}{N}\sum_{i=1}^{N} E_i^2} \tag{6.10}$$

$$\text{Error Standard Deviation} = ESD = \sqrt{\frac{1}{N-1}\sum_{i=1}^{N}(E_i - \overline{E})^2} \tag{6.11}$$

$$\text{Correlation Coefficent} = CC = \frac{\sum_{i=1}^{N}(F_i - \overline{F})(O_i - \overline{O})}{\sqrt{\sum_{i=1}^{N}(F_i - \overline{F})^2 \sum_{i=1}^{N}(O_i - \overline{O})^2}} \tag{6.12}$$

Where

N is the number of forecast–observation pairs
E_i is the difference between the forecast F_i and observation O_i
MSE_F is the Mean Square Error in the forecast
MSE_{REF} is the Mean Square Error for a reference forecast

It is important to note that different statistics measure different attributes of the forecast–observation relationship. $\overline{E}$ measures the additive bias, i.e. the difference of forecast and observation averages (this statistic only compares the marginal distributions of forecast and observation). *MSE* is a measure of the accuracy (i.e. it relates to the joint behaviour of forecast and observation, hence their joint distribution). *CC* measures the strength of the linear dependence of forecast and observation and is not affected by the bias.

These continuous evaluation scores are not independent. In fact, the *MSE*, the additive bias, forecast and observed variances and correlation are related as:

$$MSE = \overline{E}^2 + \sigma_F^2 + \sigma_O^2 - 2\sigma_F\sigma_O r_{F,O}^2 = \overline{E}^2 + ESD \tag{6.13}$$

where σ^2 denotes the sample variance. This equation decomposes the *MSE* into a bias-component, the sum of the forecast and error variances, and a correlation component (which is related to the linear association between forecast and observation). This decomposition of the *MSE* enables the different error sources to be separately assessed. Often the *MSE* is evaluated after removing the bias from the forecasts (therefore the first term in the equation becomes equal to zero): the unbiased *MSE* is the *ESD*.

Equation 6.13 shows the dependence of the *MSE* on the forecast and observed variances. High-resolution models tend to be heavily penalized by the *MSE*, because of their intrinsic higher variability with respect to coarser resolution models. In order to decrease the effect of variability on the *MSE*, the reduction of variance (*RV*) can be calculated. The *RV* is the *MSE* skill score evaluated against a reference forecast equal to the sample climatology = $\overline{O}$. The *MSE* for a constant forecast equal to the sample climatology is the observation variance, hence

$$RV = 1 - \frac{MSE}{\sigma_O^2}. \tag{6.14}$$

The *RV* removes the effect of the observation variance, however the forecast variance still affects the score. In order to remove both observed and forecast variances, the *MSE* skill score with respect to a random forecast (with the same marginal distribution as the verified forecast) can be evaluated, where $MSE_R = \overline{E}^2 + \sigma_F^2 + \sigma_O^2$.

Note that continuous evaluation statistics based on quadratic rules can be misleading if applied to variables which have a non-Gaussian distributions. For example, sea ice thickness is characterized by a right-skewed distribution and caution should be applied in the interpretation of evaluation results through *MSE* and *CC*, since they are heavily affected by large values.

6.3.2 Ice Concentration

Ice concentration is a continuous variable bounded by 0 and 1. At large scales, ice concentration distributions tend to be U-shaped with peaks near 0 and 1. The large occurrence of correct ice and water can unduly affect evaluation scores (e.g. a PCT near 1 can easily be achieved). Since large areas of the ice and water do not change over short-term forecasts, the evaluation may be chosen to focus on those areas that have changed (e.g. evaluate only those grid cells where either the reference data or model forecast has changed by some threshold). This spatial targeting method provides a focused evaluation on the most relevant areas of the domain (e.g. the marginal ice zone), resulting in a more informative assessment of forecast skill and error. This effectively amplifies the error within the evaluation signal but it also results in variation of the number and location of points verified from one time period to the next. An alternative approach that also amplifies the error signal is to verify only areas within a certain distance from the ice edge. However, if the ice edge is not the only area of importance for the APS, then this method would provide incomplete information.

Example 1: RIPS

Although CIS image analysis charts are assimilated by RIPS, an unbiased comparison between RIPS and these charts can be performed by using charts that were prepared a short time after the analysis cut-off. Errors are calculated by taking the difference between the total ice concentrations from the image analysis charts and the RIPS analysed ice concentration interpolated to the locations for which the chart concentrations are available. Since the charts are prepared for operationally active marine areas, the evaluation results are mainly representative of the Canadian northern waters in summer and Canadian East Coast in winter. Figures 6.2 a) and b) show the *ESD* and bias of the 06 UTC analysis compared with the charts valid between 09 and 15 UTC. A number of observations can be made: 1) the most general problem areas are along the coast and in narrow channels, largely because they are unresolved by passive microwave satellite observations, which are often the primary observation source; 2) larger standard deviation errors and negative bias in the Arctic Archipelago compared to the East Coast are due to underestimated retrievals related to summer melt and due to prevalence of many channels unresolved by PM observations; 3) low errors along the East Coast ice edge are likely due to averaging periods of open water when the analysis has high accuracy along with periods when there is an ice edge present and the analysis has lower accuracy. The time series results for the analysis shown in Figure 6.3 indicate an overall relatively small and stable *ESD*, although there is significant shorter-term variability. RIPS exhibits an overall negative bias but with periods of net positive bias in summer. Similarly, RIPS forecasts may be compared with similar chart data. For the 36-hour forecast results shown in Figures 6.2 c) and d), the errors are more extreme and noisy than the analysis mainly due to less frequent sampling (i.e. forecasts were only prepared once per week) but a few problematic areas are seen (72N/70W, 70N/80W, 75N/80W) and are probably related to unresolved/unmodelled tides/polynyas/land-fast ice.

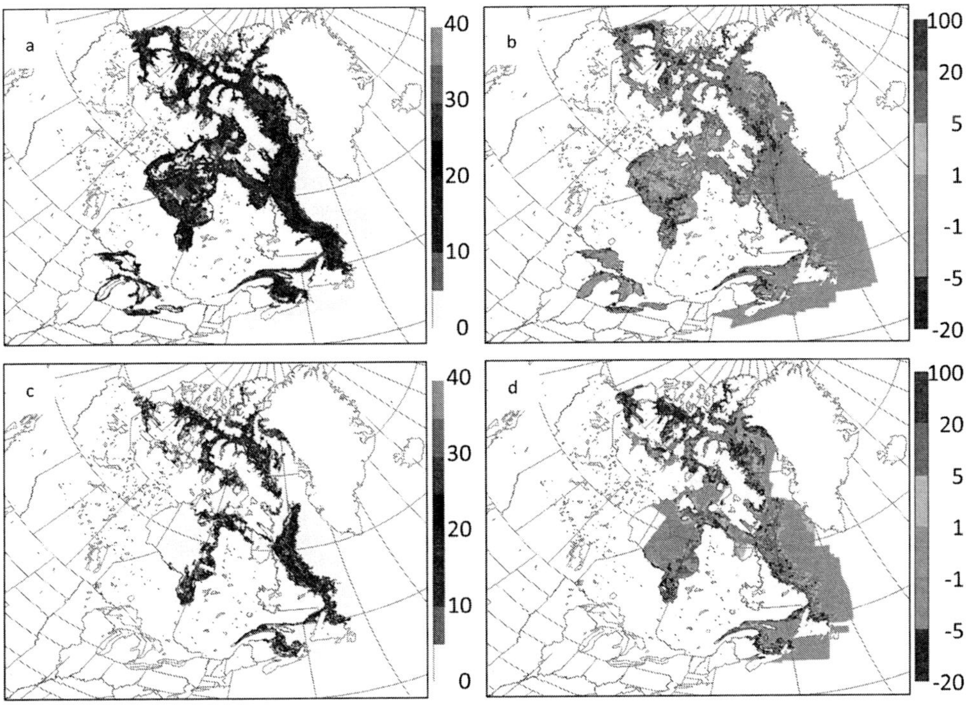

Figure 6.2: Evaluation of the RIPS ice concentration using CIS RADARSAT image analysis charts for all of 2013, (a, b) *ESD* and bias of 06 UTC analysis compared with charts valid from 09 to 15 UTC, (c, d) *ESD* and bias of 36-hour forecast valid at 12 UTC compared with charts valid from 09 to 15 UTC. (A black and white version of this figure will appear in some formats. For the colour version, please refer to the plate section.)

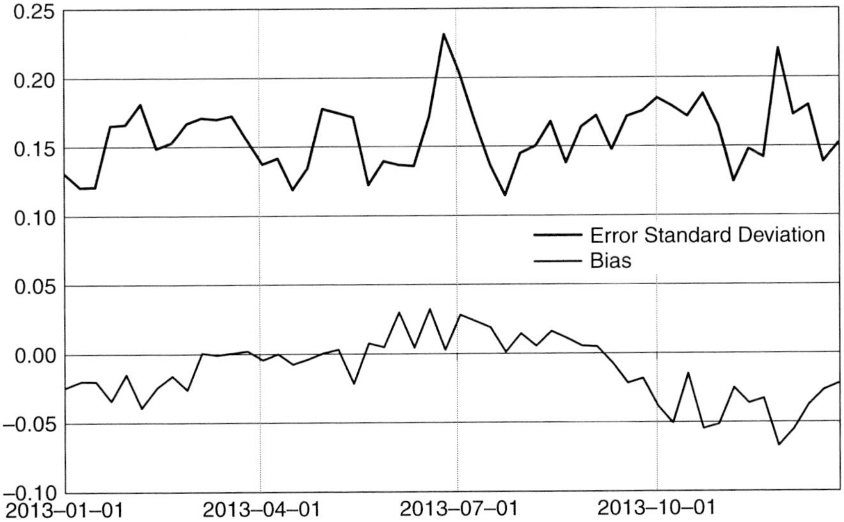

Figure 6.3: Time series comparing the RIPS 06 UTC ice concentration analysis with CIS RADARSAT image analysis charts valid from 09 to 15 UTC.

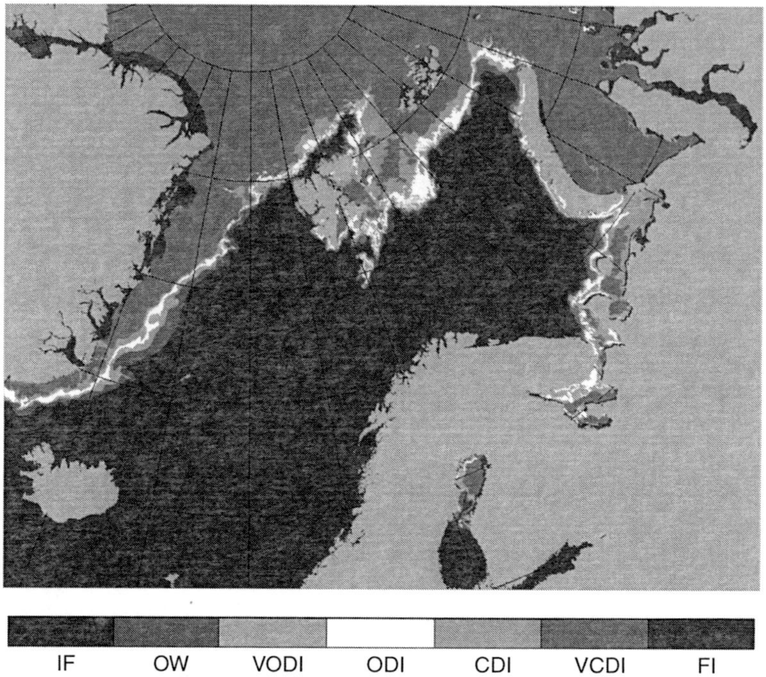

Figure 6.4: A sample map of the distribution of sea ice classes in the observational product. The entire area covered by the ice charts is displayed. Note that results for the Baltic Seas are discarded prior to evaluation. See the text for details on the definition of the various classes. (A black and white version of this figure will appear in some formats. For the colour version, please refer to the plate section.)

Example 2: Topaz4

Evaluation for TOPAZ4 (Towards an Operational Prediction system for the North Atlantic European coastal Zones) uses contingency tables similar to Table 6.1 to verify many different ice fields. For now, this approach is restricted to using Copernicus Marine Environment Monitoring Service (CMEMS) daily ice charts (see Figure 6.4), which have a horizontal resolution of 1 km and are produced for the European sector of the Arctic. For this evaluation, ice concentration categories are defined as:

- Fast Ice (FI): $A = 1.0$
- Very Close Drift (VCDI): $A = 0.95$
- Close Drift Ice (CDI): $A = 0.75$
- Open Drift Ice (ODI): $A = 0.5$
- Very Open Drift Ice (VODI): $A = 0.2$
- Open Water (OW): $A = 0.05$
- Ice Free (IF): $A = 0$

The resolutions of the TOPAZ4 product and the ice chart product have large contrasts both with respect to the horizontal representation (12.5km vs. 1km) and parameter space (continuous in the range [0,1] versus six discrete values). Moreover, the TOPAZ4 model does not distinguish between fast ice and drifting ice. Generally, interpolation from coarse resolution into a finer resolution cannot be defined uniquely, as an assumption about the distribution inside the coarse grid must be specified. On the other hand, integration from the fine resolution onto the coarse representation is a well-posed problem.

A corresponding argument leads to integration of the continuous representation of the TOPAZ4 results onto the ice chart categorical representation. Boundaries are set to the arithmetic mean of two neighbouring ice categories, so that, for example, regions with sea ice concentration in TOPAZ4 between 0.35 and 0.625 are mapped into the ODI category. Furthermore, since FI is not represented in the model results, this ice chart category is merged into the VCDI category. The very low values in the OW category are discarded by considering this region to be (truly open) water. Each grid point value then belongs to one of the four remaining categories, or (open) water. As stated above, the ice charts and the model results are both based on daily results valid at the same time, so there is no need to interpolate in time.

The entire areal evaluation of sea ice results from each day can then be provided as a 5×5 contingency table. With a forecast range of 10 days, not much change is expected at a distance into the OW and VCDI categories. Consequently, for most evaluation quantities, the analysis is restricted to a region bounded by a distance of 200 km into OW from ice chart VODI, and 200 km into VCDI from ice chart CDI. Hence, the contingency table columns for OW and VCDI are each split into two columns, representing conditions near and far (>200 km) from the respective category boundaries. As a result, the modified contingency table has a size of 7-by-5, neglecting the totals.

The results that are discarded in most of the evaluation metrics discussed below, are the contents of the top-left and bottom-right matrix cells (18025 and 0 in Table 6.3).

Sea ice concentration bias is derived from the non-diagonal entries in the contingency table. Sea ice concentration *RMSE* is derived using the same approach. Monthly average evaluation results for TOPAZ4 sea ice concentration forecasts for the period 2012–12 to 2016–02 are shown in Figure 6.5. The bias reveals that model sea ice concentration values are somewhat small during summer and a bit high during winter. The bias results for the

Table 6.3: *Sample contingency table used for TOPAZ4 evaluation.*

		Observations						
		OW	MIZ	VODI	ODI	CDI	VCDI near MIZ	VCDI
	OW	18025	5832	225	98	55	109	0
	VODI	2	466	79	87	33	19	0
Model	ODI	0	420	113	123	85	29	0
	CDI	0	87	156	222	187	153	0
	VCDI	0	27	75	242	682	6550	0

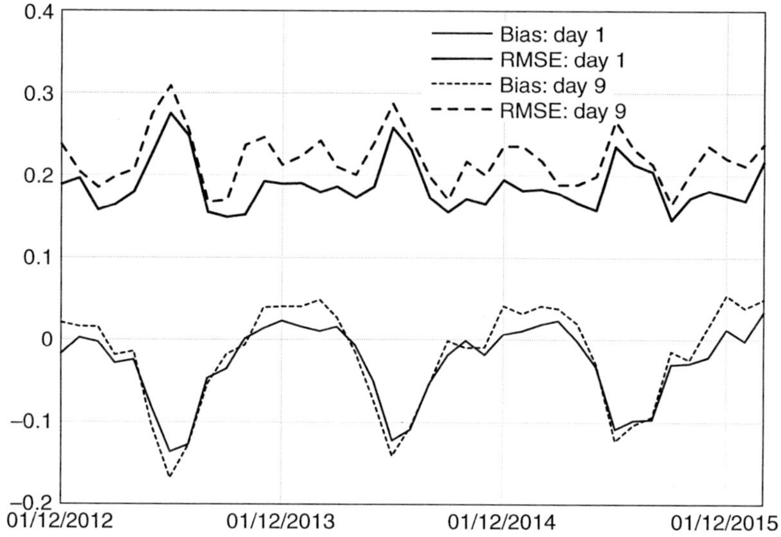

Figure 6.5: Domain average bias and root mean square difference between CMEMS observed and predicted sea ice concentrations.

one day forecast range are slightly closer to the observations than the nine-day forecast. Although there are summer maxima in the *RMSE*, the seasonal cycle in this evaluation metric is less obvious than in the bias. Again, the *RMSE* is smaller for a one-day forecast than for a nine-day forecast.

6.3.3 Ice Thickness

Ice thickness is bounded at the low end by zero and can reach up to tens of metres in extreme cases. Like ice concentration, ice thickness is a spatially discontinuous field. Uniform thermodynamic ice growth and melt may be accompanied by local strong deformations caused by dynamic processes. As described in Chapter 2, ice thickness is represented by most sea ice models using a sea ice thickness distribution (ITD). The effects of thermodynamics, deformation and advection result in an evolution of the ITD within a grid cell.

Evaluation of ice thickness usually focuses on the mean thickness within a grid cell. If observations support it, the observed and forecast ITD could be compared by looking at such things as the mean and standard deviations. Alternatively, probabilistic evaluation scores that enable the comparison of two sample distributions could be used, such as the Continuous Ranked Probability Score, or statistics that measure the distance between two sample distributions, such as the Kologomorov–Smirnoff distance.

Continuous scores can be used to verify ice thickness. Sea ice thickness, however, is characterized by a right-skewed distribution. Therefore, evaluation statistics based on a quadratic rule (e.g. *MSE* and *ESD*) will be heavily affected by even a few errors associated with large thicknesses. This might or might not be desirable, and interpretation of the

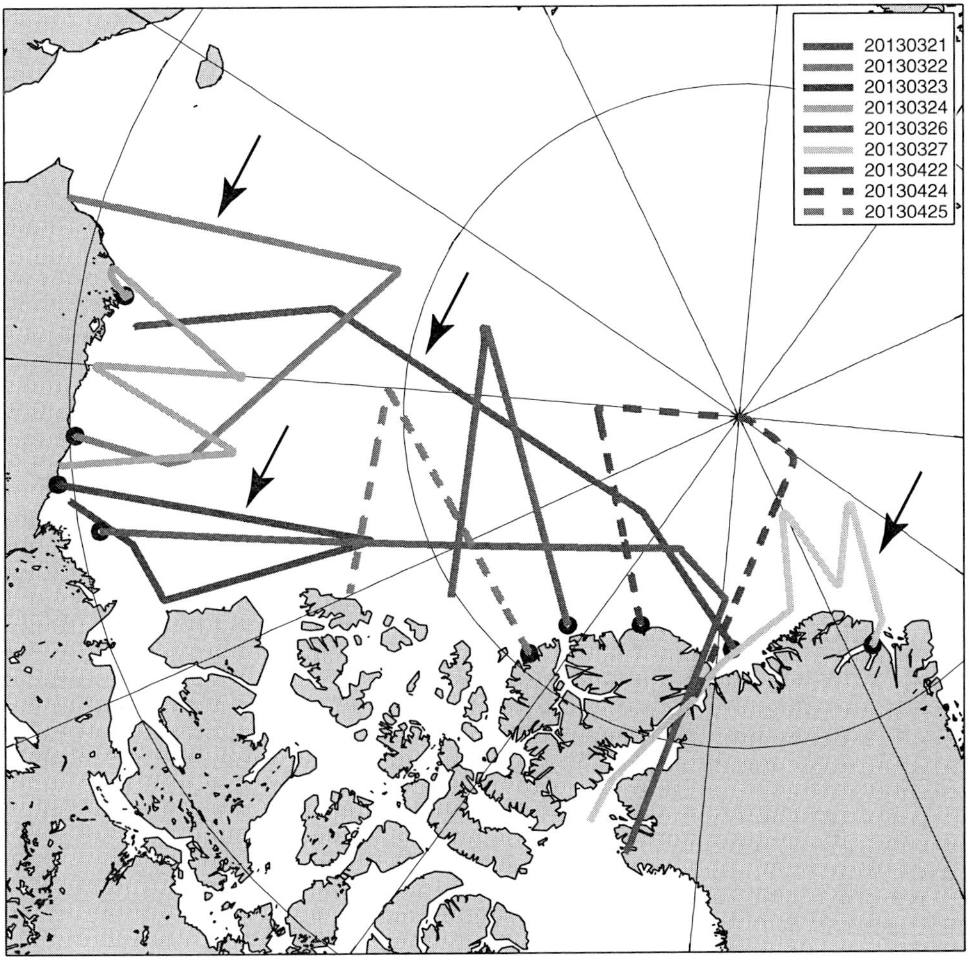

Figure 6.6: NASA Operation Ice Bridge aircraft flight paths over the Arctic Ocean in Spring 2013. The black circle indicates the starting point of each flight and the flight numbers correspond to those used in Table 6.4. (A black and white version of this figure will appear in some formats. For the colour version, please refer to the plate section.)

evaluation results needs to account for this. Moreover, sea ice thickness (like ice pressure) can be discontinuous in space, and hence evaluation results might suffer from double penalties for small errors in predicted drift and due to under-sampling.

Example: GOFS 3.1/ACNFS

To verify the Global Ocean Forecast System (GOFS) 3.1 and Arctic Cap Nowcast/Forecast System (ACNFS) ice thickness forecasts, National Aeronautics and Space Administration (NASA) Operation Ice Bridge (OIB) mission high-resolution ice thickness measurements from 2013 (see Figure 6.6) were filtered via a running mean that smoothed the time series sufficiently. Several time filters were examined, and a six-minute filter appeared reasonable. This equates to approximately 45 km of flight

Table 6.4: *Statistics comparing Ice Bridge ice thickness data (m) against GOFS 3.1 and ACNFS sampled along the same flight paths and on the same dates. Lower error is underlined.*

	Bias		Mean Absolute Error		RMS Difference	
Flight	GOFS 3.1	ACNFS	GOFS 3.1	ACNFS	GOFS 3.1	ACNFS
20130321	$\underline{-0.43}$	0.60	0.98	$\underline{0.90}$	1.22	$\underline{1.09}$
20130322	$\underline{0.39}$	0.98	$\underline{0.54}$	1.08	$\underline{0.67}$	1.33
20130323	$\underline{0.23}$	1.04	$\underline{0.55}$	1.33	$\underline{0.77}$	1.59
20130324	$\underline{0.59}$	0.82	$\underline{0.82}$	1.01	$\underline{1.05}$	1.32
20130326	−0.76	0.76	$\underline{0.96}$	1.09	$\underline{1.23}$	1.32
20130327	−1.89	$\underline{-1.11}$	1.91	$\underline{1.45}$	2.14	$\underline{1.93}$
20130422	$\underline{-0.57}$	0.80	$\underline{0.83}$	0.85	1.00	$\underline{0.99}$
20130424	−1.33	$\underline{-0.11}$	1.40	$\underline{0.62}$	1.87	$\underline{0.94}$
20130425	$\underline{-0.28}$	1.46	$\underline{0.63}$	1.47	$\underline{0.79}$	1.55

path. GOFS 3.1 and ACNFS outputs were then interpolated to these flight paths for the error statistics presented in Table 6.4. It should be noted that the Ice Bridge data exhibit significantly more spatial variability than both forecast systems and that, in general, GOFS 3.1 ice thickness is closer to the observed flight data in the regions of the Beaufort Sea and Canadian Archipelago, whereas ACNFS has lower error for the region north of Greenland.

6.3.4 Anomaly Correlation

The anomaly correlation coefficient (ACC) is equivalent to *CC* (Equation 6.12), noting that the forecast and observation means are the respective climatologies. Of importance to seasonal forecasting is the ability of the forecast to capture interannual variations as well as any longer-term trends. The example below illustrates how these separate skills can be quantified.

Example: CanSIPS

Twelve month historical forecasts produced by the Canadian Seasonal to Inter-annual Prediction System (CanSIPS) were initialized at the start of each month from January 1979 onward. These hindcasts not only serve to bias correct and calibrate operational forecasts, but also to provide an estimation of the expected forecast skill. Figure 6.7 shows the ACC for historical CanSIPS forecasts (1979–2009) of Arctic sea ice area as a function of initialization and target month. Figure 6.7 (left panel) shows that while the highest skill is found in the first few months of the forecasts, skill remains high and statistically significant up to the end of the 12-month forecasts. The right panel of Figure 6.7 reveals that most of the skill on longer lead time stems from the strong downward trend of sea ice area. It shows that the forecast skill of linearly de-trended anomalies is substantially smaller, especially for the longer lead times, and that significant trend-independent forecast skill is mostly restricted to the first 2–3 months of the forecast. A noticeable exception is skill for February forecasts, which is

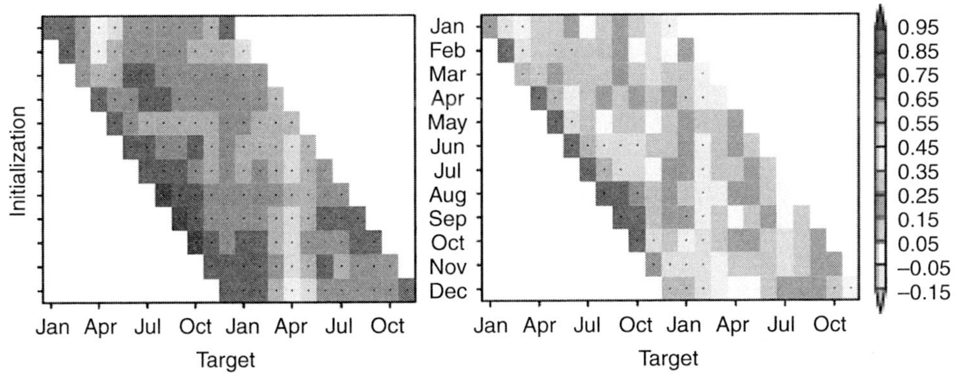

Figure 6.7: CANSIPS forecast skill (quantified by the anomaly correlation coefficient) for sea ice area as a function of initialization and target (predicted) month, evaluated against observations from the National Snow and Ice Data Center. Black dots represent statistical significance at the 95 per cent confidence level as determined by bootstrapping. The left panel is for full anomalies, and the right panel is for linearly de-trended anomalies (adapted from Sigmond *et al.*, 2013). (A black and white version of this figure will appear in some formats. For the colour version, please refer to the plate section.)

significant even when initialized on March 1 of the previous year. For most initialization and target months, trend-independent forecast skill is higher than that of a simple persistence forecast (see Sigmond *et al.*, 2013 and Merryfield *et al.*, 2013 for details).

6.4 Evaluation of Spatial and Vector Fields

6.4.1 Spatial Evaluation Approaches

Sea ice is defined over a spatial domain and is characterized by a coherent spatial structure and the presence of features. Traditional point-by-point evaluation approaches do not account for the intrinsic relation existing between nearby grid-point values. Offsets in the forecast location of an ice area can result in double penalties and these errors are typically larger as the model spatial resolution increases. Moreover, traditional point-by-point evaluation methods do not provide informative diagnostics for some of the forecast errors characterized by a spatial nature, such as distance errors.

In order to address these issues, several spatial evaluation approaches have been developed for atmospheric forecasts. These have been classified in four categories: scale-separation, neighbourhood, feature-based and field-deformation (Gilleland *et al.*, 2010), and their diagnostic capabilities have been analyzed and compared within internationally coordinated meta-verification inter-comparison projects (www.ral.ucar.edu/projects/icp). The first two classes are referred to as filtering methods since they use single-band and low-band spatial filters. The latter two classes are referred to as displacement methods since they explicitly assess distances/displacement errors.

Scale separation evaluation approaches decompose forecast and observation fields into the sum of different spatial scale components by using a single-band spatial filter (e.g.

Fourier, spherical harmonics, wavelets, etc.), and then apply traditional (continuous, categorical or probabilistic) evaluation on each scale component separately. These approaches enable the scale-dependence of the bias and error to be assessed as well as the skill–no skill transition scale, and the forecast capability in reproducing the observed scale structure.

Neighbourhood evaluation approaches (Ebert 2008) were developed to reward the enhanced resolution of NWP systems. These methods relax the strict requirement of exact forecast timing and location and consider neighbourhoods of grid-points both in space and time within which forecast and observation are matched. Within the neighbourhood, two forecast–observation matching approaches are possible: 1) if observations are available only at point locations, a single observation is compared against the distribution of the neighbourhood forecast values and 2) if spatial observations are available, the distributions of the observation and forecast values within the neighbourhood are compared. A variety of data treatment approaches within the neighbourhood define different evaluation strategies: these can range from simple up-scaling of forecast and observed values, and then evaluate standard continuous and categorical scores, to comparing the forecast and observed distribution with probabilistic scores, such as the Continuous Ranked Probability Score. In the latter case, the neighbourhood approaches enable to verify a deterministic forecast with a probabilistic approach, where the probabilities implicitly account for the uncertainties associated with small spatial and temporal scales.

Field deformation techniques use a vector field (which can be interpreted as advection or wind field) to deform the forecast field towards the observed field, up until an optimal fit is found (e.g. by maximizing a likelihood function). An amplitude (scalar) field is then applied, in order to correct the intensities of the deformed forecast field to those of the observed field. Field deformation techniques usually perform the error decomposition on different spectral components (as for the scale-separation approaches). Hence, they provide direct information about small-scale uncertainty versus large-scale errors. Feature based evaluation techniques first identify and isolate features in forecast and observation fields (e.g. by thresholding), and then separately assess the displacement and other sources of error (e.g. extent, shape, intensities) for each pair of observation and forecast features. The Method for Object-Based Diagnostic Evaluation (Davis *et al.*, 2006) is the most sophisticated feature-based evaluation approach; however, its sensitivity to the parameters which identify, merge and match the forecast and observed features makes its implementation difficult in an operational environment. Metrics associated with feature-based methods include the Hausdorff family distances (Dubuisson and Jain, 1994) and the Baddeley (1992) Delta metric. These binary image distance metrics do not need to explicitly isolate, merge and match features and hence results are more robust and suitable for operational evaluation.

Because of its enhanced diagnostic power, spatial evaluation of sea ice attributes is recommended, alongside traditional evaluation. This will become particularly useful as APS resolution increases.

6.4.2 Ice Edge

The World Meteorological Organization Manuals on Sea Ice Nomenclature[1] define ice edge as 'the demarcation at any given time between the open sea and sea ice of any kind, whether fast or drifting'. However, a more general definition used here is the contour separating ice with concentration above a certain threshold from areas with concentration lower than the threshold. The ice edge is related to ice extent because it defines the boundary between ice covered areas and open water. Ice charts generally use a threshold of 10 per cent because that is the lowest concentration level that could be reliably detected using visual observations from aircraft. Higher thresholds may be used depending on the need or the observation source.

Ice edge error statistics provide an indication of the distance offset between an observed and a predicted ice edge location. A simple first approach could be to take areal differences in ice extent and infer the ice-edge distance by dividing by the length of the ice edge. This approach requires a measurement of the predicted ice extent outside of the observed edge (IE^{+}); the observed ice extent outside the predicted ice edge (IE^{-}); and, the average of the predicted and observed ice edge length (L). The average edge error would then be $(IE^{+} + IE^{-})/L$ and the average ice edge bias is $(IE^{+} - IE^{-})/L$. An example of this technique was introduced by Goessling *et al.* (2016). This approach bridges the traditional categorical scores to spatial evaluation approaches which quantify displacement errors (such as the object-oriented methods) and provides simple yet informative evaluation results.

A second approach is to measure the average distance between a predicted and observed ice edge. In theory, points could be sampled from either edge to the other and the minimum distance would provide one measurement. Evenly sampled measurements could then be averaged to provide the average ice edge error. Keeping track of whether the measurements are inside or outside the reference edge can provide the ice edge bias.

A third approach for ice edge evaluation is the use of the distance metrics for binary images proposed in the literature of edge detection and pattern recognition analysis, such as the Hausdorff method. These metrics are sensitive to the distance between two contours and/or binary objects (e.g. forecast and observed sea ice extents) and they are capable of detecting differences/similarities in shape. These metrics rely on the evaluation of the forward distances d(a, B) for all the points $a \in A$ to the object B, and the backward distances d(b, A) for all the points $b \in B$ to the object A. Different statistics are then evaluated from the sets of forward and backward distances, such as the maximum (the Hausdorff distance), the median or a quantile (the partial Hausdorff metric) and the mean (the modified Hausdorff metric).

Relatively few studies have used the second and the third approaches for sea ice edge evaluation (e.g. Heinrichs *et al.*, 2006; Dukhovskoy *et al.*, 2015; Hebert *et al.*, 2015) described above. Although promising, general agreement has not been achieved concerning the implementation details mainly due to issues discussed below.

[1] See www.jcomm.info/index.php?option=com_oe&task=viewDoclistRecord&doclistID=160

In practice, difficulties using the second and third approaches can arise due to arrangements of the ice with respect to a coastline or with respect to the main ice pack. Some examples of different orientations of leads along a coastline and ice areas that are dissociated from the main pack are illustrated in Figure 6.8. Here are a few issues and comments related to each example:

a) The edge error calculation should be straightforward and the edge bias should be zero;
b) The edge error for the dissociated forecast ice area could be calculated as the shortest distance from the observation edge to forecast edges. Alternatively, the error could be the average radius of the dissociated ice. The forecast bias error should be positive as there is excess ice extent compared to the observation;
c) The edge error could be measured from the observation edge or from the land boundary;
d) This is a combination of b) and c) where the dissociated area of forecast ice lies along a land boundary; and
e) The observed and forecast ice edge completely overlap but the ice areas are on opposite sides of the edge. Clearly the ice edge error should not be zero.

In principle, for each of these cases, the forecast and observed ice areas should be interchangeable with similar ice edge errors obtained, other than a change in sign for the bias. Some of the above issues could be resolved by ignoring ice edges within a certain distance from a coastline and acknowledging this when interpreting the results.

Example 1: RIPS

A comparison of the difference in RIPS 48-hour lead time forecast and IMS ice edge location is show in Figure 6.9. A method similar to the second approach is used. Ice edge points within 50 km of a coastline are ignored. As a proxy for the ice edge, the RIPS 0.4 ice concentration contour location is compared to the IMS ice extent edge. The large errors in August and September are mainly related to ice areas missing in the forecast, whereas the error in mid-October is largely due to excessive forecast ice area.

6.4.3 Ice Drift

Although sea ice models predict instantaneous ice velocity, this quantity is not measured. Ice displacement over periods from hours to days can be obtained from drifting ice buoys or derived by identifying ice features in time sequential satellite images, as described in Chapter 3. One approach to compare the forecast drift to the observation is to position fictitious particles at the observed initial positions and to advect the particles in a Lagrangian way using the model ice velocity. The observed and model drifts may then be compared in a variety of ways. If the duration of all observed ice drifts is the same, the net distance travelled and the angle between the forecast and observation can be compared. Sample statistics are:

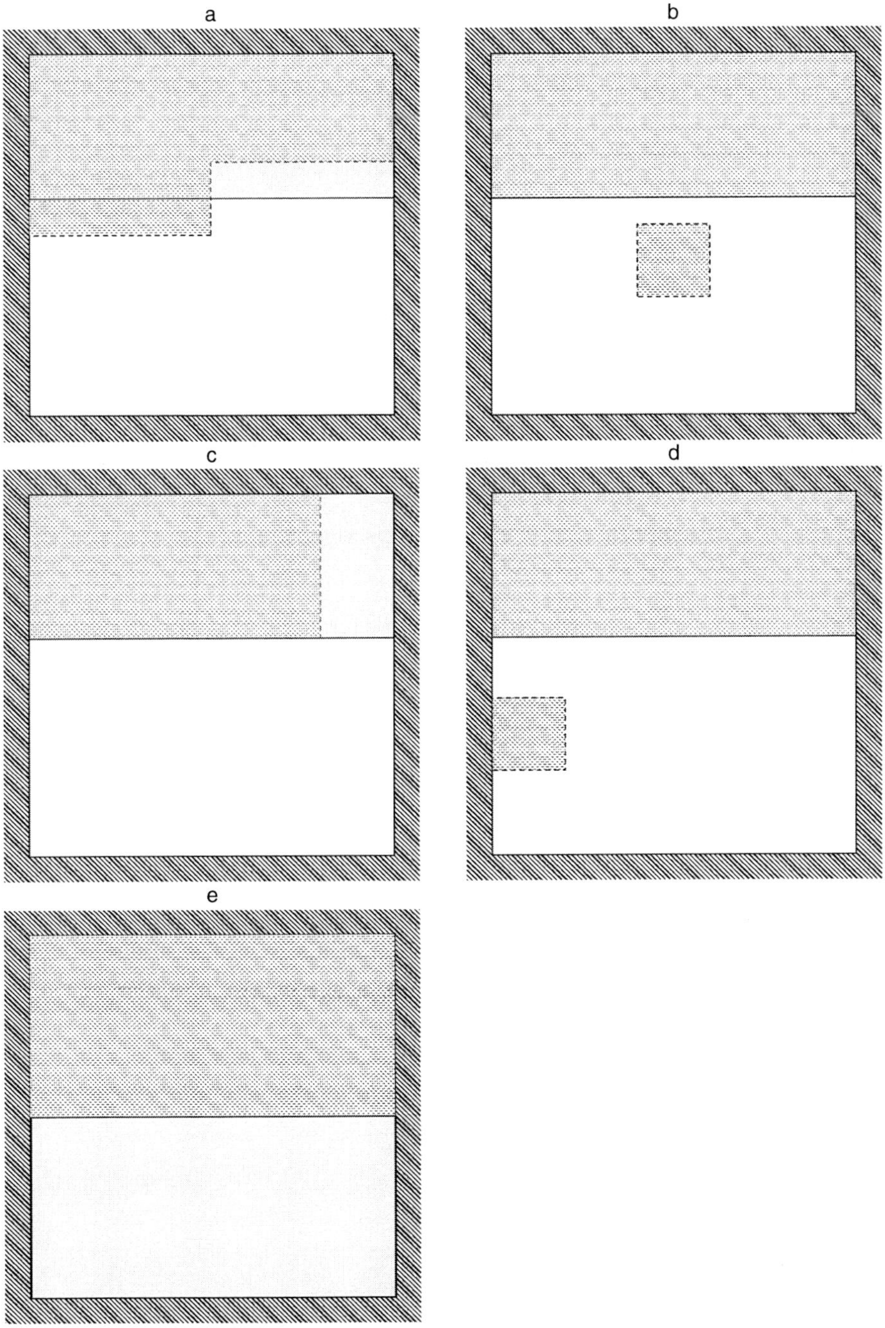

Figure 6.8: Idealized configurations of forecast ice areas (dashed), forecast ice edge (dashed line), observed ice area (shaded) and observed ice edge (solid line). Configurations illustrate: a) excess forecast edge balanced by an equivalent missing forecast area; b) forecast ice detached from the main ice area; c) forecast lead that does not exist in the observations; d) forecast ice area detached from the main overlap area and adjacent to a shore line; and, e) forecast and observed ice edges overlapping but with ice areas on opposite sides of the edges.

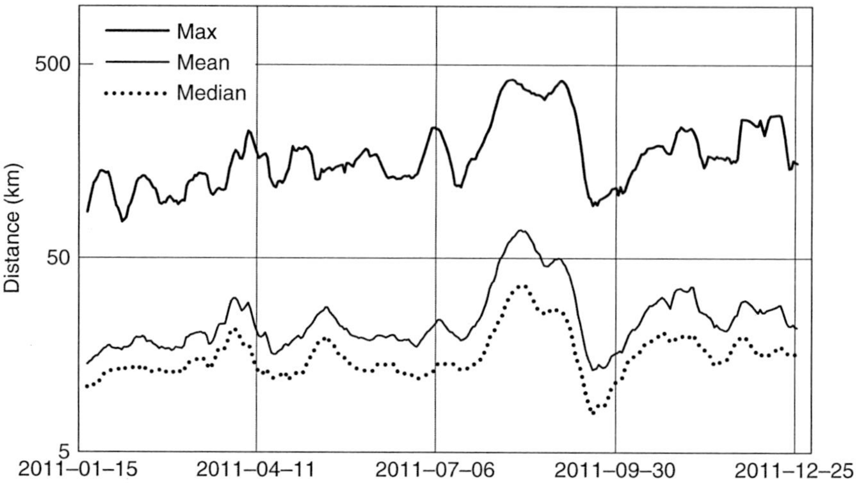

Figure 6.9: Median, mean and maximum distance to the ice edge of the RIPS 48-hour lead time forecast compared to the IMS observed ice edge. Note that an 11-day running mean filter has been applied to these results.

$$\text{Distance Error} = \frac{1}{N}\sum_{i=1}^{N}(D_{Oi} - D_{Fi}) \tag{6.15}$$

$$\text{Angle Error} = \frac{1}{N}\sum_{i=1}^{N}(\theta_{Oi} - \theta_{Fi}) \tag{6.16}$$

where D_{Oi} are the observed drift distances, D_{Fi} are the forecast drift distances, θ_{Oi} are the observed drift directions and θ_{Fi} are the forecast drift directions.

Alternatively, if the drift observations cover different time periods, average velocities may be calculated and speed and direction compared. In this case, the observed and model ice drift may be related as:

$$(V_{Fi} - \overline{V_F}) = (V_{Oi} - \overline{V_O})Ge^{i\theta} + \varepsilon, \tag{6.17}$$

where V_{Fi} and $\overline{V_F}$ are the individual and mean forecast drifts, V_{Oi} and $\overline{V_O}$ are the individual and mean observed drifts, G is a scaling factor, θ is the turning angle between the observed and forecast drift vector and ε represents the unexplained or residual drift. The latter three parameters are determined by finding the least squares fit between a series of forecast and observed ice drift pairs. A perfect forecast would have a value for 1 for G and 0 for θ and ε.

Note that the aforementioned statistics verify ice drift by using traditional continuous scores, and by matching forecast and observation point-by-point. However, sea ice drift is a vector field, and as such it should be verified spatially, in order to account

Table 6.5: *Average ice speed and component velocity (cm s^{-1}) for the IABP ice drifting buoys along with the corresponding values for GOFS 3.1 and ACNFS for the period January–August 2014. The last two columns are differences from the observations (along with the percentage). Lower error is underlined.*

Variable	Observed	GOFS 3.1	ACNFS	GOFS – Observed	ACNFS – Observed
		Statistics over the period January–August 2014			
Speed	8.78	9.97	9.59	1.19 (14%)	0.81 (9%)
U-velocity	6.09	6.85	6.41	0.77 (13%)	0.32 (5%)
V-velocity	5.07	5.89	5.85	0.83 (16%)	0.79 (16%)
		Statistics over the period January–March 2014			
Speed	7.90	9.43	9.96	1.53 (19%)	2.06 (26%)
U-velocity	5.23	5.79	6.06	0.56 (11%)	0.83 (16%)
V-velocity	4.74	6.12	6.57	1.38 (29%)	1.83 (39%)
		Statistics over the period July–August 2014			
Speed	10.41	11.20	9.87	0.79 (8%)	−0.54 (−5%)
U-velocity	7.28	7.85	6.75	0.57 (8%)	−0.53 (−7%)
V-velocity	5.95	6.49	5.91	0.54 (9%)	−0.04 (−1%)

for the intrinsic spatial correlation existing between nearby grid-point values. If satellite-derived spatial observations of sea ice drift are available, field deformation evaluation approaches may be used, since they are based on similar principles as the algorithms used to detect the ice drift from satellite image pairs. Field deformation techniques use a vector field to deform the forecast field towards the observed field, up until an optimal fit is found. In the case of sea ice drift, the vector field obtained would represent the sea ice drift error.

Example: ACNFS/GOFS 3.1

Ice drift from the GOFS 3.1 and ACNFS is compared against drifting buoys from the International Arctic Buoy Program. Over the January to August 2014 hindcast period, a total of 129 drifting buoys are used. From the daily latitude/longitude pairs of the ice-bound drifters, observed ice drift components are derived using the Haversine formula to determine the x- and y-direction distances travelled each day. Model ice velocity components are interpolated via cubic splines to the observed positions. Table 6.5 shows the ice drift error statistics that have been broken down into the winter (January–March) and summer (June–August) seasons.

6.5 Significance of the Evaluation Results

6.5.1 Robustness of the Evaluation Results

Aggregation of evaluation statistics over a sufficiently large sample of cases is sought for assuring robustness. On the other hand, aggregation should be performed only for cases with similar behaviour because signals in the error might otherwise cancel out and should be counterbalanced by adequate stratification (i.e. to separate the data sample in subsamples with similar statistical characteristics). A physically meaningful stratification (e.g. spatial or temporal) of the evaluation data may enable better interpretation of the results. However, there is a delicate balance between stratification and aggregation. Statistical inference can help resolve this balance and evaluation results should always be accompanied by significance or confidence intervals.

Significance and confidence intervals reflect the uncertainty of the evaluation statistics due to sampling (i.e. the uncertainty due to the intrinsic variability and homogeneity/inhomogeneity of the data sample). They address questions such as: a) would the same results be obtained with a sub-sample of the data?, and b) would the same evaluation results be obtained with a different sample (e.g. when verifying the same season from two different years)? Inference methods for evaluation statistics can be parametric (e.g. the traditional t-test) or non-parametric and based on re-sampling approaches (such as the bootstrapping or permutation tests). In what follows, some of the most common inference methods for evaluation scores are discussed.

6.5.2 Hypothesis Testing

One of the most common evaluation tasks is to compare a newer version of a model (APS_2) with an older one (APS_1). Evaluation statistics are then evaluated for the same period and against the same observation dataset. As an example, an ice concentration $MSE_1 = 0.20$ is obtained for APS_1 and $MSE_2 = 0.18$ is obtained for APS_2. Since $MSE_2 < MSE_1$ this implies that APS_2 performs better than APS_1. However, is this performance difference statistically significant? It is important to prove that the scores are significantly different, implying that APS_2 is really performing better than APS_1. Statistical inference through hypothesis testing can be used to address this question.

Hypothesis testing begins with the assumption that the two scores are not significantly different. This forms the null hypothesis H0 which states that MSE_1 and MSE_2 are equivalent, i.e. that their difference is close to zero. Usually parametric assumptions are made on the distribution (i.e. sample uncertainty) of MSE_1 and MSE_2, and with probabilistic argument the hypothetical (null) distribution of the difference is obtained (the null distribution then encompasses zero). The real difference of MSE_1 and MSE_2 is then placed within this hypothetical null distribution. When this difference lies outside the 95 per cent prediction interval of the null distribution, it can be concluded that the two scores are significantly different at the 95 per cent confidence level (or 5 per cent significance level). A classical example is the student t-test.

Statistical inference by using parametric tests is sometimes worrisome for some evaluation practitioners because of the hypotheses that the evaluation data ought to meet. However, these hypotheses are not as strict as they seem, and the uncertainty of most of the evaluation scores follow a normal distribution (i.e. the t-test is almost always applicable). On the other hand, non-parametric resampling methods (e.g. bootstrapping or permutation tests) are distribution free and have no constraints on probability theory.

6.5.3 Confidence Intervals by Bootstrapping

A second common task in evaluation is to relate an evaluation score to a confidence interval (CI), for a desired statistical significance level. Once the CI is provided, it can be determined whether the score is significantly different from a reference target value. For example, if the 95 per cent CI of a skill score is positive and well separated from zero, then, and only then, can it be concluded with statistical significance that the forecast is skilful. Note that the CI, which expresses the uncertainty around the score, is complementary to the prediction interval used in hypothesis testing, which express the uncertainty of the null hypothesis. For the skill score, the prediction interval would be centred on zero (i.e. no skill), and a skilful forecast would have a skill score which does not belong to the prediction interval.

Bootstrapping is a non-parametric resampling method for constructing CIs around evaluation scores. It does not require any knowledge and/or assumption on the distribution of either the variable verified or the score. In bootstrapping, the CI of a score is constructed by resampling the evaluation data. Suppose our evaluation dataset consists of a sample of 30 observation–forecast pairs. This data can be resampled using random selection and allowing repetition. If this process is repeated, multiple sets of 30 forecast–observation pairs can be obtained. The evaluation score is then computed from the original sample and from a large number of resamples. The distribution of scores from the resamples provides an indication of the uncertainty around the evaluation score of the original sample. The 95 per cent CI of the evaluation score is then simply determined by the empirical 2.5 per cent and 97.5 per cent quantile of the resampled score distribution.

If two evaluation scores have well separated, non-overlapping CIs (e.g. MSE_1 and MSE_2 in the example provided in the previous section), then it can be concluded that the scores are significantly different and that APS_2 is significantly better than APS_1. However, the reverse cannot be stated. That is, if MSE_1 and MSE_2 have overlapping CIs, APS_2 could still be better than APS_1. In fact, consider the case in which the MSE of both APS_1 and APS_2 forecasts have very large variability. The CIs of the scores will be large and possibly overlapping. However, if MSE_2 is consistently smaller than MSE_1, then the CI of the score difference won't include zero, and APS_2 would be significantly better than APS_1. When comparing two evaluation scores, bootstrapping should be performed on the differences for the two forecasts, rather than directly on the individual scores.

6.6 Chapter Summary

Sea ice APS evaluation builds on techniques developed for other applications, but presents some additional challenges. One of the challenges is that in situ measurements are sparse and might not be representative of the entire domain. Satellite observations are widely used to infer ice conditions. However, as they are not direct measurements, evaluation datasets might be affected by the assumptions behind the retrieval algorithms. Therefore, care must be taken in the interpretation of the evaluation results.

While APS predict ice conditions as continuous, sometimes bounded quantities, a number of verifying observations are only available as categorical quantities. A common approach for verifying categorical predictions such as ice extent and ice pressure is to use a (multi-categorical) contingency table. It is essential to be aware of the sensitivity of the categorical scores to the thresholds used with these approaches.

Evaluation of continuous scalars, such as ice concentration and thickness, uses common statistical measures. These statistics can be affected by large values and double penalties rising from spatial discontinuities.

Sea ice velocity and ice edge require specialized evaluation approaches which account for their spatial characteristics. Ice drift measurements are widely available with good spatial coverage and resolution but have less temporal resolution, so the tendency is to focus on ice displacement. Ice edge evaluation based on distance metrics is an area of active research.

Evaluation results provide information which must be interpreted in order to be useful. Geographical display of regional evaluation results can help with the physical interpretation of areal average results. Verification results should always be accompanied by statistical significance to ensure they are robust and meaningful.

Several aspects of sea ice evaluation practice are still unresolved. However, these should be regarded not as issues, but rather as challenges for future developments in sea ice and evaluation research. Suggestions for further developments appear in Jolliffe and Stephenson (2012) and the verification chapter in Wilks (2011). Another good reference is the verification webpage www.cawcr.gov.au/projects/verification (and references therein), maintained by the WMO World Weather Research Programme Joint Working Group in forecast Verification Research (JWGFVR). Gilleland *et al.* (2010) provides a review and classification of spatial verification methods.

References

Baddeley, A. J. (1992). An error metric for binary images. *Robust Computer Vision: Quality of Vision Algorithms*, W. Förstner and S. Ruwiedel, eds., Wichmann, 59–78.

Davis, C. A., Brown, B. G. and Bullock, R. G. (2006). Object-based verification of precipitation forecasts, Part I: Methodology and application to mesoscale rain areas. *Monthly Weather Review*, **134**, 1772–1784.

Dubuisson, M.P. and Jain, A. K. (1994). A modified Hausdorff distance for object matching. *Proceedings of the International Conference on Pattern Recognition*, Jerusalem, Israel, 566–568.

Dukhovskoy, D. S., Ubnoske, J., Blanchard-Wrigglesworth, E., Hiester, H. R. and Proshutinsky, A. (2015). Skill metrics for evaluation and comparison of sea ice models. *Journal of Geophysical Research: Oceans*, **120**, 5910–5931. doi: 10.1002/2015JC010989

Ebert, E. E. (2008). Fuzzy verification of high resolution gridded forecasts: A review and proposed framework. *Meteorological Applications*, **15**, 51–64

Gilleland, E., Ahijevych, D. A., Brown, B. G. and Ebert, E. E. (2010). Verifying forecasts spatially. *Bulletin of the American Meteorological Society*, **91**(10), 1365–1369. doi: 10.1175/2010BAMS2819.1

Goessling, H. F., Tietsche, S., Day, J. J., Hawkins, E. and Jung, T. (2016). Predictability of the Arctic sea ice edge. *Geophysical Research Letters*, **43**, 1642–1650. doi: 10.1002/2015GL067232

Hebert, D. A., Allard, R. A., Metzger, E. J., Posey, P. G., Preller, R. H., Wallcraft, A. J., Phelps, M. W. and Smedstad, O. M. (2015). Short-term sea ice forecasting: An assessment of ice concentration and ice drift forecasts using the U.S. Navy's Arctic Cap Nowcast/Forecast System. *Journal of Geophysical Research: Oceans*, **120**. doi:10.1002/2015JC011283

Heinrichs J. F., Cavalieri, D. J. and Markus, T. (2006). Assessment of the AMSR-E sea-ice concentration product at the ice edge using RADARSAT-1 and MODIS imagery. *IEEE Transactions on Geoscience and Remote Sensing*, **44**(11), 3070–3080. doi: 10.1109/TGRS.2006.880622

Jolliffe, I. T. and Stephenson, W. B. (2012). Forecast verification: A practitioner's guide in *Atmospheric Science*, 2nd edn. Wiley. doi: 10.1002/9781119960003

Merryfield, W. J., Lee, W.S., Wang, W., Chen, M. and Kumar, A. (2013). Multi-system seasonal predictions of Arctic sea ice. *Geophysical Research Letters*, **40**, 1551–1556. doi: 10.1002/grl.50317

Murphy, A. H. (1993). What is a good forecast? An essay on the nature of goodness in weather forecasting. *Weather and Forecasting*, **8**, 281–293.

Murphy, A. H. and Winkler, R. L. (1987). A general framework for forecast verification. *Monthly Weather Review*, **115**, 1330–1338

Sigmond, M., Fyfe, J. C., Flato, G. M., Kharin, V. V. and Merryfield, W. J. (2013). Seasonal forecast skill of Arctic sea ice area in a dynamical forecast system. *Geophysical Research Letters*, **40**, 529–534. doi: 10.1002/grl.50129

Smith, G. C., Roy, F., Reszka, M., Surcel Colan, D., He, Z., Deacu, D., Bélanger, J.M., Skachko, S., Liu, Y., Dupont, F., Lemieux, J.F., Beaudoin, C., Tranchant, B., Drévillon, M., Garric, G., Testut, C.E., Lellouche, J.M., Pellerin, P., Ritchie, H., Lu, Y., Davidson, F., Buehner, M., Lajoie, M. and Caya, A. (2015). Sea ice forecast verification in the Canadian Global Ice Ocean Prediction System. *Quarterly Journal of the Royal Meteorological Society*, **142**(695). doi: 10.1002/qj.2555

Wilks, D. S. (2011). *Statistical Methods in the Atmospheric Sciences*, 3rd edn, Volume 100, Academic Press, ISBN-13: 978-0123850225

7

Current Ice Services and Their Expected Evolution

Tom Carrieres, Mark Buehner, Jean-François Lemieux, Leif Toudal Pedersen, Alain Caya, Nick Hughes, Sean Helfrich, Juha Karvonen, Lars Axell

Ongoing improvements to sea ice automated prediction systems (APS) will have a positive impact on the delivery of ice services to clients. Increasing volumes and improved quality of sea ice observations (see Chapter 3) and their more effective use in sea ice data assimilation systems (see Chapter 4), combined with more accurate sea ice models (see Chapter 2), will be incorporated into sea ice APS (such as those described in Chapter 5) as they become available. This chapter outlines our view of how a National Ice Information Service (NIIS) could evolve to increasingly make use of output from automated systems and the benefits that would be realized from this change. Chapters 2 to 6 provided an objective description of existing theory and practice, whereas this chapter is more speculative.

In order to identify how the analysis and forecast products from a sea ice APS can provide numerical guidance to improve operational ice services, it is instructive to draw parallels with the evolution of operational weather prediction. The first numerical weather prediction (NWP) systems, circa 1960, had poor spatial resolution (due to a lack of computing resources), inadequate initialization (due to a lack of both observations and effective data assimilation methods), and incomplete or poorly modelled physics. Following this, these systems underwent a period of evaluation and refinement, resulting in a gradual improvement in forecast quality. By the year 2000, forecasts for days three to five were routinely performing better than human forecasts. Currently, automated weather forecasts are used to generate forecasts for day two and beyond, allowing human forecasters to focus on very short-range forecasts of high-impact weather.

Where available, the output from a sea ice APS is used currently to varying degrees in the delivery of operational ice services according to its perceived accuracy and utility. To identify the reasons why automated systems are not used more extensively, an assessment of the current capabilities of manual techniques and numerical ice guidance is presented. Given this assessment, and taking into account external factors, a scenario is proposed for the evolution of future operational ice services towards an increased reliance on APS. As in the case for NWP, this evolution should result in improving the quality of these services and increasing the efficiency with which they are delivered.

7.1 An Assessment of Current Capabilities

Ideally, an assessment of current capabilities would objectively evaluate how well the available operational products meet client needs. However, since client needs tend to be subjective and specified in general terms, an objective assessment may not be feasible. The components of the process currently used to construct operational products are first reviewed. Then, some of the evaluation techniques outlined in Chapter 6 are applied to a selection of typical operational products and relevant results used to guide the assessment where feasible.

7.1.1 Current Process

A schematic of the current, largely manual process for generating operational ice products is shown in Figure 7.1. Only the process leading up to the generation of products is considered here, while the actual product delivery and use by clients and any interactions between clients and the NIIS are not. An overview of the essential elements of Figure 7.1 follows:

- Ice Analyst (IA): the current workflow is designed to allow an IA with significant experience to visually examine geographical maps of key information and to prepare manual operational products;
- Climatology experience: a knowledge of historical ice conditions prevalent in a given area allows the IA to better interpret satellite imagery;
- Observations: satellite images provide the bulk of the information while in situ measurements provide very limited coverage;
- Atmosphere and ocean APS predictions: these provide important guidance on evolving environmental conditions;

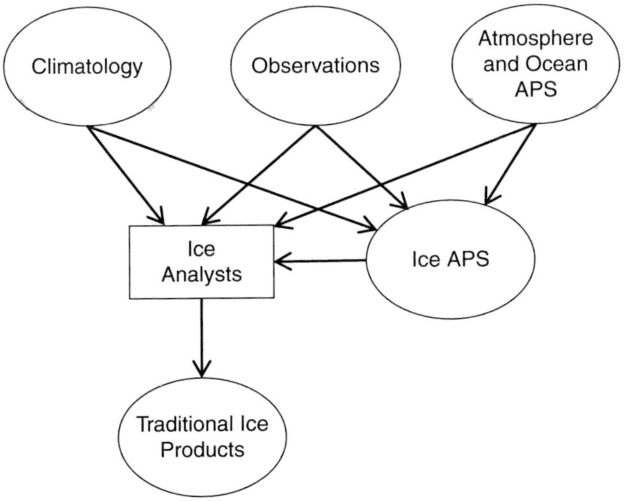

Figure 7.1: A schematic of the current ice product preparation process.

- Ice APS predictions: these provide more direct guidance on evolving ice conditions; and
- Traditional Ice Products: the resulting products are generated by the IA through a manual synthesis of all available information listed above.

7.1.2 Analysis Functions

The IA routinely prepares ice analysis charts, such as those depicted in Figures 1.1 and 3.15, by assessing and incorporating a large variety of relevant information, as well as considering ice information from a previous analysis, when available. In-depth knowledge of local ice climatology and of the relevant predominant physical processes provides a guide to expected ice stages of development and structure, which aids the interpretation of in situ observations and satellite imagery. Ice buoys, satellite imagery and sea ice model drift predictions may be used to track ice from the previous analysis to the current analysis. To infer ice coverage, satellite synthetic aperture radar (SAR) imagery is preferred due to its all-weather and day/night capabilities, but this is often complemented by imagery mainly from satellites with visual and infrared sensors. In situ measurements of ice thickness may be available from shore stations for land-fast ice and from ships for mobile ice floes. Ship-based and aerial observations of concentration, ice edge location, navigability and possibly ice stage of development are often available. Meteorological and oceanographic information is also used to estimate new ice growth, to infer surface melt conditions and as a guide to the net ice drift that has occurred since a previous analysis time. The process of assessing and synthesizing all of this information into a single definitive product is somewhat subjective. Given operational time constraints, it is not practical for ice analysts to provide detailed confidence estimates in these products, although it is possible to indicate what sources of data were used.

The correct interpretation of satellite imagery is crucial to the above process. For example, for the analysis of SAR imagery, an IA infers ice concentration, floe size category and ice stage of development based on the relative strength of the backscatter return and on the apparent shape of ice floes and structure of ice deformation. Areas with highly deformed, very thin or smooth land-fast ice and areas with surface melt may present challenges. SAR incidence angle and noise floor, wind effects and other ocean features also complicate the analysis. An IA can also manually derive an estimate of ice concentration and possibly even the ice stage of development from visual and infrared imagery in cloud-free areas.

The automated process of data assimilation, as outlined in Chapter 4, is similar in many respects to the manual process just described. A short-term model forecast is used as the initial estimate or background state, and a correction to this is computed in an optimal way using numerous diverse types of observations that are weighted according to their estimated accuracy. The development of retrieval algorithms, observation operators and quality control for satellite data is a major endeavour. For example, significant ambiguity in the SAR backscatter between some ice types and water in certain wind situations restricts assimilation of automated retrievals to only portions of a SAR image. Similarly, even though ice concentration retrieval algorithms from passive microwave imagery are well established, the data in areas with surface melt, thin or low concentration ice and high wind speeds require quality control. Retrieval of ice information from optical satellites is hampered by difficulties in accurate cloud masking. Despite these challenges, the strengths of APS ice analysis are the ability to assimilate large quantities of observations simultaneously and to quantify the uncertainty in the analysis objectively. Extensive evaluation of APS ice analyses has been conducted and many aspects of the systems are continually being improved.

It is expected that manual analyses will still usually provide better estimates of the ice conditions than automated systems. This is especially true in areas with significant operational marine activity, for which the extra time and effort required for the IA to synthesize a larger number of data sources manually is justified. However, automated analyses can provide significant benefits in terms of being able to provide consistent estimates over very large areas efficiently and are approaching the point where they could be used for areas of more limited marine activity or for clients who consider the current information content and quality adequate. For many clients navigating in and around sea ice, the analyses must approach 1 km resolution, and this is beyond the capability of most current operational automated systems.

7.1.3 Forecast Functions

NIIS forecasts are limited mainly to very short range and seasonal products. At some NIIS, the IA evolves ice conditions to a common valid time for daily ice charts. An IA may also produce text forecasts of up to 24 hours for quantities such as ice drift, edge location or pressure. Seasonal forecasts are mainly descriptive bulletins. In contrast, an APS produces forecasts spanning hours to seasons of a large variety of fields over the entire system domain. As a result, the following qualitative comparison between IA and APS forecasts is informative but certainly is not an exhaustive assessment of the potential use of sea ice APS output for future NIIS products.

7.1.3.1 Ice Drift

A general rule of thumb used by the IA is that freely drifting ice moves at about 2 per cent of the near-surface wind speed. Other factors, such as the Coriolis effect, ocean currents, ice surface roughness, internal ice stress and coastal effects, are subjectively accounted for. An early study (Carrieres, 1996) of Canadian Ice Service (CIS) 24-hour ice drift forecasts using this simple technique revealed a root mean square error (RMSE) of 11 km for displacement (85 per cent of the mean observed displacement) and an RMSE of about 50° for drift direction. It is expected that improvements have been made since that study, but no more recent studies are available.

As outlined in Chapters 2 and 5, APS sea ice models account for the effects described above although very few – if any – observations of ice thickness and surface roughness are available or assimilated yet. Evaluation of numerical ice drift forecasts is quite extensive with representative mean displacement errors of 11 km for a 30-hour forecast reported (Hebert *et al.*, 2015) for ACNFS (Arctic Cap Nowcast/Forecast System). In fact, APS forecasts of ice drift are already widely used.

7.1.3.2 Ice Edge Location

Ice drift, growth and melt can result in an expansion or contraction of the ice edge. Very little or no observation-based guidance has been developed for thermodynamic effects, so the IA usually relies on his or her experience. Figure 7.2 compares the location of an ice edge derived from the Interactive Multisensor Snow and Ice Mapping System (IMS) product with the location of the 0.4 ice concentration contour from CIS regional charts for a three-month period in 2015. It is worth noting that there is no specific attempt to provide contours of ice concentration in the ice charts, but rather polygons of fairly uniform ice conditions. Similarly, IMS is designed to provide an ice (and snow) mask for NWP with a 0.4 concentration threshold and it is not designed to help mariners locate the ice edge. These results show that the difference in ice edge location ranges from 10 to 30 km. If the errors in each analysis can be assumed to be uncorrelated, then this difference can be interpreted as an upper limit for the error in each product.

In an APS, sea ice is represented as an average concentration over a grid cell and therefore an ice edge cannot be defined with more precision than the grid cell size. This becomes less of an issue with increasing model resolution. As an example, evaluations against IMS indicate mean errors of 14 to 16 km for RIPS (Regional Ice Prediction System) analyses and forecasts up to 48 hours. This is an area where APS could provide useful

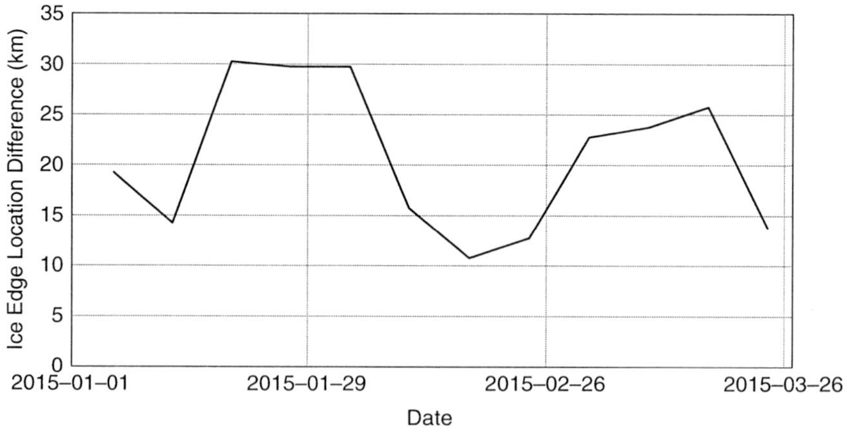

Figure 7.2: Median distance between locations of the 0.4 ice concentration edge from CIS regional ice charts and the ice extent boundary derived from the IMS product over a three-month time frame.

guidance. It is important to assess the accuracy of IA forecasts more formally and to assess numerical guidance subjectively in a variety of situations.

7.1.3.3 Ice Concentration

The IA uses personal experience and APS output as guidance to account for future changes in ice concentration due to advection, convergence, divergence, estimates of new ice formation and melt. Enhanced melt is also estimated near the ice edge, polynyas and warm water. A comparison of ice concentration products from the Norwegian and Danish Meteorological Institutes (Eastwood, 2016) indicated standard deviations between the two operational products in the range of 10–30 per cent. Figure 7.3 compares CIS Daily Chart products valid at 18 UTC with CIS RADARSAT Image Analysis Chart products that are valid around 22 UTC. The differences in the products are partly due to the evolution of ice conditions over the time offset between the two products. However, changes in the overall ice conditions over such a short period should be quite small. Therefore, assuming the errors in each are uncorrelated with the other, these results are an indication of an upper limit on each product's uncertainty.

Extensive comparison of APS forecasts with ice analysis charts and other products has been made for various prediction systems. A typical root mean square difference is around 20 per cent but this is highly dependent on the location and time of year. This indicates that APS output is approaching the point where it could provide a useful background estimate with corrections applied by an IA for areas of high priority or high uncertainty as required.

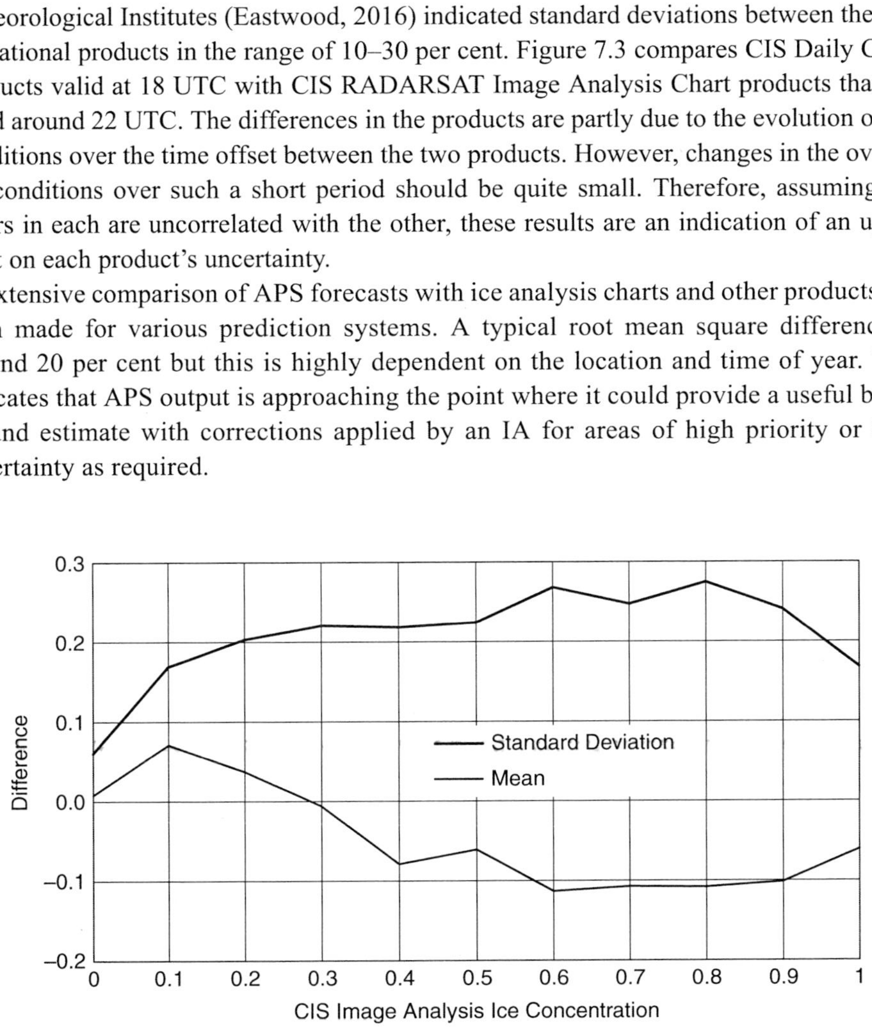

Figure 7.3: Mean and standard deviation of the ice concentration difference between CIS daily ice charts valid at 18 UTC compared with CIS RADARSAT Image Analysis Charts valid on the same day at approximately 22 UTC calculated for all of 2010.

Table 7.1: *Empirical ice pressure guidance for grey-white or thicker ice for different wind speeds and corresponding durations (source: Canadian Ice Service).*

Wind	<12 hours	12–24 hours	24–36 hours	36–48 hours
Strong (20–33 knots)	Light	Moderate	Strong	Strong
Gale (34–47 knots)	Moderate	Strong	Strong	Strong
Storm (48–63 knots)	Strong	Strong	Strong	Strong
Hurricane (>63 knots)	Strong	Strong	Strong	Strong

7.1.3.4 Ice Pressure

NIIS ice pressure forecasts are issued as light, moderate or strong pressure categories. Since a given category or value of ice pressure may have significantly different impact on different ship classes, ship masters must interpret the expected impact on their vessel. Although some empirical guidance exists (see Table 7.1), no systematic evaluation of subjective ice pressure forecasts has been conducted.

Assessing the quality of APS ice pressure forecasts and providing useful information to clients remain difficult. Indeed, the forecast pressure depends on various assumptions in the formulation of the sea ice rheology (including the ice strength parameterization) that are still not well constrained by observations. Moreover, APS simulated ice pressure is representative of the mean pressure over the grid cell and may not be consistent with in situ measurements. A downscaling approach could be developed by relating APS forecast pressure with ship reports. A related pressure severity index could also be defined.

7.1.3.5 Ice Thickness

Air temperature provides a guide to IA for forecasting ice thickness using a variety of available empirical relations. These use accumulated freezing degree days to predict landfast ice thickness, which in turn can be used as a proxy for nearby mobile ice thickness. Forecasts of initial ice formation are based primarily on IA subjective experience.

As reviewed in Chapter 2, APS models represent many thermodynamic and dynamic processes for calculating the evolution of the sea ice thickness. In an operational setting, an APS often involves an analysis of ice concentration for initialization while other fields, such as snow coverage and thickness, vertical temperature, sea ice thickness and salinity profiles, are usually obtained from the previous forecasts. Evaluation studies of APS ice thickness forecasts are rare, but the mean absolute error has been reported to be around 1.2 m in the relatively thick ice in the Beaufort Sea (see Section 6.3.3). Smaller absolute errors would be expected in the marginal ice zone where sea ice thickness is generally lower. Clearly, APS ice thickness predictions require careful assessment and further evaluation.

7.2 Expected Changes in External Influences

Operational ice services have naturally evolved over the years due to the availability of new sources of remotely sensed observations and specialized ice analysis workstations. This has resulted in more accurate and flexible digital products. However, making effective use of the many planned satellite systems will also place new demands on NIIS and necessitate the adoption of more efficient and more consistent procedures. Additionally, more demanding client requirements and improvements in the accuracy and diversity of sea ice APS outputs will result in new types of NIIS products.

7.2.1 Current and Planned Satellite Systems

Both the Canadian Space Agency and the European Space Agency have made ongoing commitments to support satellite systems with improving SAR sensors. The RADARSAT Constellation and Sentinel-1 missions will provide high-volume operational streams of SAR data for many years to come. It is unlikely that all of the data can be analysed manually, and therefore a variety of automated systems will need to be employed to supplement human analysts. When these missions become fully operational, it is reasonable to expect that ice information from SAR for virtually all ice-covered areas of the globe will be available on a daily or more frequent basis.

Optical satellite systems have been operational for many decades, largely in support of weather applications, and this will continue with the Suomi mission of the United States and the European Sentinel-2 and -3 and MetOp missions. The data will become even more useful for NIIS with higher spatial and spectral resolutions.

Passive microwave missions have provided continuous global coverage since the 1970s, and this is expected to continue with even more applications under development. In addition to ice concentration and some ice typing, ice thickness information for thin ice may be estimated from these data, albeit at relatively coarse spatial resolution and only for relatively thin ice.

Altimetry sensors (e.g. Cryosat-2) now provide ice thickness information with sufficiently short latencies as to be useful for operational ice analysis as well as model validation. This information complements the thin ice retrievals from passive microwave (e.g. SMOS and SMAP) and infrared sensors.

7.2.2 Client Requirements

There is a large variety of NIIS clients with many diverse and evolving needs. A common request from clients is for more complete and accurate information and for customized products. Specifically, some areas that have been identified as new requirements include:

- Improved accuracy of ice edge location, ice concentration, thickness, pressure and drift;
- Client-dependent spatial and temporal resolution;

- Probabilistic forecasts and characterization of uncertainty; and
- Locations of leads and polynyas in sea ice.

In essence, some clients would like to effectively access continuous spatial and temporal ice information, selecting the particular variables that are crucial to their activities. Other clients prefer products that present essential ice information in a simpler manner. In all cases, clients would benefit from knowledge of the uncertainty associated with these products so they can incorporate the associated risk into their decision making. Some clients may require additional documentation and training so that they can use products appropriately. Some clients will also want to be involved in the development and evaluation of new products.

7.2.3 Automated Prediction System Developments

Ongoing improvements to existing sea ice APS (some of which were outlined in Chapter 5) and NWP and environmental prediction systems will result in improved numerical guidance for NIIS. In general, NWP is moving towards more comprehensive environmental prediction that incorporates coupling between atmosphere, ice, ocean (including waves) and land surface components. This coupling will include interactions between the system components within both the forecast and the data assimilation procedures. For shorter-term forecasts, this will improve forecasts locally while longer forecasts will likely have a smaller but more widespread improvement. There are also increasing efforts towards ensemble systems that will provide information on forecast uncertainty, including forecasts of the probability of extreme events. Continuing efforts towards higher-resolution systems will allow better simulation of small-scale processes and a more defined topography for better representing orographic effects (e.g. wind channelling in narrow passages).

The continuing development of APS specifically designed for the use of NIIS will parallel many of the advances being made for atmospheric APS, although with a slightly different focus. In particular, research will concentrate on improving sea ice data assimilation systems to make more effective use of the numerous types of satellite observations, including those which will allow higher-resolution analyses. These systems will also advance through an improved characterization of uncertainty within both the data assimilation and forecasting procedures by using ensemble methods. Sea ice models will introduce more complete physics (e.g. wave–ice interactions, floe size distribution, blowing snow) with more efficient and robust numerical schemes. It is expected that large-scale continuum-based sea ice models will be used together with limited area models representing sea ice as a collection of particles. The diversity of ice APS output fields including ensemble predictions will allow a large variety of tailored products to be generated.

7.3 Evolution of Ice Services Through Increased Use of Numerical Guidance

Given the changes presented in the previous section and the assessments made in Section 7.1, a schematic is shown in Figure 7.4 of how the NIIS production could evolve in the future from the current state, as shown in Figure 7.1. The most crucial changes will be:

- Increased volume of observations and fully coupled (atmosphere–ice–ocean–wave) APS output will be available to the IA;
- Given this, it is expected that the IA will use only the most relevant data while the APS will assimilate the bulk of the least ambiguous observations;
- Increased volume and more accurate gridded ice APS output will be the main source of information for many new and existing products; and
- IA will use their knowledge of marine activity and APS uncertainty information to focus their attention on areas where their contributions will have the greatest impacts by refining and applying a final quality check to the APS output, with less time spent on traditional products and more time devoted to new, more tailored products.

7.3.1 Role of Ice Analysts

The IA will continue to serve a central role in ensuring that the best available ice information and products are provided to clients in a timely fashion. As APS improve and automation increases, this will allow the IA to focus their efforts more on areas that

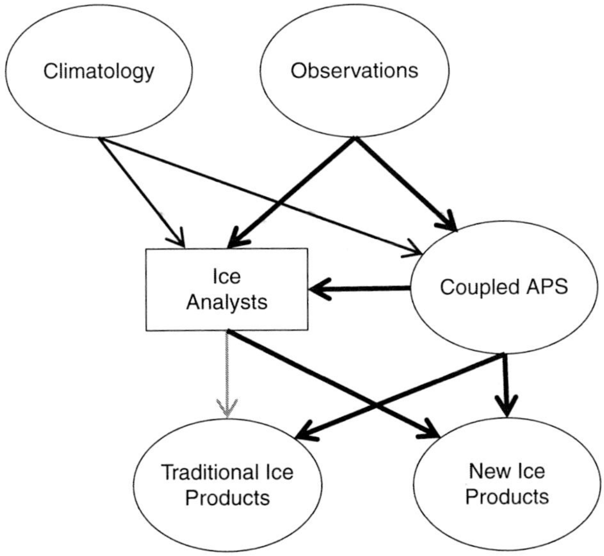

Figure 7.4: A schematic of ice product preparation in the future. To emphasize changes compared to Figure 7.1, increased information flow is shown by thicker lines while decreased information flow is shown by grey lines.

would have the greatest benefit, such as site or route-specific products or areas where APS output uncertainty exceeds an acceptable level. Good two-way communication with clients for personal consultation and training will ensure clients are aware of available products and services and can use them effectively.

Increasing integration of automated products generated from coupled APS output into operational services will be gradual and the IA will need to be an integral part of the transition. Initially, metrics will be established to determine the accuracy of existing products, both to document this information for clients and to evaluate when and where APS-based products are ready for operational dissemination to clients. It should be expected that some APS products may continue to require human intervention while others may be totally automated. New APS-based products will also be developed in consultation with clients and will require their evaluation through field trials.

Although it is not yet clear how NIIS staff will use APS data, good examples are available now for meteorological applications. APS output could in many cases become the backbone of product generation as it is not feasible for NIIS staff to prepare forecasts or even analysis products that cover all areas of potential interest or for all required time periods. The spatial resolution of APS output will continue to increase in response to client requirements, computer capabilities, increased observations and the developments of accurate representation of sea ice processes in forecast models.

There will be an ongoing need for IA staff to work with researchers to identify APS strengths and weaknesses in order to make improvements on an ongoing basis. As a direct link to clients, operational staff will help researchers understand ongoing and evolving client needs. Similarly, clients will need to be aware of APS limitations and how to use outputs effectively. Close linkages will be developed between operational staff, APS developers and end users.

7.3.2 Clients and Development of New Products

In addition to numerical guidance being available to NIIS, the direct output from coupled APS should be made available to clients. Ultimately, clients should have the capability to define products that are most suitable to their needs, including the required spatial and temporal resolution. For some, there will be limitations in terms of bandwidth for communications, end user processing capabilities and user expertise. However, improvements in technology should reduce these obstacles over time.

It is not straightforward to produce traditional ice charts from APS output and it may not be necessary to do so. Output from a sea ice APS currently provides valuable guidance for IA staff in terms of ice drift and pressure predictions, and products based on this output could be made directly available to clients. New products should therefore take advantage of APS strengths: (1) they can operate 24 hours per day and 7 days per week with very little additional cost; (2) they can produce fairly high-resolution output at high frequency intervals, including analyses of current conditions and forecasts ranging from hours to seasons; (3) improved methodologies are being developed to characterize analysis and forecast uncertainty; (4) sea ice models accurately solve the equations related to sea ice dynamics and thermodynamics with completely reproducible results; (5) some fields that are not currently well observed or forecasted with manual procedures can be obtained from APS output.

Using APS output could lead to an exponential growth in available products. A few examples are provided below.

7.3.2.1 Direct Outputs and Derived Products

Most sea ice models incorporate sufficient physics to provide forecasts of a variety of fields, including instantaneous ice velocity, ice concentration, thickness distribution, edge location and pressure. Some of these direct outputs have potential utility and others may be more useful after applying additional processing, such as by generating ice trajectories from the predicted ice velocity fields. In many ways, these numerical forecasts of continuous quantities are more flexible than categorical forecasts because they allow clients to identify their own critical thresholds that are most relevant for their activities. As an example, a potentially useful product is shown in Figure 7.5 that depicts the number of hours that ice pressure (vertically integrated) is forecast to be above an arbitrarily chosen value of 2.5×10^3 N m^{-1}. Expert clients may identify a critical pressure value and duration that is most relevant to their activities.

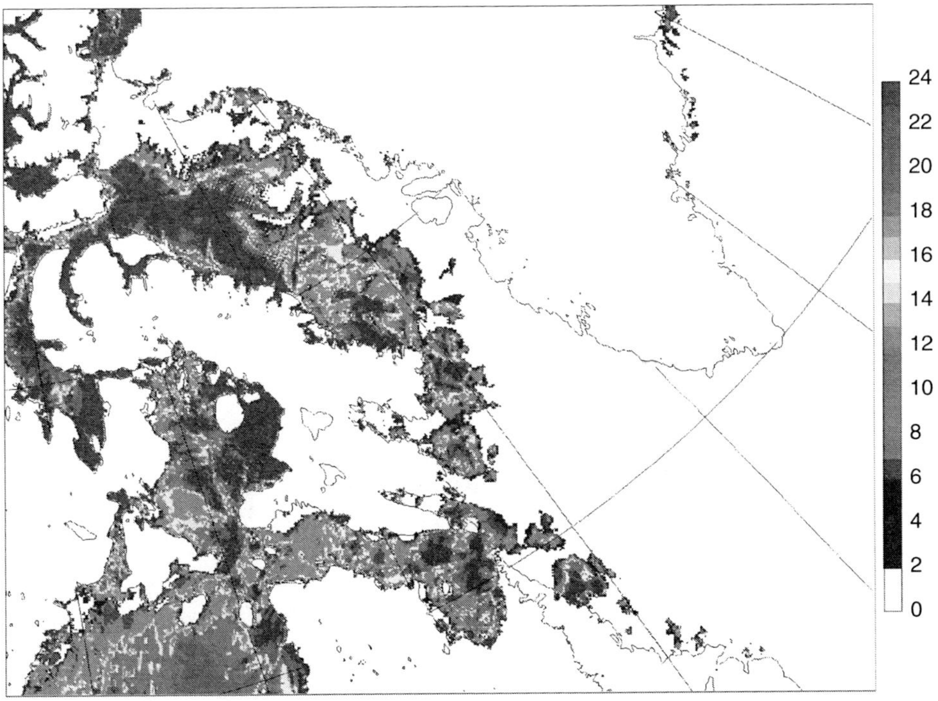

Figure 7.5: 24-hour forecast of the duration in hours of ice pressure greater than 2.5×10^3 Nm^{-1} from RIPS, valid 1200 UTC 11 March 2016. (A black and white version of this figure will appear in some formats. For the colour version, please refer to the plate section.)

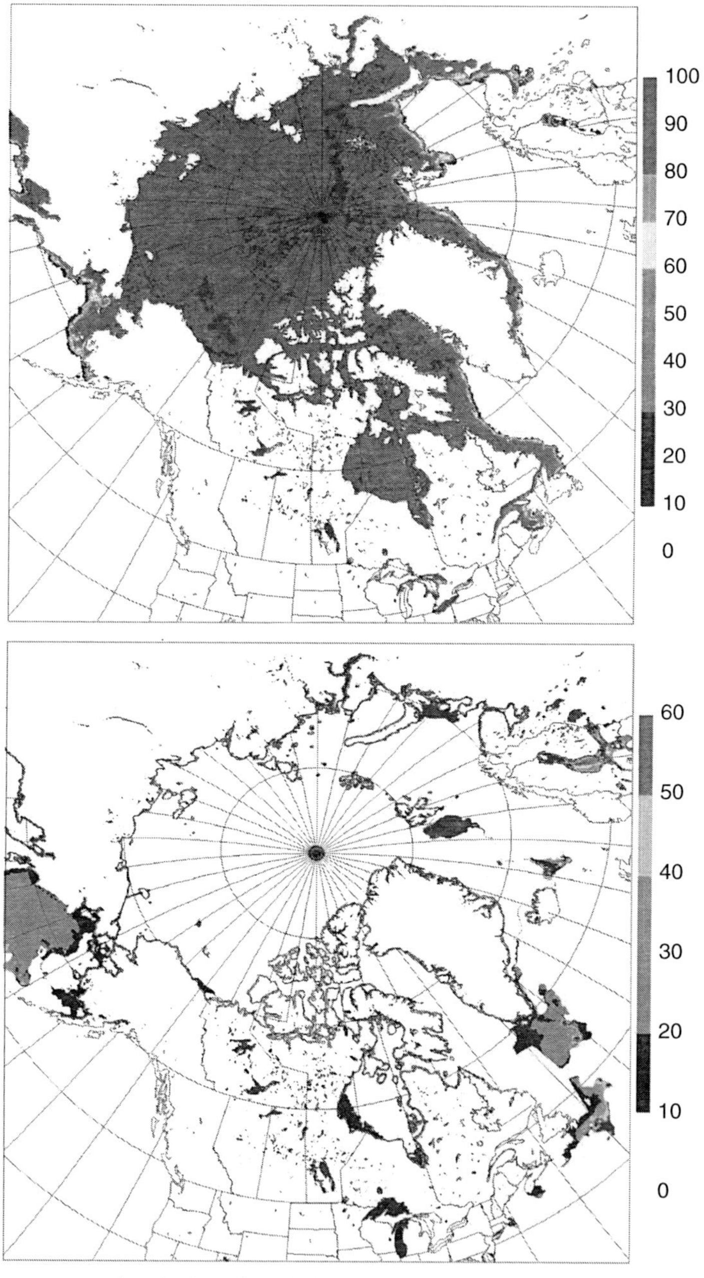

Figure 7.6: Ice concentration (top) and analysis uncertainty (bottom) from RIPS, valid 1800 UTC 31 January 2015. Units as percentages. (A black and white version of this figure will appear in some formats. For the colour version, please refer to the plate section.)

A further example of derived products could be to use post-processing techniques to enhance APS output or derive forecasts of quantities not directly predicted by the APS. Downscaling techniques (such as described in Section 5.5) can be employed when the APS resolution is too coarse. Alternatively, APS output can be used as a predictor within a statistical regression system to provide forecasts of derived quantities, such as the ice breakup date for a specific body of water.

7.3.2.2 Probabilistic Analyses and Forecasts

It has been shown that, for a variety of consumers of weather forecasts, introducing the probability or uncertainty in forecasts adds value to the forecasts. For NIIS clients, these types of products could support risk-based decisions. There are various ways of presenting probability in numerical guidance. Some current automated analysis systems already provide an estimate of the uncertainty in the analysis obtained with a relatively simple approach. Initial attempts at generating ensembles of forecasts also have demonstrated their potential to generate useful information. An example product is shown in Figure 7.6. The ice concentration analysis field available from RIPS is complemented by a fairly sophisticated estimate of the uncertainty in the analysis. This uncertainty is based on the quality and number of assimilated observations over the past several days and the proximity to the ice edge (see Buehner *et al.*, 2016 for further details).

Another forecast product that could help clients is shown in Figure 7.7. This product illustrates the probability of encountering ice with a concentration greater than 10 per cent. In effect, this product informs clients about the most likely location of the ice edge and also indicates the level of forecast uncertainty in a given area. This type of processing could be further expanded to include multiple conditions such as shown in Figure 7.8, where the probability of mean ice thickness in a grid cell greater than 1 m and concentration greater than 90 per cent is depicted. This product could be an indication where only the more powerful icebreakers should venture.

7.3.3 ***Evaluation***

Compared to NWP, evaluation methods for sea ice forecasts are still in their infancy. More advanced and tailored evaluation techniques will be developed to guide the future research aimed at improving sea ice APS and to provide information to clients. Spatial verification techniques adapted from what is done in NWP (e.g. methods for spatial precipitation forecasts) are expected to bring valuable information to clients and to the developers of sea ice APS. As APS are foreseen to evolve toward ensemble prediction, probabilistic evaluation methods will be developed. The advent of particle-based sea ice models in the future will also represent a challenge for evaluating forecasts made with such novel systems.

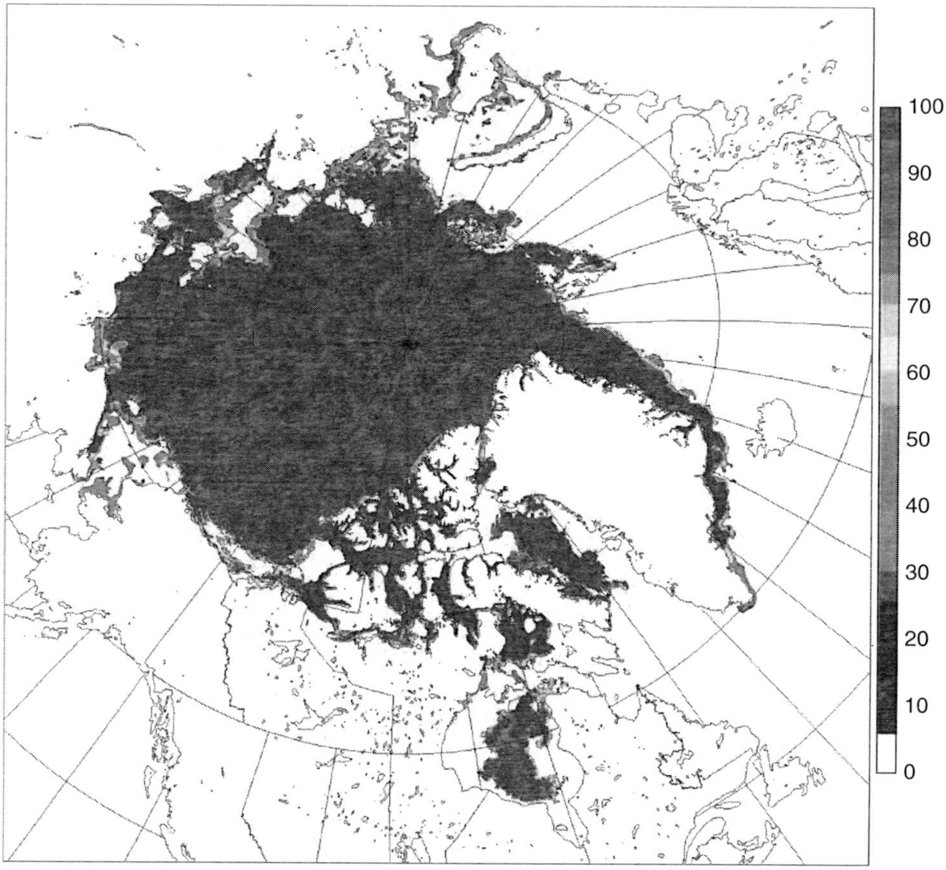

Figure 7.7: Forecast of the probability of encountering ice with concentration greater than 10 per cent, valid 0600 UTC 10 July 2011. This forecast is based on a 21-member ensemble system (Shlyaeva *et al.*, 2016). (A black and white version of this figure will appear in some formats. For the colour version, please refer to the plate section.)

7.3.4 Ice Analysis and Forecast Workstations and Delivery Systems

Current ice analysis and forecast workstations are mainly geographic information system (GIS)-based with an emphasis on vector analysis and raster images. This allows for the preparation of high-resolution and high-quality products. Currently, APS output is mainly limited to passive display of predetermined APS products derived from the gridded output. For local ice forecasts, a future system would include all required forecast fields at the APS resolution. Whether the IA has the ability to adjust output spatially, or only for individual or groups of grid points, remains to be determined. It will also be feasible to pre-process the APS output to either downscale the results or to apply bias corrections, for example. Once the IA completes the necessary adjustments to the gridded data, forecast bulletins and other products could be generated automatically.

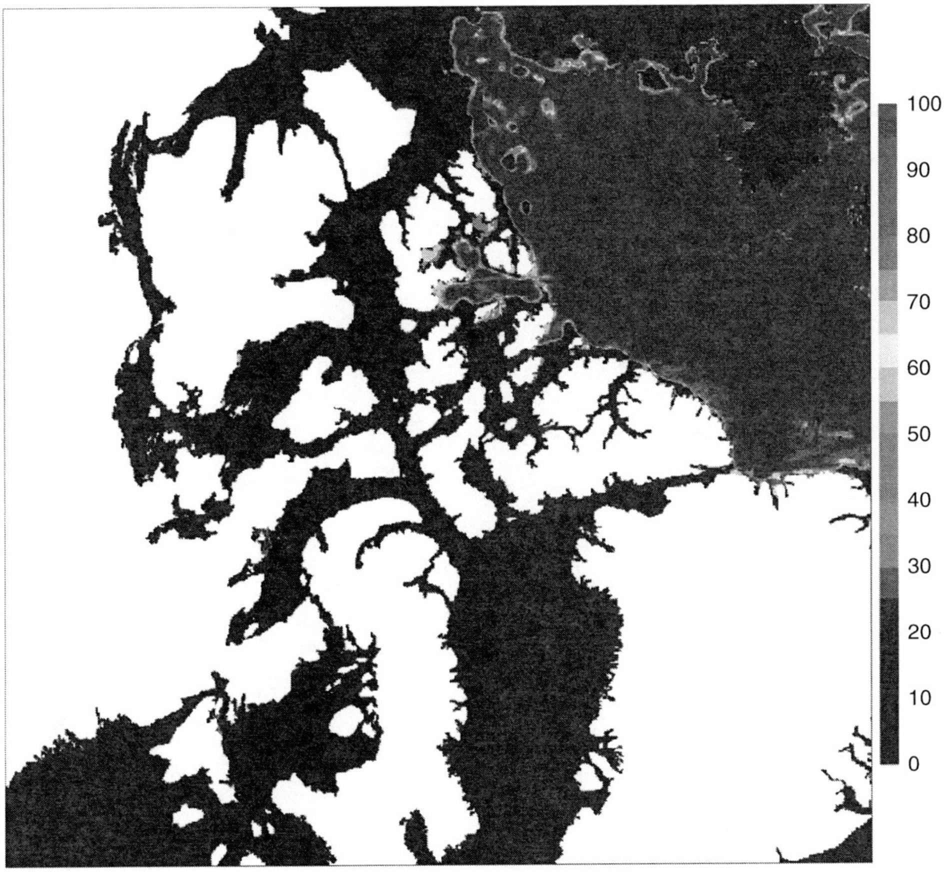

Figure 7.8: Forecast of the probability of encountering ice with concentration greater than 90 per cent with mean ice thickness in a grid cell greater than 1 m, valid 0600 UTC 31 August 2011. This forecast is based on a 21-member ensemble system (Shlyaeva *et al.*, 2016). (A black and white version of this figure will appear in some formats. For the colour version, please refer to the plate section.)

Product servers will expand from their current emphasis on delivering predetermined products to allow either the IA or clients to derive new products based on their needs. Using the corrected or uncorrected APS output database, new products such as those described above will be made available either through an interactive user system or as part of a client request system.

7.4 Chapter Summary

Techniques used at NIIS are well established, but remain largely manual processes that have benefited from the greatly improved availability of satellite imagery and developments in workstation design. Very little quantitative information is available concerning the accuracy of the information provided to clients, and a continued emphasis remains on current conditions as opposed to forecasts.

A number of external factors will affect the evolution of ice services over the coming years. Even though there has already been a revolution in product generation due to the wide availability of satellite SAR imagery, much more satellite information will be available in the near future. Clients are also making greater demands for NIIS products that better suit their individual needs. Standalone sea ice APS developments and sea ice predictions from the evolution of NWP into coupled environmental prediction systems will provide support for both improved and expanded ice analysis and forecast products.

It is expected that the role of IA will evolve as a result of these external factors. The use of APS output should evolve from providing supporting information to becoming the backbone of an expanded suite of ice information products. This will allow the IA to focus on areas where they may have the most effective impact while also enabling clients to access much more diverse information that is better suited to their individual needs.

7.5 References

Buehner, M, Caya, A., Carrieres, T. and Pogson, L. (2016). Assimilation of SSMIS and ASCAT data and the replacement of highly uncertain estimates in the Environment Canada Regional Ice Prediction System. *Quarterly Journal of the Royal Meteorological Society*, **142**, 562–573. doi: 10.1002/qj.2408

Carrieres, T., Greenan, B., Prinsenberg, S. and Peterson, I.K. (1996). Comparison of Canadian daily ice charts with surface observations off Newfoundland, winter 1992. *Atmosphere-Ocean*, **34**(1), 207–226. doi: 10.1080/07055900.1996.9649563

Eastwood, S., Karvonen, J., Dinessen, F., Fleming, A., Hamre, T., Toudal Pedersen, L., Saldo, R., Buus-Hinkler, J., Hackett, B., Ardhuin, F. and Kreiner, M.B. (2016). *Quality Information Document for OSI TAC Sea Ice Products* 011-001, -002, -003, -004, -005, -006, -007, -009, -010, -011, -012. System documentation report for the Copernicus Marine Environment Monitoring Service http://cmems-resources.cls.fr/documents/QUID/CMEMS-OSI-QUID-011-001to007-009to012.pdf

Hebert, D.A., Allard, R.A., Metzger, E.J., Posey, P.G., Preller, R.H., Wallcraft, A.J., Phelps, M.W. and Smedstad, O.M. (2015). Short-term sea ice forecasting: an assessment of ice concentration and ice drift forecasts using the U.S. Navy's Arctic Cap Nowcast/Forecast System. *Journal of Geophysical Research: Oceans*, **120**. doi: 10.1002/2015JC011283

Shlyaeva, A., Buehner, M., Caya, A., Lemieux, J.F., Smith, G.C., Roy, F., Dupont, F. and Carrieres, T. (2016). Towards ensemble data assimilation for the Environment Canada Regional Ice Prediction System. *Quarterly Journal of the Royal Meteorological Society*, **142**, 1090–1099, doi: 10.1002/qj.2712

Index